Electroanalytical Methods

Fritz Scholz
Editor

Electroanalytical Methods

Guide to Experiments and Applications

Second, Revised and Extended Edition

With contributions by
A.M. Bond, R.G. Compton, D.A. Fiedler, G. Inzelt,
H. Kahlert, Š. Komorsky-Lovrić, H. Lohse, M. Lovrić,
F. Marken, A. Neudeck, U. Retter, F. Scholz, Z. Stojek

Springer

Editor
Prof. Dr. Fritz Scholz
University of Greifswald
Inst. of Biochemistry
Felix-Hausdorff-Str. 4
17487 Greifswald
Germany
fscholz@uni-greifswald.de

ISBN 978-3-642-42531-8 ISBN 978-3-642-02915-8 (eBook)
DOI 10.1007/978-3-642-02915-8
Springer Heidelberg Dordrecht London New York

Cover design: WMXDesign GmbH, Heidelberg

Printed on acid-free paper

Springer is part of Springer Science+Business Media (www.springer.com)

Electroanalytical Methods

Fritz Scholz dedicates this book to the memory of his late parents Anneliese and Herbert Scholz

Preface to the Second Edition

The authors are pleased to present here the second edition of the book "Electroanalytical Methods. Guide to Experiments and Applications." All chapters have been carefully revised and updated, and a new chapter, devoted to the use of the electrochemical quartz crystal nanobalance, has been added.

Greifswald, Germany, October 2009 *Fritz Scholz*

Preface to the First Edition

Electroanalytical techniques offer a unique access to information on chemical, biochemical, and physical systems. Both the instrumental basis and the theoretical fundamentals have been developed such that non-specialists can easily apply them. However, there is still a considerable lack in acceptance of these techniques by others except those who have experience and training in electrochemistry. The authors of this volume felt that it was time to write a volume that concentrates on the really important techniques together with the most essential information to make them applicable for potential users who do not possess specialist knowledge of electrochemistry. All the authors have a long experience in teaching and know the most frequent experimental pitfalls as well as theoretical misunderstandings and misinterpretations. This book has been written to become a bench book used in the laboratory. The "Electroanalytical Methods" addresses chemists and biochemists who are interested in using electroanalytical techniques to supplement spectroscopic and perhaps theoretical calculations. It also addresses biologists, environmental and material scientists, physicists, medical scientists, and, most importantly, students in all branches of science, who are confronted with the need to apply electroanalytical techniques. In the short first part of the book, entitled "Basic Electrochemistry," the essentials of electrochemical thermodynamics and kinetics are given. The second part, entitled "Electroanalytical Techniques," contains the most frequently utilized techniques, i.e., cyclic voltammetry, pulse and square-wave voltammetry, chronocoulometry, etc. The third part is devoted to electrodes and electrolytes, which are the major constituents of an electrochemical cell. Throughout the book, special attention is given to guide the user to successful laboratory experiments and a sound data evaluation. This book focuses only on modern and widespread techniques. To give the interested reader a key to the historic background, a short list of seminal publications in electrochemistry and electroanalysis is provided in Chap. IV.1.

There are many carefully written monographs on special electroanalytical techniques and textbooks on fundamental electrochemistry available. We refer to this fundamental literature (see Chap. IV.2) for a deeper insight into the subject. The World Wide Web is of steadily growing importance also for electrochemical information. Although it is constantly changing and renewing, some key addresses are provided to make access easier and the search more successful.

Greifswald, Germany, 2001 *Fritz Scholz*

Contents

Part I Basic Electrochemistry

Part II Electroanalytical Techniques

Contributors

Alan M. Bond Department of Chemistry, Monash University, Clayton VIC 3168, Australia, alan.bond@sci.monash.edu.au

Richard G. Compton Physical Chemistry Laboratory, Oxford University, South Parks Road, Oxford OX1 3QZ, UK, richard.compton@chem.ox.ac.uk

Dirk A. Fiedler Principal Systems Engineer, Cochlear, Limited, 14 Mars Road, Lane Cove, NSW 2066, Australia, dirkfiedler1@optusnet.com.au

György Inzelt Department of Physical Chemistry, Eötvös Loránd University, 1518 Budapest 112, P.O. Box 32, Hungary, inzeltgy@chem.elte.hu

Heike Kahlert Institut für Biochemie, Universität Greifswald, Felix-Hausdorff-Str. 4, 17487 Greifswald, Germany, hkahlert@uni-greifswald.de

Heinz Lohse Peter-Hille-Str. 127, 12587 Berlin, Germany, u.h.lohse@arcor.de

Milivoj Lovrić Rudjer Boskovic Institute, POB 1016, 10001 Zagreb, Croatia, mlovric@irb.hr

Šebojka Komorsky-Lovrić Rudjer Boskovic Institute, POB 1016, 10001 Zagreb, Croatia, slovric@irb.hr

Frank Marken Department of Chemistry, University of Bath, Bath BA2 7AY, UK, F.Marken@bath.ac.uk

Andreas Neudeck Textilforschungsinstitut Thüringen-Vogtland e.V., Stellv. Bereichsleiter Forschung und Entwicklung –Vorlaufforschung, Zeulenrodaer Str. 42, 07973 Greiz, Germany, andreas.neudeck@t.online.de

Utz Retter Balzerstr. 45, 12683 Berlin, Germany, Utz.Retter@gmx.de

Fritz Scholz Institut für Biochemie, Universität Greifswald, Felix-Hausdorff-Str. 4, 17487 Greifswald, Germany, fscholz@uni-greifswald.de

Zbigniew Stojek Department of Chemistry, University of Warsaw, Pasteura 1, 02-093 Warsaw, Poland, stojek@chem.uw.edu.pl

Index of Symbols and their SI Dimensions

I. Roman Symbols

a	activity
$a_{\mathrm{B}},\ a_{\mathrm{i}}$	relative activity of component B, i
$a_{\pm}$	mean ionic activity
A	absorbance
A	area, m^2, cm^2
A	coefficient of Debye–Hückel equation, $dm^{3/2}\ mol^{-1/2}$
A_{real}	real surface area, m^2
b	Tafel slope, V
$b_{\mathrm{a}},\ b_{\mathrm{c}}$	anodic, cathodic Tafel slope, V
B	coefficient of Debye–Hückel equation, $dm^{1/2}\ mol^{-1/2}$
c	concentration, mol m^{-3} or mol L^{-1}
$c_{\mathrm{B}},\ c_{\mathrm{B}}^{*}$	bulk concentration of component B in solution, mol dm^{-3}
$c_{\mathrm{B}}\ (\infty, t)$	bulk concentration of component B at time t , mol dm^{-3}
$c_{\mathrm{B}}\ (x = 0)$	concentration of species B at the electrode surface, mol dm^{-3}
$c_{\mathrm{B}}\ (0, t)$	concentration of species B at the electrode surface at time t, mol dm^{-3}
c_{f}	differential sensitivity of EQCM, Hz g^{-1}
c_{O}	concentration of species O (oxidized form), mol dm^{-3}
c_{R}	concentration of species R (reduced form), mol dm^{-3}
C	capacitance, F m^{-2}
C	differential capacitance, F m^{-2}
C_{a}	adsorption capacitance, F m^{-2}
C_{d}	differential capacitance of the electrochemical double layer, F m^{-2}
C_{D}	capacitance of the diffuse double layer, F m^{-2}
C_{f}	integral sensitivity of EQCM, Hz cm^2 g^{-1}
C_{g}	geometric capacitance, F
C_{H}	capacitance of the Helmholtz layer, F m^{-2}
C_{HF}	high-frequency capacitance, F m^{-2}

C_{LF}, C_L	low-frequency capacitance, F m^{-2}
d	thickness, m
D	diffusion coefficient, m^2 s^{-1}, cm^2 s^{-1}
D_B	diffusion coefficient of species B, m^2 s^{-1}
D_{KA}	diffusion coefficient of electrolyte KA, m^2 s^{-1}
D_O, D_R	diffusion coefficient of species O, R, m^2 s^{-1}
e	quantity of charge on the electron (elementary charge), C ($e = 1.60217653(14) \times 10^{-19}$ C)
ΔE	electric potential difference, V m^{-1}
E	electromotive force (E_{mf}, emf), V
E	electrode potential, V
$E^{\ominus}$	standard[1] potential of electrochemical half-cell reaction, V
$E^{\ominus}$	standard potential of electrode reaction (standard electrode potential), V
$E_c^{\ominus\prime}$	formal potential, V
E_{cell}	potential of electrochemical cell reaction, V
E_e	equilibrium electrode potential, V
E_F	Fermi energy (Fermi level), eV
E_i	initial potential, V
E_p	polarization potential, V
E_{pa}, E_{pc}	peak potential (anodic, cathodic in cyclic voltammetry), V
$E_{p/2}$	half-peak potential, i.e., potential at 1/2 of the peak current (in cyclic voltammetry), V
$E_{3/4}$	potential at 3/4 of the peak current (in cyclic voltammetry), V
$E_{1/4}$	potential at 1/4 of the peak current (in cyclic voltammetry), V
E_{rv}	reversal potential in cyclic voltammetry, V
$E_{1/2}$	half-wave potential, V
E_r	rest potential, V
$E_{\sigma=0}$	potential of zero charge, V
E_{ref}	potential of the reference electrode, V
f	F/RT
f	frequency, Hz
f	activity coefficient (molar fraction bases)
f_B, f_i	rational activity coefficient of species B, i
$f_\pm$	mean rational activity coefficient
f_0	fundamental frequency, Hz
f_r	roughness factor
F	force, N
F	Faraday constant, C mol^{-1} ($F = 9.64853383(83) \times 10^4$ C mol^{-1})

[1] In agreement with the IUPAC suggestions, we are using here the Plimsoll symbol $\ominus$ to indicate standard values. The Plimsoll is a circle with a vertical line, named after Samuel Plimsoll (1824–1898), and originally meant to mark the limit to which a ship may be loaded.

g	interaction parameter
G	Gibbs free energy, J
ΔG	change of Gibbs free energy, J
ΔG_{ads}	Gibbs free energy of adsorption, J
ΔG_s	Gibbs free energy of solvation, J
$\Delta^{\ddagger} G$	Gibbs free energy of activation, J
$\Delta^{\ddagger} G^{\ominus}$	Standard Gibbs free energy of activation, J
h	hydration number
h	Planck constant, J s
H	enthalpy, J
H_0	Hammett-acidity function
ΔH	enthalpy change, J
ΔH_s	solvation enthalpy change, J
I	electric current, A
I	ionic strength, mol kg^{-1}, mol dm^{-3}
I_a	anodic current, A
I_c	cathodic current, A
I_d	diffusion current, A
I_L	limiting current, A
I_m	migration current, A
I_m	amplitude of sinusoidal current, A
I_0	exchange current, A
j	electric current density, A m^{-2}
j_0	exchange current density, A m^{-2}
j_a, j_c	anodic, cathodic current density, A m^{-2}
j_D	diffusion current density, A m^{-2}
j_L	limiting current density, A m^{-2}
J	flux, mol m^{-2} s^{-1}
J_B	flux of species B, mol m^{-2} s^{-1}
k	rate constant for reaction of the n -th order $\left(m^3\ mol^{-1}\right)^{n-1}$ s^{-1}($n = 1,2,\ldots$)
k	rate constant of heterogeneous reaction, m s^{-1}
k_a	rate constant of adsorption, m s^{-1}
k_{an}, k_{cn}	anodic, cathodic rate constants for n -th order reactions, A m$^{(3n\text{-}2)}$ mol^{-n}
k_B	Boltzmann constant, J K^{-1} (k_B= 1.3806504(24) $\times$ 10^{-23} J K^{-1})
k_d	rate constant of desorption, mol s^{-1} cm^{-2}
k_d	rate constant of diffusion mass transport, m s^{-1}
k_m	mass transport rate constant, m s^{-1}
k_{ox}	heterogeneous rate constant for oxidation, m s^{-1}
k_{red}	heterogeneous rate constant for reduction, m s^{-1}
k_s	standard heterogeneous rate constant, m s^{-1}

K	equilibrium constant
K_a	thermodynamical equilibrium constant
K_c	conditional equilibrium constant
L	thickness, length, distance, m
m	mass, g, kg
m_e	mass of the electron, kg
m_i	molarity of species i, mol kg^{-1}
M	molar mass, kg mol^{-1}
M_i	relative molar mass of species i
n	charge number of electrochemical cell reaction
n	reaction order
n	refractive index
n_i	number of density of species i, mol
N_A	Avogadro constant, mol^{-1} ($6.02214179(30)\times 10^{23} mol^{-1}$)
p	pressure, Pa
pH	negative decadic logarithm of the relative activity of H_3O^+ ions
pK_a, pK_b	negative decadic logarithm of acidity and base constant resp.
P	pressure, Pa
Q	charge, C
Q	constant phase element (CPE) coefficient
r	radius, m
r	inhomogeneity factor
r_i	radius of species i, m
r_{kr}	crystallographic radius of species i, m
r_0	radius of a disc microelectrode, m
r_{st}	Stokes radius, m
R	gas constant, J mol^{-1} K^{-1} (8.314472(15) J mol^{-1} K^{-1})
R_a	adsorption resistance, Ωm^2
R_b	cell resistance, Ω
R_{ct}	charge transfer resistance, Ω
R_d	diffusion resistance, Ω
R_e	Reynolds number
R_m	metal wire resistance, Ω
R_p	polarization resistance, Ω
R_s	solution resistance, Ω
R_{sa}	relative surface area ($R_{sa} = A_{3D}/A_{2D}$), A_{3D} is the "real surface area, "A_{2D} is the projected surface area
R_u	uncompensated ohmic resistance, Ω
R_Ω	ohmic resistance, Ω
S	entropy, J K^{-1}
t	time, s
t_i	transport number of species i

T	(thermodynamical or absolute) temperature, K
u_i	mobility of ion i, $m^2\ V^{-1}\ s^{-1}$
u_0	linear velocity of the liquid, $m\ s^{-1}$
U	internal energy, J
U	electric potential, potential difference, V
U_m	amplitude of sinusoidal voltage, V
v	potential scan rate, $V\ s^{-1}$
v	rate of a chemical reaction, $mol\ s^{-1}$, $mol\ s^{-1}\ cm^{-3}$
v_a	rate of adsorption, $mol\ s^{-1}\ cm^{-2}$ (variable)
v_b	rate of homogeneous reaction, $mol\ s^{-1}\ cm^{-3}$
v_d	rate of desorption, $mol\ s^{-1}\ cm^{-2}$ (variable)
$\bar{v}_i$	mean velocity of species i, $m\ s^{-1}$
v_{ox}	rate of oxidation, $m\ s^{-1}\ cm^{-2}$
v_{red}	rate of reduction, $m\ s^{-1}\ cm^{-2}$
$v(x)$	velocity of the volume element of a fluid in x direction $m^3\ s^{-1}$, $kg\ s^{-1}$
V	volume, m^3
w_{el}	electrical work, J
W	Warburg impedance, Ω
x_h	thickness of the Prandtl layer, m
x_H	thickness of the Helmholtz layer, m
x_i	molar fraction of species i
y_i	activity coefficient of component i (concentration basis)
$y_\pm$	mean ionic activity coefficient
Y	admittance, $S\ m^{-2}$
z_i	charge number of an ion i
Z	impedance, Ω, Ωm^2
Z'	real part (in-phase) impedance Ω, Ωm^2
Z''	imaginary part (out of phase) impedance, Ω, Ωm^2
Z_{dl}	impedance of the double layer, Ωm^2
Z_{el}	electrode impedance, Ωm^2
Z_F	Faraday impedance, Ωm^2
Z_R	solution impedance, Ωm^2

II. Greek Symbols

α	degree of dissociation
α	transfer coefficient (electrochemical)
α	real potential, V
α_a, α_c	anodic, cathodic transfer coefficient
α_f	constant phase element (CPE) exponent
δ	thickness of diffusion layer
δ_i	thickness of diffusion layer related to species i, m
γ	surface tension, J m^{-2}
γ_i	activity coefficient of species i (molality basis)
$\gamma_\pm$	mean ionic activity coefficient
Γ_i	surface (excess) concentration of species i, mol m^{-2}
$\Gamma_{i(H_2O)}$	surface (excess) concentration of species i related to water (solvent), mol m^{-2}
ε	molar (decadic) absorption coefficient, m^2 mol^{-1}, or L mol^{-1} cm^{-1}
ε	dielectric permittivity, Fm^{-1} $(C\ V^{-1}\ m^{-1})$
ε_0	permittivity of vacuum, F m^{-1} ($8.8541878176 \times 10^{-12}$ F m^{-1})
$\varepsilon_{0(\omega \to 0)}$	static relative permittivity
$\varepsilon_{\infty(\omega \to \infty)}$	optical relative permittivity
ε_r	relative permittivity
ϑ	phase angle, °, rad
η	overpotential, V
η	dynamic viscosity, P (g $cm^{-1}s^{-1}$)
η_{diff}	diffusion overpotential, V
Θ	surface coverage
Θ_i	surface coverage of species i
χ	conductivity, S m^{-1}
χ	surface electric potential, V
χ	transmission coefficient
χ^{-1}	radius of ionic atmosphere, Debye length, m
λ_i	molar ionic conductivity of ion i, S m^2 mol^{-1}
λ_B	absolute activity of species B
Λ	molar conductivity of an electrolyte, S m^2 mol^{-1}
Λ^0	molar conductivity at infinite dilution, S m^2 mol^{-1}
μ	electric dipole moment, C m
μ_i	chemical potential of species i, J mol^{-1}
$\mu_i^{\ominus}$	standard chemical potential of species i, J mol^{-1}
$\tilde{\mu}_i$	electrochemical potential of species i, J mol^{-1}
ν	kinematic viscosity, m^2 s^{-1}
ν_i	stoichiometric number of species i

ρ	density, kg dm^{-3}, g cm^{-3}
ρ	charge density, C m^{-3}
ρ	resistivity, Ω m
σ	surface charge density, C m^{-2}
σ	Warburg coefficient, Ω cm^2 s$^{-1/2}$
τ	characteristic time, relaxation time, s
φ	potential, V
$\Delta\varphi_{\text{diff}}$	junction (diffusion) potential, V
ϕ^{α}	inner electric potential of phase α, V
$\Delta\phi$	Galvani-potential difference, V
Φ	work function $\Phi = E - E_{\text{F}}$, eV
Ψ^{α}	outer electric potential of phase α, V
$\Delta\Psi$	Volta potential difference (contact potential), V
ω	angular frequency $\omega = 2\pi f$, Hz
ω_{r}	angular velocity, rad s^{-1}

Part I
Basic Electrochemistry

Chapter I.1
The Electrical Double Layer and Its Structure

Zbigniew Stojek

I.1.1 Introduction

At any electrode immersed in an electrolyte solution, a specific interfacial region is formed. This region is called the double layer. The electrical properties of such a layer are important, since they significantly affect the electrochemical measurements. In an electrical circuit used to measure the current that flows at a particular working electrode, the double layer can be viewed as a capacitor. Figure I.1.1 depicts this situation where the electrochemical cell is represented by an electrical circuit and capacitor C_d corresponds to the differential capacity of the double layer. To obtain a desired potential at the working electrodes, the double-layer capacitor must be first appropriately charged, which means that a capacitive current, not related to the reduction or oxidation of the substrates, flows in the electrical circuit. While this capacitive current carries some information concerning the double layer and its structure, and in some cases can be used for analytical purposes, in general, it interferes with electrochemical investigations. A variety of methods are used in electrochemistry to depress, isolate, or filter the capacitive current.

Although many models for the double layer have been published in the literature, there is no general model that can be used in all experimental situations. This is because the double-layer structure and its capacity depend on several parameters such as electrode material (metals, types of carbon, semiconductors, material porosity, the presence of layers of either oxides or polymeric films or other solid materials at the surface), type of solvent, type of supporting electrolyte, extent of specific adsorption of ions and molecules, and temperature.

The composition of the double layer influences the electron transfer rate (see Sect. I.3.1.5). Some ions and molecules specifically adsorbed at the electrode surface enhance the rate of the electrode process. In such a situation, we talk about heterogeneous electrocatalysis. On the other hand, there are numerous compounds that, after adsorption, decrease the electron transfer rate and therefore are simply

Z. Stojek (✉)
Department of Chemistry, University of Warsaw, Pasteura 1, 02-093 Warsaw, Poland
e-mail: stojek@chem.uw.edu.pl

F. Scholz (ed.), *Electroanalytical Methods*, 2nd ed.,
DOI 10.1007/978-3-642-02915-8_1,

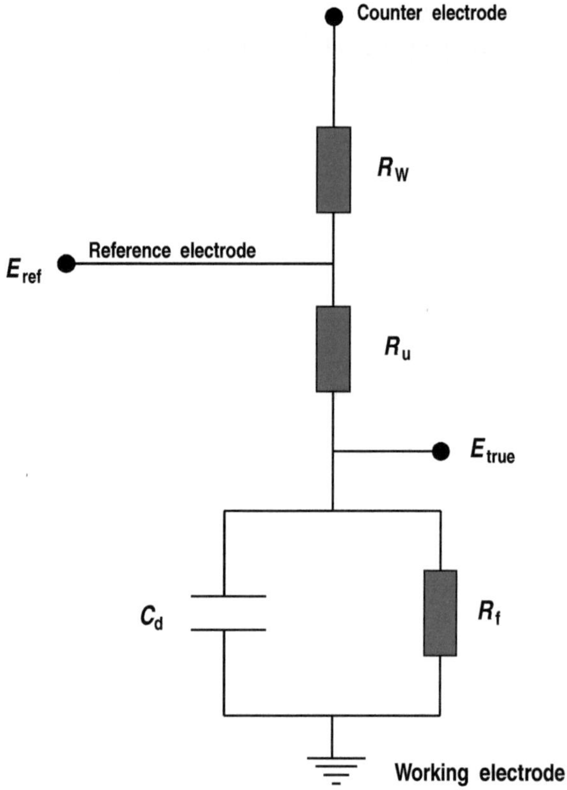

Fig. I.1.1 A simple electronic scheme equivalent to the electrochemical cell. R_u, resistance uncompensated in the regular three-electrode system; C_d, differential capacity of the double layer; R_f, resistance to faradaic current at the electrode surface; R_Ω, solution resistance compensated in the three-electrode system

inhibitors. Some surface-active compounds can be very strongly adsorbed. This may lead to the total isolation of the electrode surface and, finally, to the disappearance, or substantial decrease, of the voltammetric peaks or waves.

An imposition of a potential from an external source (potentiostat/voltammograph) to a metallic electrode results in generation of a charge, σ_M, on the metal, and a charge σ_S in the solution. The charge at the electrode is related directly to the interfacial (double-layer) capacity or capacitance. There are two ways to describe the capacity of an electrode:

- the differential capacitance, C_d, which naturally is at the minimum for the potential of zero charge and which is represented by Eq. (I.1.1):

$$C_d = \frac{\partial \sigma_M}{\partial E} \tag{I.1.1}$$

- and the integral capacitance, C_i, described by Eq. (I.1.2).

$$C_{\mathrm{i}} = \frac{\sigma_{\mathrm{M}}}{E - E_{\sigma=0}} \tag{I.1.2}$$

The excess charge on the metallic electrode, σ_{M}, is a function of the electrode potential. The simplest equation that describes the charge on the metal is given for mercury electrodes. This is because the excess charge strongly affects the surface tension of mercury, and the latter can be easily measured experimentally. One simple method to measure the surface tension vs. potential is to measure the drop time of a mercury-dropping electrode immersed in an electrolyte solution. The surface tension of mercury plotted vs. potential usually gives a parabolic curve. The maximum of this curve is located at the potential of zero charge, $E_{\sigma = 0}$, since

$$-\sigma_{\mathrm{M}} = \frac{\partial \gamma}{\partial E} \tag{I.1.3}$$

and the derivative of the surface tension equals 0 at the maximum. The differential capacity, C_{d}, reaches its minimum also at the potential of zero charge, a fact that can be concluded from a simple inspection of Eq. (I.1.1).

I.1.2 Double-Layer Models

The concept of the existence of the double layer at the surface of a metal being in contact with an electrolyte appeared in 1879 (Helmholtz). That first theoretical model assumed the presence of a *compact layer* of ions in contact with the charged metal surface. The next model, of Gouy and Chapman, involves *a diffuse double layer* in which the accumulated ions, due to the Boltzmann distribution, extend to some distance from the solid surface. In further developments, Stern (1924) suggested that the electrified solid–liquid interface includes both the rigid Helmholtz layer and the diffuse one of Gouy and Chapman. Specific adsorption of ions at the metal surface was pointed out by Graham in 1947. In consecutive developments, the role of the solvent has been taken into account (Parsons, 1954; Bockris, 1963). It soon became clear that in dipolar solvents, such as water, the dipoles must interact with the charged metal surface. It is also worth noting here that these interactions are promoted by the high concentration of the solvent, which is usually at least several moles per liter, and, in particular, for water it is around 55.5 M. In his theory, Parsons recognized that the dielectric constant of the solvent in the compact layer of adsorbed molecules is much lower compared to the outer region and approaches the limiting Maxwell value. A detailed description of the double-layer models mentioned above can be found in the literature [1–4].

A classic, simplified model of the double layer formed at the metal electrode surface is presented in Fig. I.1.2. There is a layer of adsorbed water molecules on the electrode surface. Since it has been assumed that there is excess of negative charge at the metal phase, the hydrogen atoms of adsorbed water molecules are oriented toward the metal surface. However, it is not a prerequisite that all water molecules

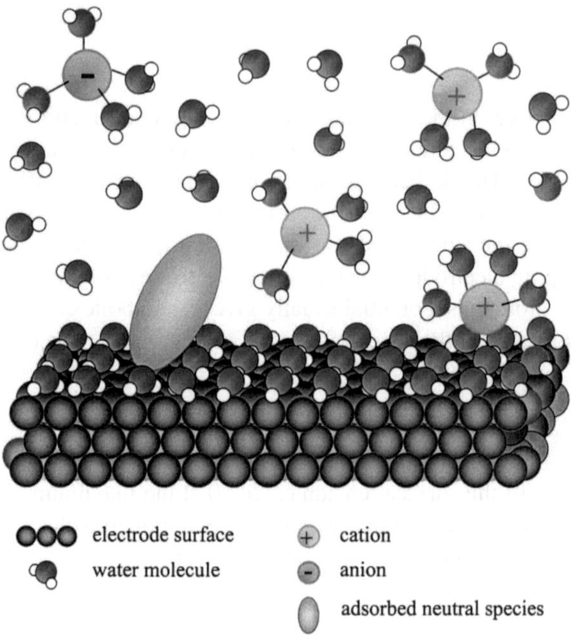

Fig. I.1.2 General representation of the double layer formed at the metal–electrolyte interface

at a particular electrode potential and the corresponding excess charge have the same orientation. For excess of positive charge at the metal surface, the dipoles of water will have a different orientation: the oxygen atoms of adsorbed water molecules will be oriented toward the metal surface. A specifically adsorbed large neutral molecule is also shown in Fig. I.1.2. Note that this molecule has removed some water molecules from the surface. The situation is different when a hydrated cation is present at the surface: it has not removed surface water and therefore cannot be considered as specifically adsorbed.

Two planes are usually associated with the double layer. The first one, the inner Helmholtz plane (IHP), passes through the centers of specifically adsorbed ions (compact layer in the Helmholtz model), or is simply located just behind the layer of adsorbed water. The second plane is called the outer Helmholtz plane (OHP) and passes through the centers of the hydrated ions that are in contact with the metal surface. The electric potentials linked to the IHP and OHP are usually written as Ψ_2 and Ψ_1, respectively The diffuse layer develops outside the OHP. The concentration of cations in the diffuse layer decreases exponentially vs. the distance from the electrode surface. The hydrated ions in the solution are most often octahedral complexes; however, in Fig. I.1.2. they are shown as tetrahedral structures for simplification.

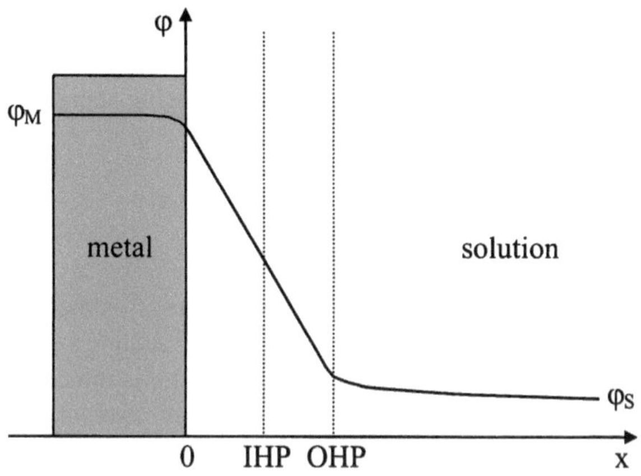

Fig. I.1.3 Potential profile in the double layer formed at a metallic electrode charged negatively

The change in the electric potential within the double layer is illustrated in Fig. I.1.3. It is assumed that the electrode is charged negatively. The electric potential, ϕ_M, is virtually constant throughout the metallic phase except for the layers of metal atoms located next to the solution, where a discontinuity in the metal structure takes place and the wave properties of the electron are exposed (the jellium model [1, 3]). This effect is much stronger in semiconductor electrodes, where the accessible electronic levels are more restricted [5].

At carbon electrodes, which are widely used in electrochemistry, the double layer develops too; however, these electrodes have some specific interfacial properties. The two main types of carbon electrodes, glassy carbon and highly oriented pyrolitic graphite (HOPG) and the recently introduced boron-doped diamond, differ much in the bulk and the surface structure. They also differ in electrochemical activity. Particularly big differences exist for the two surfaces of highly oriented pyrolitic graphite: the basal (hexagonal) plane and the edge one. At the edge plane, the electrode processes are usually much faster. An additional important factor for the electron transfer rate is the presence of oxygen at the surface. Oxygen easily chemisorbs on sp^2 carbon present in graphitic materials. This leads to the formation of many functional groups, mainly carbonyl, phenolic, and carboxylate, to an increase in the electrode differential capacity, and to an increase in the rate of the electrode processes. To reverse the chemisorption of oxygen, and to obtain, reproducibly, oxygen-free surfaces, is not easy. Neither is it easy to keep the surface oxygen-to-carbon ratio constant in the consecutive experiments. A positive aspect of the presence of the functional groups at the graphitic surfaces is that they make the chemical modification of the electrodes easier.

Details of the properties of carbon electrodes can be found in the literature [6]. For the effect of the double-layer structure on the electrode reaction rate see Sect. I.3.2.5.

I.1.3 Thickness of the Electric Double Layer

The thickness of the double layer is usually given as being approximately 1.5 κ^{-1}, where κ^{-1} is the Debye–Hückel length:

$$\kappa^{-1} = \left(\varepsilon_{\mathrm{r}}\varepsilon_0 k_{\mathrm{B}} T / 2c^0 z_{\mathrm{i}}^2 e^2\right)^{1/2} \tag{I.1.4}$$

where c^0 is the bulk concentration of the z:z electrolyte, ε_{r} is the relative dielectric permittivity of the solvent, ε_0 is the permittivity of the vacuum, k_{B} is the Boltzmann constant, T is the temperature, z is the ion charge, and e is the elementary charge. For $z = 1$, the approximate κ^{-1} values calculated for electrolyte concentrations of 1×10^{-3}, 1×10^{-5}, and 1×10^{-7} M are 10 nm, 100 nm, and 1 μm, respectively. The thickness of the double layer also depends on the potential: the larger the difference between the electrode potential and the potential of zero charge (the potential at which the excess charge on the electrode equals zero), the smaller the Debye–Hückel length.

I.1.4 Recent Developments

There is still much to do to be able to predict the behavior and the capacitance of the double layer in the entire available potential window and under all conditions. The introduction of rigorous theories that can take into account the various chemical and electrochemical properties of electrode materials, the specific adsorption of ions and neutral molecules, and the dynamics of adsorbed solvent molecules and other adsorbates is not trivial. In consequence, there is still no satisfactory agreement between the experimental and theoretical data regarding capacitance of the double layer. Hopefully, the new experimental techniques, such as atomic force and scanning tunneling microscopies [7], and scanning electrochemical microscopy [8], will allow electrochemists to learn more about the structure of the double layer at the atomic level. On the theoretical side, the new digital methods of calculations provide a possibility to simulate, in time, all the changes within the double layer. The recent progress in the research on the solid–liquid electrochemical interfaces is given in, e.g., [9] and [10].

References

1. Brett CMA, Brett AMO (1993) Electrochemistry: principles, methods, and applications. Oxford University Press, Oxford
2. Bard AJ, Faulkner RF (2000) Electrochemical methods, 2nd edn. John Wiley, New York
3. Parsons R (1990) Electrical double layer: recent experimental and theoretical developments. Chem Rev 90: 813
4. Trasatti S (1985) In: Silva AF (ed) Trends in interfacial electrochemistry. Proceedings of NATO ASI (1984), Reidel, Dordrecht, pp 25–48

5. Morrison SR (1980) Electrochemistry at semiconductor and oxidised metal electrodes. Plenum, New York
6. McCreery RL (1999) Electrochemical properties of carbon surfaces. In: Wieckowski A (ed) Interfacial electrochemistry. Theory, experiment, and applications. Marcel Dekker, New York, pp 631–647
7. Moffat TP (1999) Scanning tunnelling microscopy studies of metal electrodes. In: Bard AJ, Rubinstein I (eds) Electroanalytical chemistry, vol 21. Marcel Dekker, New York
8. Bard AJ, Fan Fu-Ren, Mirkin MV (1994) Scanning electrochemical microscopy. In: Bard AJ, Rubinstein I (eds) Electroanalytical chemistry, vol 18. Marcel Dekker, New York
9. Jerkiewicz G (1997) From electrochemistry to molecular-level research on the solid-liquid electrochemical interface. An overview. In: Jerkiewicz G, Soriaga MP, Uosaki K, Wieckowski A (eds) Solid-liquid electrochemical interfaces. American Chemical Society, Washington
10. Philpott MR, Glosli JN (1997) Molecular dynamics simulation of interfacial electrochemical processes: electric double layer screening. In: Jerkiewicz G, Soriaga MP, Uosaki K, Wieckowski A (eds) Solid-liquid electrochemical interfaces. American Chemical Society, Washington

Chapter I.2
Thermodynamics of Electrochemical Reactions

Fritz Scholz

I.2.1 Introduction

The wish to determine thermodynamic data of electrochemical reactions and of the involved compounds is one of the most important motivations to perform electrochemical measurements. After calorimetry, electrochemistry is the second most important tool to determine thermodynamic data. Although ab initio quantum chemical calculations can be used for the calculation of thermodynamic data of small molecules, the day is not yet foreseeable when electrochemical experiments will be replaced by such calculations. In this chapter we provide the essential information as to what thermodynamic information can be extracted from electrochemical experiments and what the necessary prerequisites are to do so.

The first step in this discussion is to distinguish between the thermodynamics and kinetics of an electrochemical reaction. Thermodynamics only describes the changes in energy and entropy during a reaction. The interplay between these two fundamental state functions determines to what extent a reaction will proceed, i.e., what the equilibrium constant is. Nothing can be said about the rate at which this equilibrium state can be reached, and nothing can be said about the mechanism of the proceeding reaction. In general, thermodynamic information can only be obtained about systems that are in equilibrium, or at least very near to equilibrium. Since electrochemical reactions always involve the passage of current, it is in many cases easy to let a reaction proceed near to the equilibrium by limiting the current, i.e., the passage of charge per time, which is nothing else but the reaction rate.

In this chapter, no attempt is made to provide a comprehensive account of electrochemical thermodynamics; but rather a survey of what is essential to understand the thermodynamic information provided by electroanalytical techniques. The fundamentals of electrochemical thermodynamics are available elsewhere [1].

F. Scholz (✉)
Institut für Biochemie, Universität Greifswald, 17487 Greifswald, Germany
e-mail: fscholz@uni-greifswald.de

F. Scholz (ed.), *Electroanalytical Methods*, 2nd ed.,
DOI 10.1007/978-3-642-02915-8_2,

I.2.2 The Standard Potential

The electroanalytical techniques considered in this volume are such that one always measures an electrode potential–current relationship, which is determined by the electrochemical reaction proceeding at one electrode only, i.e., the so-called working electrode. Of course, the same current must flow through the counter, or auxiliary, electrode as well; however, the experiments are designed in such a way that the process at the counter electrode is not rate determining. To give an example, when a platinum disc electrode of 1 mm diameter is used as the working electrode and the counter electrode is a sheet of platinum with a surface area of 4 cm^2, and the solution contains 10^{-3} mol L^{-1} $K_4[Fe(CN)_6]$ and 0.1 mol L^{-1} KNO_3, the dependence of current on electrode potential will be determined by the following electrochemical reaction only:

$$\left[\mathrm{Fe}\,(\mathrm{CN})_6\right]^{4-} \rightleftarrows \left[\mathrm{Fe}\,(\mathrm{CN})_6\right]^{3-} + \mathrm{e}^- \qquad (\mathrm{I.2.1})$$

Of course, on the counter electrode, another electrochemical reaction proceeds, the nature of which remains usually unknown. Let us assume that we measure a cyclic voltammogram (Fig. I.2.1), so that, in the first potential scan going in the positive direction, the hexacyanoferrate(II) ions are oxidized at the working electrode to hexacyanoferrate(III). The counterbalancing reaction at the second (auxiliary) electrode is not known; however, it is probable that hydronium ions of the water are reduced to hydrogen. In the following scan to negative potentials, the hexacyanoferrate(III) ions formed in the first scan are reduced back to hexacyanoferrate(II). Here the counterbalancing reaction on the auxiliary electrode may be an oxidation of the adsorbed hydrogen or the oxidation of the hexacyanoferrate(II) ions, which are also present at the auxiliary electrode (provided that this electrode is, as normally, in the same solution as the working electrode). The fact that we do not know what happens at the counter electrode, and, even worse, the fact that *different* processes may occur on the counter electrode, would make it very hard to use such electrode potential–current relationships for the determination of thermodynamic data, if we could not provide conditions where only the current at the working electrode is determining the measured response: this is achieved by using a working electrode having a surface area which is much smaller than that of the auxiliary electrode. Such ratio of electrode surface areas assures that the electrode reaction at the working electrodes limits the measured currents. Further, it is advantageous to control the electrode potential of the working electrode always versus the potential of an electrode having a fixed and stable reference potential; therefore, such measurements are nowadays always performed in a three-electrode arrangement: A third electrode, the so-called reference electrode (see Chap. III.2), is in electrolytic contact with the solution to be studied, only for the purpose to control the potential of the working electrode throughout the experiment. Practically no current is allowed to flow through the reference electrode and its construction is such that its potential is constant (equilibrium potential) under all conditions, in particular, independent of the composition of the solution being studied. Since

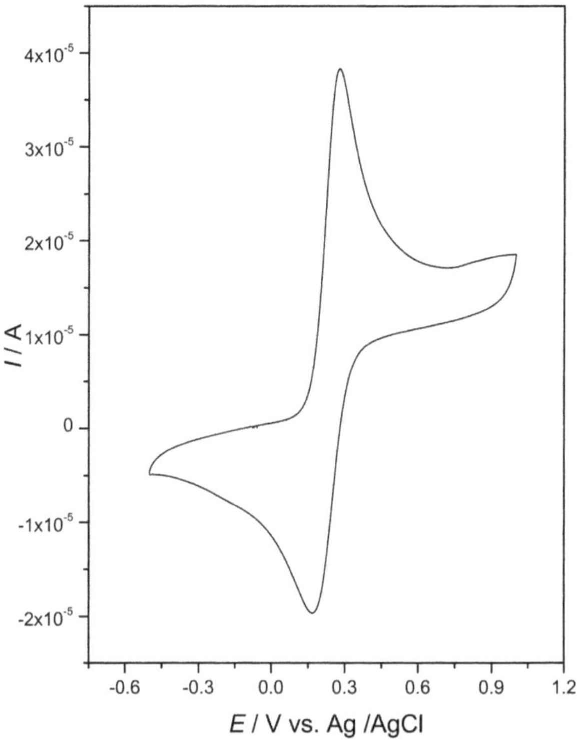

Fig. I.2.1 Cyclic voltammogram (fifth cycle) of a solution containing 6.25×10^{-3} mol L^{-1} $K_4Fe(CN)_6$. Pt working electrode with 3 mm diameter; electrolyte: 0.5 mol L^{-1} KNO_3; scan rate: 20 mV s^{-1}

the potential of the working electrode is always referred to that of the reference electrode, one has to recognize that the electrochemical reaction at the reference electrode is the theoretically counterbalancing reaction for the process studied. This means that the cyclic voltammogram shown in Fig. I.2.1 corresponds to the following cell reaction, provided that the reference electrode is a silver/silver chloride electrode:

$$\left[\mathrm{Fe\,(CN)_6}\right]^{3-} + \mathrm{Ag_{met}} + \mathrm{Cl^-} \rightleftarrows \left[\mathrm{Fe\,(CN)_6}\right]^{4-} + \mathrm{AgCl} \qquad (I.2.2)$$

Usually, reference electrodes are chosen for convenience, and the potentials may be recalculated versus the standard hydrogen electrode (SHE), which was selected as the zero point of the potential scale. When this is done for the given example, the following reaction is considered:

$$\left[\mathrm{Fe\,(CN)_6}\right]^{3-} + \tfrac{1}{2}\mathrm{H_2} \rightleftarrows \left[\mathrm{Fe\,(CN)_6}\right]^{4-} + \mathrm{H^+} \qquad (I.2.3)$$

This, of course, is also a chemical reaction and it could proceed without any electrodes in a solution. However, in our experiment, the oxidation and reduction are proceeding at separate electrodes, which have the task of transferring the electrons. The electrical work w_{el} that can be done by this system is

$$w_{el} = -Q\Delta E = -nF\Delta E \quad \text{(I.2.4)}$$

where ΔE is the potential difference between the electrodes, Q is the transported charge, which is n times the Faraday constant (96,484.6 C mol^{-1}), and n is the number of electrons transferred within the reaction ($n = 1$ in reaction (I.2.3)). Fundamental thermodynamics tell that the electrical work equals the change in Gibbs free energy (this quantity is also called free energy, or Gibbs energy):

$$w_{el} = \Delta G_{T,p} \quad \text{(I.2.5)}$$

The subscripts T and p indicate that this holds true for constant temperature and pressure, a condition which can be realized in electrochemical reactions. Conventional electrochemistry treatise would now discuss a cell in which the reaction (I.2.3) takes place. This could be a cell where equilibrium has been established, which in the example means that the species on the right side of Eq. (I.2.3) are strongly predominating, and, by application of a potential difference, the reaction is driven to the left side. This case is called electrolysis. The other possibility is realized when a hexacyanoferrate(III) solution is in one electrode compartment and the other compartment contains a platinum electrode around which hydrogen gas is bubbled. In that case a current flow will be observed to establish equilibrium conditions, i.e., to drive the reaction to the right side. This case is called a galvanic cell. In a cyclic voltammetric experiment (the recorded voltammogram is shown in Fig. I.2.1), the potential of the working electrode is changed in a controlled manner, first from left to right (to positive potentials) and later from right to left (to negative potentials) and the current response is measured. The current flow is the consequence of a fundamental dependence of the ratio of the activities of the hexacyanoferrate(III) and hexacyanoferrate(II) ions on the potential of the electrode:

$$E = E^{\ominus}_{[Fe(CN)_6]^{3-/4-}} + \frac{RT}{nF}\ln\frac{a_{[Fe(CN)_6]^{3-}}}{a_{[Fe(CN)_6]^{4-}}} \qquad (n = 1) \quad \text{(I.2.6)}$$

This equation is referred to as the Nernst equation. This equation requires that at each potential of the working electrode there is a specific value of the ratio:

$$\frac{a_{[Fe(CN)_6]^{3-}}}{a_{[Fe(CN)_6]^{4-}}}$$

To establish this ratio it is necessary to interconvert the involved ions, which is only possible by a flow of current. The Nernst equation follows from the requirement that reaction (I.2.1) is at equilibrium when the electrochemical potentials of reactants and products are equal:

$$\tilde{\mu}_{[\mathrm{Fe(CN)_6}]^{3-}} + \tilde{\mu}_{\mathrm{e}^-}^{\mathrm{solution}} = \tilde{\mu}_{[\mathrm{Fe(CN)_6}]^{4-}} \tag{I.2.7}$$

Since $\tilde{\mu}_{\mathrm{e}^-}^{\mathrm{metal}}$ is equal to $\tilde{\mu}_{\mathrm{e}^-}^{\mathrm{solution}}$ it follows that the second term on the left side of Eq. (I.2.7) is the electrochemical potential of the electrons in the inert metal electrode. The electrochemical potentials are connected with the chemical potentials according to

$$\tilde{\mu}_i^{\alpha} = \mu_i^{\ominus,\alpha} + RT \ln a_i^{\alpha} + zF\phi^{\alpha} \tag{I.2.8}$$

(ϕ^{α} is the inner electric potential of the phase α in which the species i are present (cf. Fig. I.2.2), $\mu_i^{\ominus,\alpha}$ is the standard chemical potential of the species i, and z is its charge). The electrochemical potential differs from the chemical potential only by the electric work, i.e., by the product 'charge times voltage.' The chemical potential of i is

$$\mu_i = \mu_i^{\ominus} + RT \ln a_i = \left(\frac{\partial G}{\partial n_i}\right)_{p,T,n_j} \tag{I.2.9}$$

i.e., the partial derivative of the Gibbs free energy over the change in the number of ions i. Writing Eq. (I.2.8) for all species of reaction (I.2.1) and introducing it into Eq. (I.2.7) yields

$$\begin{aligned} &\mu^{\ominus}_{[\mathrm{Fe(CN)_6}]^{4-}} + RT \ln a_{[\mathrm{Fe(CN)_6}]^{4-}} + z_{[\mathrm{Fe(CN)_6}]^{4-}} F\phi^{\mathrm{solution}} \\ &= \mu^{\ominus}_{[\mathrm{Fe(CN)_6}]^{3-}} + RT \ln a_{[\mathrm{Fe(CN)_6}]^{3-}} + z_{[\mathrm{Fe(CN)_6}]^{3-}} F\phi^{\mathrm{solution}} \\ &\quad + \mu_{\mathrm{e}^-}^{\ominus,\mathrm{metal}} + RT \ln a_{\mathrm{e}^-}^{\mathrm{metal}} + z_{\mathrm{e}^-} F\phi^{\mathrm{metal}} \end{aligned} \tag{I.2.10}$$

The standard potentials and activities without a phase index all relate to the solution phase. Since the following relation holds $z_{[\mathrm{Fe(CN)_6}]^{4-}} - z_{[\mathrm{Fe(CN)_6}]^{3-}} = z_{\mathrm{e}^-} = -n$, where n is the number of exchanged electrons, one can rearrange Eq. (I.2.10) as follows:

$$\begin{aligned} &z_{\mathrm{e}^-} F\phi^{\mathrm{metal}} + z_{[\mathrm{Fe(CN)_6}]^{3-}} F\phi^{\mathrm{solution}} - z_{[\mathrm{Fe(CN)_6}]^{4-}} F\phi^{\mathrm{solution}} \\ &= z_{\mathrm{e}^-} F\left(\phi^{\mathrm{metal}} - \phi^{\mathrm{solution}}\right) = -nF\left(\phi^{\mathrm{metal}} - \phi^{\mathrm{solution}}\right) \\ &= \mu^{\ominus}_{[\mathrm{Fe(CN)_6}]^{4-}} - \mu^{\ominus}_{[\mathrm{Fe(CN)_6}]^{3-}} + RT \ln a_{[\mathrm{Fe(CN)_6}]^{4-}} - RT \ln a_{[\mathrm{Fe(CN)_6}]^{3-}} \\ &\quad - \mu_{\mathrm{e}^-}^{\ominus,\mathrm{metal}} - RT \ln a_{\mathrm{e}^-}^{\mathrm{metal}} \end{aligned} \tag{I.2.11a}$$

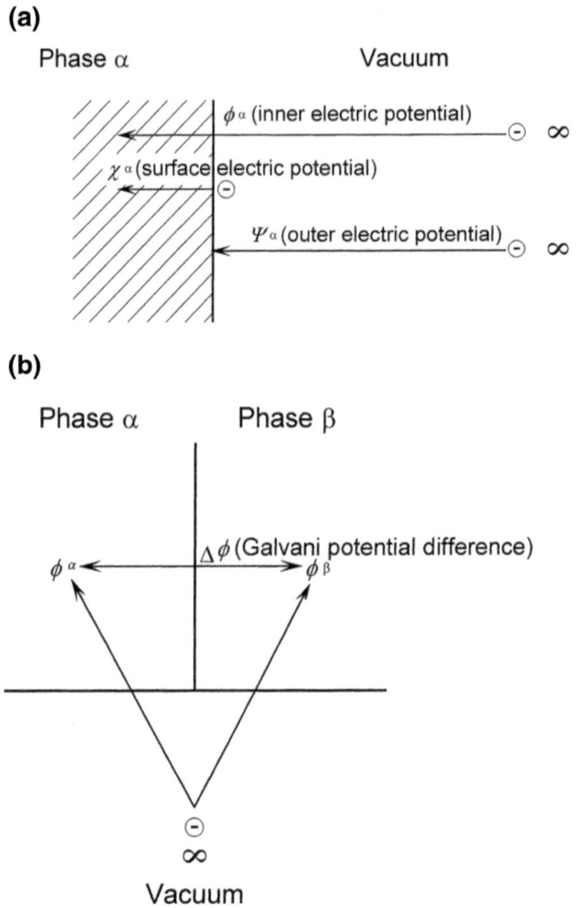

Fig. I.2.2 (**a**) Schematic situation at the border of a phase with vacuum. Ψ^{α} is the outer electric potential of phase α, i.e., the work that must be done when a unit charge is transferred from infinity (in the vacuum) to the surface of phase α. (The difference in the two outer electric potentials of two different phases is called the Volta potential difference.) χ^{α} is the surface electric potential of phase α, i.e., the work to be done when a unit charge is transferred from the surface into phase α, and ϕ^{α} is the inner electric potential of phase α, i.e., the work to be done when a unit charge is transferred from infinity (in vacuum) into the inner of phase α. ϕ^{α} is a nonmeasurable quantity, whereas Ψ^{α} can be calculated and measured. The three potentials are interrelated as follows: $\phi^{\alpha} = \Psi^{\alpha} + \chi^{\alpha}$. (**b**) Schematic situation at the interface of two phases α and β: The difference in inner electric potentials is called the Galvani potential difference $\Delta\phi$

The activity of electrons in the metal phase is 1 because they are in their standard state and $z_{e^-} = -1$. From Eq. (I.2.11a) follows the Nernst equation in the form

$$\Delta\phi = \phi^{\text{metal}} - \phi^{\text{solution}} = \frac{\mu^{\ominus}_{[\text{Fe(CN)}_6]^{3-}} - \mu^{\ominus}_{[\text{Fe(CN)}_6]^{4-}} + \mu^{\ominus,\text{metal}}_{e^-}}{nF} + \frac{RT}{nF}\ln\frac{a_{[\text{Fe(CN)}_6]^{3-}}}{a_{[\text{Fe(CN)}_6]^{4-}}} \qquad \text{(I.2.11b)}$$

Whereas a direct measurement of the inner electric potential of a single phase is impossible, the difference, i.e., the Galvani potential difference of two phases $\Delta\phi$ having identical composition or its variation for two phases having a common interface, is accessible when a proper reference electrode is used, i.e., a metal/electrolyte system, which should guarantee that the chemical potential of the species i is the same in both electrolytes, i.e., the two electrolytes contacting the metal phases I and II. In addition, the absence of a junction potential between the two electrolytes is required. Under such circumstances it is possible to measure a potential difference, ΔE, that is related to $\Delta\phi$; however, it always includes the $\Delta\phi$ of the reference electrode. The latter is set to zero for the Standard Hydrogen Electrode (see below). In fact, the standard chemical potential of the formation of solvated protons is zero by convention.

$$\Delta\phi = \phi^{\mathrm{metal}} - \phi^{\mathrm{solution}} = \Delta E = E^{\ominus}_{[\mathrm{Fe(CN)_6}]^{3-/4-}} + \frac{RT}{nF}\ln\frac{a_{[\mathrm{Fe(CN)_6}]^{3-}}}{a_{[\mathrm{Fe(CN)_6}]^{4-}}} \qquad \text{(I.2.12)}$$

The standard potential $E^{\ominus}$ is an important value as it is related to the standard Gibbs free energy of the reaction $\Delta G^{\ominus}_{T,p}$ and also to the equilibrium constant K according to

$$-nFE^{\ominus} = \Delta G^{\ominus}_{T,p} = -RT\ln K \qquad \text{(I.2.13)}$$

When one wants to calculate the equilibrium constant of reaction (I.2.3) from the standard potentials of the system hexacyanoferrate(II/III) and $2H^+/H_2$, it is essential that one writes this equation with the oxidized form of the system and hydrogen on the left side and the reduced form and protons on the right side. Only then does the sign convention hold true and Eq. (I.2.13) yields the equilibrium constant for the reaction when the tabulated standard potentials are used. Note also that the standard potential of the hydrogen electrode is 0 V for the reaction written as: $2H^+ + 2e^- \rightleftarrows H_2$, or written as $H^+ + e^- \rightleftarrows \frac{1}{2}H_2$. Table I.2.1 gives a compilation of standard potentials of electrode reactions. (Standard potentials are available from many different sources [2].) Although only single redox couples are listed, the standard potentials of each system always refer to the reaction:

$$\text{Oxidised form} + \text{hydrogen} \rightleftarrows \text{reduced form} + \text{hydronium ions} \qquad \text{(I.2.14)}$$

In many cases, standard potentials of electrode reactions can be determined by electrochemical measurements. However, this is not trivial for the following reasons: according to the Nernst equation, one will measure $E = E^{\ominus}$ when the activities of all species are 1, and, of course, at 25°C and 1 bar pressure. However, the activity condition is hard to realize as, at the high concentrations which would be necessary to realize it, the activity *coefficients* strongly deviate from 1. Therefore, one measures the potentials at concentrations orders of magnitude lower and extrapolates the linear part of the dependence to unit activities. The standard potential can also be calculated from the standard enthalpies and entropies of the involved species,

Table I.2.1 Standard potentials of electrode reactions[a]

Half cell	Electrode reaction	$E^{\ominus}$ (V vs. SHE)
$Li \mid Li^+$	$Li^+ + e^- \rightleftarrows Li$	−3.040
$Rb\mid Rb^+$	$Rb^+ + e^- \rightleftarrows Rb$	−2.924
$K\mid K^+$	$K^+ + e^- \rightleftarrows K$	−2.924
$Cs\mid Cs^+$	$Cs^+ + e^- \rightleftarrows Cs$	−2.923
$Ca\mid Ca^{2+}$	$Ca^{2+} + 2\,e^- \rightleftarrows Ca$	−2.76
$Na\mid Na^+$	$Na^+ + e^- \rightleftarrows Na$	−2.713
$Mg\mid Mg^{2+}$	$Mg^{2+} + 2\,e^- \rightleftarrows Mg$	−2.375
$Al\mid Al^{3+}$	$Al^{3+} + 3\,e^- \rightleftarrows Al$	−1.706
$Zn\mid Zn^{2+}$	$Zn^{2+} + 2\,e^- \rightleftarrows Zn$	−0.7628
$Pt\mid Cr^{2+}, Cr^{3+}$	$Cr^{3+} + e^- \rightleftarrows Cr^{2+}$	−0.41
$Fe\mid Fe^{2+}$	$Fe^{2+} + 2\,e^- \rightleftarrows Fe$	−0.409
$Cd\mid Cd^{2+}$	$Cd^{2+} + 2\,e^- \rightleftarrows Ni$	−0.4026
$Ni\mid Ni^{2+}$	$Ni^{2+} + 2\,e^- \rightleftarrows Ni$	−0.23
$Pb\mid Pb^{2+}$	$Pb^{2+} + 2\,e^- \rightleftarrows Pb$	−0.1263
$Pt\mid H_2, H^+_{aq}$	$2H^+ + 2\,e^- \rightleftarrows H_2$	0.0000
$Pt\mid Cu^{2+}, Cu^+$	$Cu^{2+} + e^- \rightleftarrows Cu^+$	+0.167
$Cu^{2+}\mid Cu$	$Cu^{2+} + 2\,e^- \rightleftarrows Cu$	+0.3402
$Pt\mid [Fe(CN)_6]^{3-}, [Fe(CN)_6]^{4-}$	$[Fe(CN)_6]^{3-} + e^- \rightleftarrows [Fe(CN)_6]^{4-}$	+0.356
$Pt\mid [W(CN)_8]^{3-}, [W(CN)_8]^{4-}$	$[W(CN)_8]^{3-} + e^- \rightleftarrows [W(CN)_8]^{4-}$	+0.457
$Pt\mid [Mo(CN)_8]^{3-}, [Mo(CN)_8]^{4-}$	$[Mo(CN)_8]^{3-} + e^- \rightleftarrows [Mo(CN)_8]^{4-}$	+0.725
$Ag\mid Ag^+$	$Ag^+ + e^- \rightleftarrows Ag$	+0.7996
$2\,Hg\mid Hg_2^{2+}$	$Hg_2^{2+} + 2\,e^- \rightleftarrows 2\,Hg$	+0.7961
$Pt\mid Cr_2O_7^{2-}, Cr^{3+}, H^+$	$Cr_2O_7^{2-} + 14\,H^+ + 6\,e^- \rightleftarrows 2\,Cr^{3+} + 7H_2O$	+1.36
$Pt\mid O_2, H_2O, H^+$	$1/2\,O_2 + 2H^+ + 2\,e^- \rightleftarrows H_2O$	+1.229
$Au^+\mid Au$	$Au^+ + e^+ \rightleftarrows Au$	+1.83
$Pt\mid MnO_4^-, Mn^{2+}, H^+$	$MnO_4^- + 8H^+ + 5\,e^- \rightleftarrows Mn^{2+} + 4H_2O$	+1.491
$Pt\mid H_4XeO_6, XeO_3$	$H_4XeO_6 + 2\,H^+ + 2\,e^- \rightleftarrows XeO_3 + 3\,H_2O$	+2.42
$Pt\mid F_2, F^-$	$F_2 + 2\,e^- \rightleftarrows 2\,F$	+2.866

[a] Some half-cells are given with platinum as the inert electrode; however, this is only taken as an example for an inert electrode and it does not mean that there is any dependence of the standard potentials on the electrode material. The standard potentials of dissolved redox systems are independent of the electrode material. This is opposite to the standard rate constants of electron transfer, which are very dependent on the electrode material. Please note also that many of the given standard potentials cannot be obtained by electrochemical measurements. They are calculated from thermodynamic data obtained, e.g., from calorimetry. The system $Pt\mid MnO_4^-$, Mn^{2+}, H^+ is irreversible not only on platinum but also on all other electrode materials. When a platinum wire is introduced into an acidic solution containing permanganate and manganese(II) ions, the measured potential is a so-called *mixed potential*. This term refers to the fact that it is the result of two different electrode reactions, the reduction of permanganate to some intermediate redox state (+6 or +5) and the oxidation of water to oxygen. Both processes occur with a certain exchange current density and the electrode attains a potential at which the cathodic and anodic current densities are equal, so that no net current flows. Hence the mixed potential depends on the kinetics of these two processes and it will more or less strongly deviate from the standard and formal potential of the two redox species constituting a possible redox pair.

which is the only possibility for such systems were a reversible electrochemical measurement is impossible (see footnote of Table I.2.1).

When hydronium or hydroxide ions are involved in redox equilibria without being themselves reduced or oxidized, it is essential to define standard potentials for the overall reaction, not only for the electron transfer equilibrium. An example is the following reaction:

$$Cr_2O_7^{2-} + 14\,H^+ + 6e^- \rightleftarrows 2Cr^{3+} + 7\,H_2O \tag{I.2.15}$$

with $E^{\ominus} = +1.35$ V vs. SHE. *The splitting of this composite equilibrium into a pure redox equilibrium and a pure acid–base equilibrium is senseless because the two are not experimentally feasible.* When protons or hydroxide ions are involved in a redox equilibrium only via acid–base equilibria, their activities are also defined as 1 for the standard potential. For biochemists, standard potentials which relate to pH 0 or 14 are not very useful, as biochemical reactions proceed at pH values around 7 (± 5 at most). Therefore, in biochemistry, another set of standard potentials E' was introduced, the so-called biochemical standard potentials, which refer to the standard state of H^+ and OH^- as 10^{-7} mol L^{-1}. The E' values of a reaction

$$\mathrm{Ox} + n\mathrm{H}^+ + me^- \rightleftarrows \mathrm{H}_n\mathrm{Red}^{(n-m)^+} \tag{I.2.16}$$

can be calculated from the standard potential $E^{\ominus}$ (defined for $a_{H^+} = 1$) of this reaction with the help of the following relationship:

$$E' = E^{\ominus} - 0.414[\mathrm{V}]\frac{n}{m} \tag{I.2.17}$$

This relationship holds true for 25°C (for details see [1a]).

I.2.3 The Formal Potential

Although the standard potentials are the fundamental values for all thermodynamic calculations, in practice, one has more frequently to deal with the so-called *formal potentials*. The formal potentials are *conditional* constants, very similar to the conditional stability constants of complexes and conditional solubility products of sparingly soluble salts (see [2c]). The term *conditional* indicates that these constants relate to *specific conditions*, which deviate from the usual *standard conditions*. Formal potentials deviate from standard potentials for two reasons, i.e., because of nonunity activity coefficients and because of chemical 'side reactions'. The latter should better be termed 'side equilibria'; however, this term is not in common use. Let us consider the redox system iron(II/III) in water:

$$Fe^{3+} + e^- \rightleftarrows Fe^{2+} \tag{I.2.18}$$

By common agreement this notation means that both iron(II) and iron(III) are present in the form of aqua complexes. The Nernst equation for reaction (I.2.18) is

$$E = E^{\ominus}_{[\mathrm{Fe(H_2O)_6}]^{3+/2+}} + \frac{RT}{F}\ln\frac{a_{[\mathrm{Fe(H_2O)_6}]^{3+}}}{a_{[\mathrm{Fe(H_2O)_6}]^{2+}}} \tag{I.2.19}$$

(Note that $n = 1$ in this case.) For the activities we can write $a_i = \gamma_i\left(c_i/c_i^{\mathrm{o}}\right)$ where γ_i is a dimensionless activity coefficient, c_i is the concentration of the species i in mol L^{-1}, and c° is the unit concentration 1 mol L^{-1}. Introducing this into the Nernst equation yields

$$E = E^{\ominus}_{[\mathrm{Fe(H_2O)_6}]^{3+/2+}} + \frac{RT}{F}\ln\frac{c_{[\mathrm{Fe(H_2O)_6}]^{3+}}}{c_{[\mathrm{Fe(H_2O)_6}]^{2+}}} + \frac{RT}{F}\ln\frac{\gamma_{[\mathrm{Fe(H_2O)_6}]^{3+}}}{\gamma_{[\mathrm{Fe(H_2O)_6}]^{2+}}} \tag{I.2.20}$$

It is easy to imagine that only in acidic solutions can both iron(II) and iron(III) be present as aqua complexes: the pK_{a1} ($= -\log K_{\mathrm{a1}}$; K_{a1} being the first acidity constant) of the iron(III) hexaqua complex $[\mathrm{Fe(H_2O)_6}]^{3+}$ is 3.1. Hence this is an acid almost two orders of magnitude stronger than acetic acid! $[\mathrm{Fe(H_2O)_6}]^{3+}$ strongly tends to transfer a proton to the solvent water and to become a $[\mathrm{Fe(H_2O)_5(OH)}]^{2+}$ ion. Other protons, although less acidic, may subsequently be transferred and the resulting hydroxo complexes will further tend to form polynuclear complexes. This reaction cascade may easily go on for hours and days and all the time there is no equilibrium established. If that happens, the Nernst equation cannot be applied at all. However, if this reaction cascade comes to a quick end, perhaps because the solution is rather acidic, a number of different iron(III) species may coexist in equilibrium with the iron(III) hexaqua complex $[\mathrm{Fe(H_2O)_6}]^{3+}$. In such cases it is useful to define a so-called side reaction coefficient $\alpha_{\mathrm{Fe(III)}}$ according to the following equation:

$$\alpha_{\mathrm{Fe(III)}} = \frac{c_{[\mathrm{Fe(H_2O)_6}]^{3+}}}{c_{\mathrm{Fe(III)_{total}}}} \tag{I.2.21}$$

Where $c_{\mathrm{Fe(III)_{total}}}$ is the sum of the concentrations of all iron(III) species. Formulating for the iron(II) species a similar equation to Eq. (I.2.21) and introducing both into Eq. (I.2.20) yields

$$\begin{aligned} E = E^{\ominus}_{[\mathrm{Fe(H_2O)_6}]^{3+/2+}} &+ \frac{RT}{F}\ln\frac{\alpha_{\mathrm{Fe(III)}}}{\alpha_{\mathrm{Fe(II)}}} + \frac{RT}{F}\ln\frac{c_{\mathrm{Fe(III)_{total}}}}{c_{\mathrm{Fe(II)_{total}}}} \\ &+ \frac{RT}{F}\ln\frac{\gamma_{[\mathrm{Fe(H_2O)_6}]^{3+}}}{\gamma_{[\mathrm{Fe(H_2O)_6}]^{2+}}} \end{aligned} \tag{I.2.22}$$

Because the total concentrations of iron(III) and iron(II) are analytically accessible values, and because the second and fourth term on the right side of Eq. (I.2.22) are constant under well-defined experimental conditions (i.e., when the solution has a constant composition), it is convenient to define a new constant, the *formal potential* $E^{\ominus\prime}_{\mathrm{c,Fe^{3+/2+}}}$, as follows:

$$E^{\ominus\prime}_{\mathrm{c,Fe^{3+/2+}}} = E^{\ominus}_{[\mathrm{Fe(H_2O)_6}]^{3+/2+}} + \frac{RT}{F}\ln\frac{\alpha_{\mathrm{Fe(III)}}}{\alpha_{\mathrm{Fe(II)}}} + \frac{RT}{F}\ln\frac{\gamma_{[\mathrm{Fe(H_2O)_6}]^{3+}}}{\gamma_{[\mathrm{Fe(H_2O)_6}]^{2+}}} \tag{I.2.23}$$

Equation (I.2.22) can now be written in the following way:

$$E = E^{\ominus\prime}_{\mathrm{c,Fe^{3+/2+}}} + \frac{RT}{F}\ln\frac{c_{\mathrm{Fe(III)_{total}}}}{c_{\mathrm{Fe(II)_{total}}}} \tag{I.2.24}$$

(The subscript c of $E^{\ominus\prime}_{\mathrm{c,Fe^{3+/2+}}}$ indicates that this is a conditional constant.) A formal potential characterizes an equilibrium between two redox states; however, one should never forget that it is strongly dependent on the solution composition, as side reactions (equilibria) and activity coefficients strongly influence it. If the solution would be so acidic that both Fe(II) and Fe(III) are present as hexaqua complexes, it may still be that the activity coefficients of these two species strongly deviate from 1 because either the concentrations of the two iron forms are rather high, or also because there are other electrolytes in high concentration present. In such case it would make sense to define a formal potential on the basis of Eq. (I.2.20) as follows:

$$E^{\ominus\prime}_{\mathrm{c,Fe^{3+/2+}}} = E^{\ominus}_{[\mathrm{Fe(H_2O)_6}]^{3+/2+}} + \frac{RT}{F}\ln\frac{\gamma_{[\mathrm{Fe(H_2O)_6}]^{3+}}}{\gamma_{[\mathrm{Fe(H_2O)_6}]^{2+}}},$$

i.e., only taking into account the deviations of the standard potential caused by the activity coefficients.

In defining the formal potential in Eq. (I.2.23), the side reactions are acid–base equilibria. Of course, all other kinds of chemical equilibria, e.g., complex formation and precipitation, have similar consequences. In the case of an electrode of the second kind, e.g., a calomel electrode, the so-called 'standard potential' of the calomel electrode is nothing but the formal potential of the electrode at $a_{\mathrm{chloride}} = 1$. The potential of the calomel electrode at various KCl concentrations is always the formal potential of this electrode at the specified concentration (see Chaps. II.9 and III.2).

The concept of formal potentials has been developed for the mathematical treatment of redox titrations, because it was quickly realized that the standard potentials cannot be used to explain potentiometric titration curves. Generally, formal potentials are experimentally determined using equations similar to Eq. (I.2.24) because it is easy to control the overall concentrations of species in the two redox states. For calculating formal potentials it would be necessary to know the standard potential, all equilibrium constants of 'side reactions', and the concentrations of all solution constituents. In many cases this is still impossible as many equilibrium constants and the underlying chemical equilibria are still unknown. It is the great advantage of the concept of formal potentials to enable a quantitative description of the redox equilibrium without the exact knowledge of the side reactions. Formal potentials are tabulated for strictly defined experimental conditions.

I.2.4 Characteristic Potentials of Electroanalytical Techniques

Each electroanalytical technique has certain characteristic potentials, which can be derived from the measured curves. These are the *half-wave potential* in direct current polarography (DCP), the *peak potentials* in cyclic voltammetry (CV), the *mid-peak potential* in cyclic voltammetry, and the *peak potential* in differential pulse voltammetry (DPV) and square-wave voltammetry. In the case of electrochemical reversibility (see Chap. I.3) all these characteristic potentials are interrelated and it is important to know their relationship to the standard and formal potential of the redox system. Here follows a brief summary of the most important characteristic potentials.

I.2.4.1 Direct Current Polarography (Employing a Dropping-Mercury Electrode)

I.2.4.1.1 The Half-Wave Potential $E_{1/2}$

There are four fundamentally different factors that will lead to a deviation between half-wave and standard potentials. The first one is related to the diffusion of the species towards the electrode and within the mercury drop. These diffusion processes are also influenced by the sphericity of the mercury drop. The second factor is due to any amalgamation reaction, and the third factor is due to solution equilibria. The fourth factor, which will force the half-wave potential to deviate from the standard potential, is a possible irreversibility of the electrode system. Before discussing these four effects, the general equation relating the half-wave potential to the standard potential is given as

$$E_{1/2} = E^{\ominus} + \frac{RT}{nF} \ln \left[\frac{\gamma_{\mathrm{Ox}} D_{\mathrm{Red}}^{1/2}}{\gamma_{\mathrm{Red}} D_{\mathrm{Ox}}^{1/2}} \right] \tag{I.2.25}$$

I.2.4.1.2 Influence of Diffusion

The deviation expected as a result of (i) unequal diffusion coefficients of the oxidized and reduced species, and (ii) electrode sphericity can be described as follows [3]:

$$E_{1/2} = E^{\ominus} - 3.4 \frac{RT}{nF} \frac{t_1^{1/6}}{m^{1/3}} \left(D_{\mathrm{Ox}}^{1/2} + D_{\mathrm{Red}}^{1/2} \right) - \frac{RT}{nF} \ln \frac{D_{\mathrm{Ox}}^{1/2}}{D_{\mathrm{Red}}^{1/2}} \tag{I.2.26}$$

(t_1 is the drop time, m is the flow rate of mercury, and D are diffusion coefficients of the involved species. In Eq. (I.2.25), activity coefficients are not taken into account.) For a typical set of parameters, i.e., for equal diffusion coefficients (this is frequently a good approximation), $D_{\mathrm{Ox}} = D_{\mathrm{Red}} = 10^{-9}\ \mathrm{m^2 s^{-1}}$, a drop time of 1 s, a mercury flow rate of 1 mg s^{-1}, and $n = 1$ the result is that $E_{1/2} = E^{\ominus} - 0.0093\ \mathrm{V}$. This is indeed a rather small deviation. Equation (I.2.25) was derived taking into account the sphericity of the mercury drop. When the sphericity is neglected, the second term in Eq. (I.2.26) may be omitted.

I.2.4.1.3 Influence by Amalgamation

When the deviation caused by unequal diffusion coefficients can be neglected, it generally holds true that the half-wave potential equals the formal potential, i.e., $E_{1/2} = E_c^{\ominus'}$; however, the chemical system has a significant influence on how much the formal potential really deviates from the standard potential. Even in the case of very simple systems, the thermodynamic deviation between the measured half-wave potential and the standard potential can be quite large, as, e.g., for $Ba^{2+}/Ba_{amalgam}$ where the polarographic half-wave potential is −1.94 V vs. SCE and the standard potential of the system Ba^{2+}/Ba is −2.90 V vs. SCE. The reason for that deviation is the amalgamation of metallic barium. For amalgam-forming metals, the relationship between the half-wave potential and the standard potential is as follows:

$$E_{1/2} = E^{\ominus} - \frac{\Delta G^{\ominus}_{amal}}{nF} + \frac{RT}{nF}\ln a_{sat} + \frac{RT}{nF}\ln\frac{\gamma_{Ox}D^{1/2}_{Red}}{\gamma_{Red}D^{1/2}_{Ox}} \tag{I.2.27}$$

Where $\Delta G^{\ominus}_{amal}$ is the standard Gibbs free energy of amalgam formation and a_{sat} is the activity of the metal in the mercury at saturation. (In the derivation of this equation it has been assumed that the activity of mercury is not altered by the amalgam formation [4].) Note that the very negative standard Gibbs free energy of amalgam formation of barium shifts the half-wave potential by almost 1 V, fortunately to more positive values, so that barium becomes accessible in polarography.

I.2.4.1.4 Influence by Solution Equilibria

(i) **Acid–Base Equilibria**

In the polarography of organic compounds in protic solvents such as water, the electron transfer is frequently accompanied by a proton transfer:

$$Ox + 2e^- + 2H^+ \rightleftarrows H_2Red \tag{I.2.28}$$

The pure redox equilibrium is

$$Ox + 2e^- \rightleftarrows Red^{2-} \tag{I.2.29}$$

As Red^{2-} is a Brønsted base it will be stepwise protonated and these equilibria can be described as follows:

$$H_2Red \rightleftarrows HRed^- + H^+ \qquad K_{a,1} \tag{I.2.30}$$

$$HRed^- \rightleftarrows Red^{2-} + H^+ \qquad K_{a,2} \tag{I.2.31}$$

Provided that the system is reversible and not complicated by side reactions, the half-wave potential will be equal to the formal potential and the relation to the standard potential is as follows:

$$E_{1/2} = E_c^{\ominus'} = E^{\ominus} + \frac{RT}{2F}\ln\frac{\gamma_{Ox}}{\gamma_{Red^{2-}}} + \frac{RT}{2F}\ln\left(\frac{a_{H^+}^2}{K_{a,1}K_{a,2}} + \frac{a_{H^+}}{K_{a,2}} + 1\right) \quad (I.2.32)$$

As expected, the half-wave potential will depend on the pH and, from a plot of $E_{1/2}$ vs. pH, one can determine the pK_a values of the system, provided that they are within the pH range. Whenever the solution pH equals a pK_a value, the slope of the plot $E_{1/2}$ vs. pH changes. More information on the influence of pH on half-wave potentials of more complex systems is available from a publication by Heyrovský [5].

(ii) **Complex Formation**

A very frequent case in inorganic chemistry is the formation of metal complexes according to the general reaction

$$Me_{aq}^{n+} + pAn^{m-} \rightleftarrows MeAn_p^{(mp-n)-} \quad (I.2.33)$$

If the metal ions can be reduced to the metal, which means that all ligands will be stripped off during this reduction, the following equation can be derived [6] for the dependence of the half-wave potential (which is equal to $E_c^{\ominus'}$) on the activity of ligands An^{m-} and the stability constant of the complex K:

$$E_{1/2} = E^{\ominus}_{Me_{aq}^{n+}/Me_{amal}^0} - \frac{3.4RTt_1^{1/6}}{nFm^{1/3}}\left(D^{1/2}_{Me_{aq}^{n+}} + D^{1/2}_{Me_{amal}^0}\right) - \frac{RT}{nF}\ln\frac{D^{1/2}_{Me_{aq}^{n+}}}{D^{1/2}_{Me_{amal}^0}}$$
$$-\frac{RT}{nF}\ln\frac{D^{1/2}_{Me_{aq}^{n+}}}{D^{1/2}_{[MeAn_p]^{(mp-n)-}}} - \frac{RT}{nF}\ln K - \frac{RT}{nF}\ln a^p_{An^{m-}} \quad (I.2.34)$$

The subscript on the diffusion coefficient D indicates the species, i.e., the aqua metal ion Me_{aq}^{n+}, the complex metal ion $[MeAn_p]^{(mp-n)-}$ and the metal atoms in the liquid mercury Me^0_{amal}. Because of the small contributions from the second, third and fourth term on the right-hand side of Eq. (I.2.34), the following simplified equation is often used to determine the stoichiometric coefficient p and the stability constant K:

$$E_{1/2} = E^{\ominus}_{Me_{aq}^{n+}/Me_{amal}^0} - \frac{RT}{nF}\ln K - \frac{RT}{nF}\ln a^p_{An^{m-}} \quad (I.2.35)$$

To determine p and K a plot of $E_{1/2}$ vs. the logarithm of the concentration of the ligand is useful. The slope gives p and the intercept gives K. The following prerequisites have to be fulfilled: (i) reduction of the metal ions to the metal, (ii) ligand concentration must exceed that of the metal, and (iii) the reduction must be reversible in dc polarography.

When a metal ion is not reduced to the metal but instead to a lower oxidation state, the dependence of $E_{1/2}$ on ligand concentration gives only the difference in p values and the ratio of K values of the two complexes of the metal in the two oxidation states.

From the preceding it follows that the half-wave potential measured in DCP will only in rare cases approximately equal the standard potential. The requirements for this are (i) no side reactions (equilibria) of the reduced or oxidized form (esp. no protonation reactions), (ii) no amalgamation, or a dissolution in mercury with negligible Gibbs free energy of amalgamation, and (iii) no strong deviation of the activity coefficient ratio from unity.

I.2.4.1.5 Influence by Irreversibility of the Electrode System

In the case of irreversible reactions, the polarographic half-wave potential also depends on the standard potential (formal potential); however, the kinetics of the electrode reaction lead to strong deviation as an overpotential has to be applied to overcome the activation barrier of the slow electron transfer reaction. In the case of a totally irreversible electrode reaction, the half-wave potential depends on the standard rate constant k_s of the electrode reaction, the transfer coefficient α, the number n_{e-} of transferred electrons, the diffusion coefficient D_{ox}, and the drop time t_1 [7] as follows:

$$E_{1/2} = E_c^{\ominus\prime} + \frac{RT}{\alpha nF} \ln \left(\frac{2.31 k_s t_1^{1/2}}{D_{Ox}^{1/2}} \right) \tag{I.2.36}$$

I.2.4.2 Cyclic Voltammetry

I.2.4.2.1 The Peak Potentials

In the case of a reversible electrode reaction, the cathodic and anodic peak potentials depend in the following way on the formal potential:

$$E_{pc} = E_c^{\ominus\prime} - 1.109\frac{RT}{nF} - \frac{RT}{nF} \ln \frac{D_{Ox}^{1/2}}{D_{Red}^{1/2}} \tag{I.2.37}$$

$$E_{pa} = E_c^{\ominus\prime} + 1.109\frac{RT}{nF} - \frac{RT}{nF} \ln \frac{D_{Ox}^{1/2}}{D_{Red}^{1/2}} \tag{I.2.38}$$

Assuming equal diffusion coefficients, the difference between the anodic and cathodic peak potentials is

$$E_{pa} - E_{pc} = 2\left(1.109\frac{RT}{nF}\right) \approx \frac{57}{n} \text{mV} \tag{I.2.39}$$

at 25°C.

The latter relationship is a good indication of the reversibility of the electrode reaction, although some caution is necessary because a more complex electrode reaction may give the same difference (see Chap. II.1).

Provided that the diffusion coefficients of the oxidized and reduced forms are equal, a simple relation between the peak potentials and the formal potential follows:

$$E_{c}^{\ominus'} = \frac{E_{pa} + E_{pc}}{2} \tag{I.2.40}$$

Equation (I.2.40) is very frequently used to determine the formal potential of a redox system with the help of cyclic voltammetry; however, one should never forget that it holds true only for reversible systems, and provided that the symmetry coefficient $\alpha = 0.5$ (see Chap. I.3). To be cautious, it is better to refer to the value determined by Eq. (I.2.40) as the *mid-peak potential determined by cyclic voltammetry*. The formal potential $E_{c}^{\ominus'}$ has the same meaning as discussed above for direct current polarography. Hence Eqs. (I.2.32) and (I.2.34) can be applied accordingly.

When cyclic voltammetry is performed with microelectrodes it is possible to record wave-shaped steady-state voltammograms at not too high scan rates, similar to dc polarograms. Ideally, there is almost no hysteresis and the half-wave potential is equal to the mid-peak potential of the cyclic voltammograms at macroelectrodes (see Chap. II.1).

In the case of totally irreversible electrode reactions, only one peak, e.g., the reduction peak when the oxidized form is present in the solution, will be visible. The cathodic peak potential depends on the formal potential as follows:

$$E_{pc} = E_{c}^{\ominus'} - \frac{RT}{\alpha nF}\left(0.780 + 0.5\ln\frac{\alpha n D_{Ox} F v}{RT} - \ln k_{s}\right) \tag{I.2.41}$$

It is impossible to disentangle $E_{c}^{\ominus'}$ and k values for totally irreversible reactions.

For quasi-reversible electrode reactions it is not easy to say how much the peak potential difference can be to still allow a fairly reliable determination of the formal potential with the help of Eq. (I.2.40); however, differences up to 120 mV can be tolerated if α and β are near to 0.5.

I.2.4.3 Differential Pulse Voltammetry (DPV), Alternating Current Voltammetry (ACV), and Square-Wave Voltammetry (SWV)

For reversible systems there is no special reason to use these techniques, unless the concentration of the electrochemical active species is too low to allow application of DCP or cyclic voltammetry. For a reversible electrochemical system, the peak potentials in alternating current voltammetry (superimposed sinusoidal voltage perturbation) and in square-wave voltammetry (superimposed square-wave voltage perturbation) will be equal to the formal potential, i.e., $E_p = E_c^{\ominus'}$. However, in differential pulse voltammetry, there is a systematic deviation according to

$$E_p = E_c^{\ominus'} - \frac{\Delta E_{pulse}}{2} \tag{I.2.42}$$

for a reduction. ΔE_{pulse} is the amplitude of the pulse. In the case of oxidation, the deviation is positive. As in the previous methods, in the case of irreversible electrode systems, the deviation of the peak potential from the formal potential will also depend on the kinetic parameters. Whereas it is easy to detect irreversibility in ACV and SWV (see Chap. II.2), this is not trivial in DPV as the peak width of a totally irreversible system is almost as for a reversible system.

I.2.5 Thermodynamics of the Transfer of Ions Between Two Phases

Reactions in which electrons are transferred from one phase to another are of electrochemical nature, because a *charged* particle, the electron, is transferred by an applied electric field. However, it would not be reasonable to confine electrochemistry to electron transfer only. There is no difference in principle when other charged species, i.e., ions, are transferred under the action of an electric field. The driving force for an ion transfer between two phases I and II is the establishment of equal electrochemical potentials $\tilde{\mu}_i$ in both phases. The electrochemical potential of a charged species in phase I is

$$\tilde{\mu}_i^{I} = \tilde{\mu}_i^{\ominus,I} + RT \ln a_i^{I} + z_i F \phi^{I} \tag{I.2.43}$$

($\tilde{\mu}_i^{\ominus,I}$ is the standard chemical potential of the ion i in I, ϕ^{I} is the inner electric potential of phase I, z_i is the charge of the species i). An ion can be driven into phase I by two different forces, either by chemical forces, due to $\tilde{\mu}_i^{\ominus,I}$, or by the electric potential ϕ^{I}. When an ion has a high chemical affinity toward a certain phase II, it will not cross the interface from phase I to phase II until the electrochemical potentials are equal. This will create a potential difference between the two phases, which counterbalances the chemical affinity. This process is the basis of all ion-selective electrodes, e.g., a glass electrode. It is also possible to force ions deliberately from one phase into the other when a potential difference is applied across the interface. Imagine that two immiscible liquid phases are filled into a tube so that they build up a common interface in the middle (Fig. I.2.3). When each of the two liquids contains an electrolyte, which is dissociated (this needs dipolar liquids), and two inert metal electrodes are inserted into the two liquids, it is possible to apply a potential difference across the liquid–liquid interface. For exact measurements one will further introduce into each liquid a reference electrode to control the potential of each of the metal electrodes separately. Upon application of a voltage between the two working

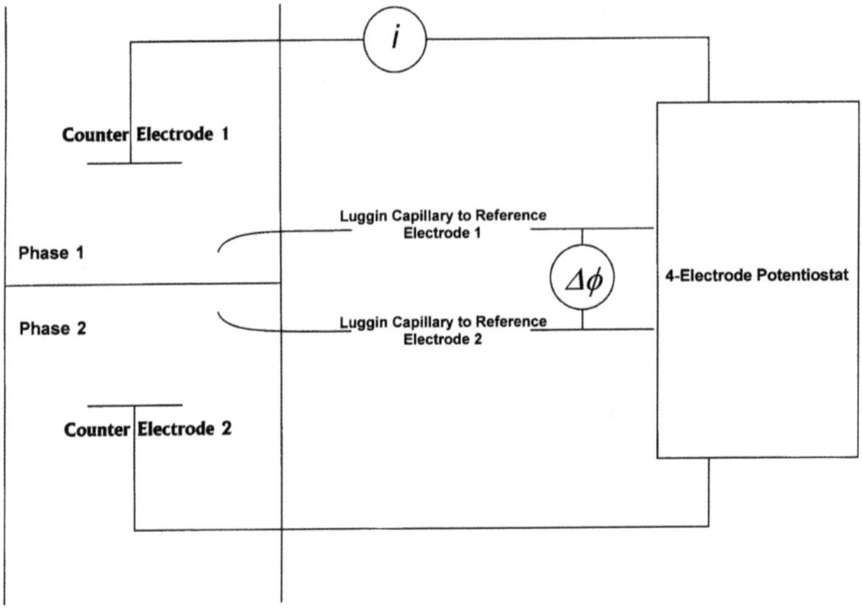

Fig. I.2.3 Experimental arrangement for measuring the transfer of ions between two immiscible liquid electrolyte solutions

electrodes, a current may flow. At the two metal electrodes unknown faradaic reactions will occur (electron transfer reactions). However, the overall current has also to cross the liquid–liquid interface. Since the electrolyte solutions on both sides are ion conductors only, passage of current can occur only when ions are transferred from one liquid to the other. The ion transfer at the interface is the rate-determining process of the entire current flow. Indeed, it is possible to record a cyclic voltammogram which shows current peaks due to the transfer of, e.g., an anion from water to nitrobenzene and back (Fig. I.2.4). The mid-peak potential of such cyclic voltammograms also has a thermodynamic meaning. The difference between the two standard chemical potentials of *i* in the two phases I and II is called the standard Gibbs free energy of ion transfer:

$$\tilde{\mu}_i^{\ominus,\mathrm{I}} - \tilde{\mu}_i^{\ominus,\mathrm{II}} = \Delta G_{\mathrm{transfer},i}^{\ominus,\mathrm{I}\rightarrow\mathrm{II}} \tag{I.2.44}$$

Hence, the difference of the standard Galvani potentials of the two phases is related to the standard Gibbs free energy of ion transfer:

$$\Delta_{\mathrm{I}}^{\mathrm{II}}\phi_i^{\ominus} = -\frac{\Delta G_{\mathrm{transfer},i}^{\ominus,\mathrm{I}\rightarrow\mathrm{II}}}{z_i F} \tag{I.2.45}$$

The *mid-peak potential* $E_{1/2}$ of the cyclic voltammogram is equal to the standard Galvani potential $\Delta_{\mathrm{I}}^{\mathrm{II}}\phi_i^{\ominus}$. Again there is a similar problem as encountered for the electron transfer reaction, i.e., the impossibility to determine a single-electron

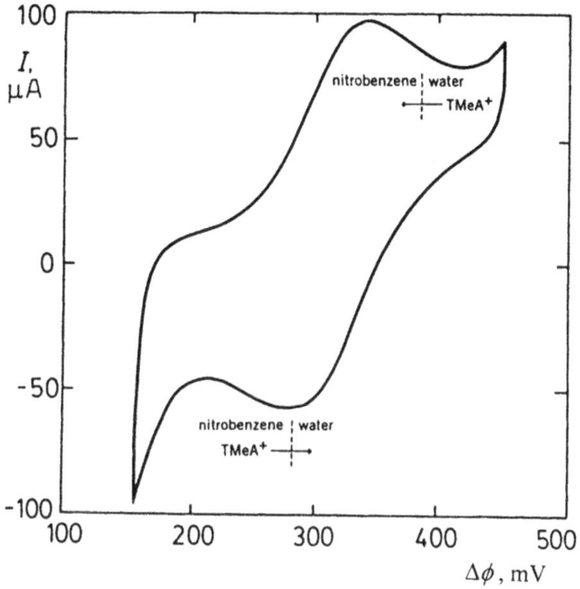

Fig. I.2.4 Cyclic voltammogram of the transfer of tetramethylammonium ions between water and nitrobenzene. $c(\mathrm{TmeA}^+) = 4.7 \times 10^{-4}$ mol L^{-1}; the supporting electrolyte is in the aqueous phase 0.1 mol L^{-1} LiCl, and in nitrobenzene 0.1 mol L^{-1} tetrabutylammonium tetraphenylborate; the scan rate is 20 mV s^{-1}. (Adapted from [11], with permission)

transfer equilibrium. All electron-transfer equilibria have been referred to that of the hydrogen electrode. To build up a thermodynamic scale of standard Galvani potentials, an *extra thermodynamic assumption* has to be made. One such assumption is that the standard Gibbs free energies of ion transfer of the anions and cations of tetraphenylarsonium tetraphenylborate are equal for all pairs of immiscible liquids. It may be generally stated that the standard Galvani potentials of ion transfer are much less accurately known than the standard potentials of electron transfer.

It is interesting that the transfer of ions from one phase to another can also result from the creation of a potential difference by electron transfer. Imagine that a solid phase contains immobile electroactive ions like Fe^{3+}. These ions can be reduced; however, this would violate the charge balance, unless other cations can diffuse into the solid, or anions could leave the solid. This is a very frequently encountered case in solid-state electrochemistry (see Chap. II.8). One can understand this insertion electrochemistry as resulting from the creation of an electric field due to the electron transfer. The same phenomena can also be observed when droplets of an immiscible liquid contain electroactive species, and these droplets are deposited onto an electrode surface, which is introduced into an aqueous solution [8].

The transfer of ions between phases is still a minor field in electrochemical studies and therefore this very brief introduction should suffice. Detailed information is available elsewhere [8–11]. A constantly updated listing of standard Galvani potentials of ion transfer is available on the Internet [12].

I.2.6 Thermodynamic Data Derived from Standard and Formal Potentials

I.2.6.1 Data Derived from Standard Potentials

Equation (I.2.13) has shown us that the standard potential gives access to the standard Gibbs free energy of the electrochemical reaction. Since the Gibbs free energy, the enthalpy and the entropy are connected via the relationship $G = H - TS$, the following equation holds true for the standard values:

$$-nFE^{\ominus} = \Delta_r G^{\ominus}_{T,p} = \Delta_r H^{\ominus} - T\Delta_r S^{\ominus} \quad \text{(I.2.46)}$$

By differentiation one can easily obtain the standard entropy as

$$\left(\frac{\partial E^{\ominus}}{\partial T}\right)_p = -\frac{1}{nF}\left(\frac{\partial \Delta_r G^{\ominus}}{\partial T}\right) = \frac{\Delta_r S^{\ominus}}{nF} \quad \text{(I.2.47)}$$

Equation (I.2.46) means that by plotting the standard potential vs. the temperature, a straight line will result from the slope of which the standard entropy can be calculated. However, the experiment is not as easy to perform when high-precision data are aimed at. Usually, one thermostats the voltammetric cell and keeps the reference electrode at a constant reference temperature. Of course, the temperature gradient between the working electrode and the reference electrode gives rise to an extra potential difference, which will be rather small for small temperature differences (up to 30–50°C). Only for high-precision data has this to be taken into account. The alternative is to bring the reference electrode to the same temperature as the working electrode. This can be done when the temperature coefficient of the reference electrode is known. Once the standard entropy of a reaction has been determined it is trivial to calculate the standard enthalpy. Often it is desirable to compare electrochemically determined standard potentials with those calculated from tabulated thermodynamic data. When these data are available, the standard values of the thermodynamic functions G, H, and S of a reaction can be calculated from the standard values of formation (index f) of the products (index P) and reactants (index R), as exemplified for H as follows:

$$\Delta_r H^{\ominus} = \sum_{\mathrm{R,P}} \left(\nu_{\mathrm{P}} \Delta_f H^{\ominus}_{\mathrm{P}} - \nu_{\mathrm{R}} \Delta_f H^{\ominus}_{\mathrm{R}}\right) \quad \text{(I.2.48)}$$

Obviously, it will be possible to determine standard formation values of thermodynamic functions from standard potentials, and of course vice versa (see oxygen electrode).

I.2.6.2 Data Derived from Formal Potentials

The measurement of formal potentials allows the determination of the Gibbs free energy of amalgamation (cf. Eq. I.2.27), acidity constants (pK_a values) (cf. Eq. I.2.32), stability constants of complexes (cf. Eq. I.2.34), solubility constants, and all other equilibrium constants, provided that there is a definite relationship between the activity of the reactants and the activity of the electrochemical active species, and provided that the electrochemical system is reversible. Today, the most frequently applied technique is cyclic voltammetry. The equations derived for the half-wave potentials in dc polarography can also be used when the mid-peak potentials derived from cyclic voltammograms are used instead. Provided that the mechanism of the electrode system is clear and the same as used for the derivation of the equations in dc polarography, and provided that the electrode kinetics is not fully different in differential pulse or square-wave voltammetry, the latter methods can also be used to measure the formal potentials. However, extreme care is advisable to first establish these prerequisites, as otherwise erroneous results will be obtained.

References

1. (a) Rieger PH (1987) Electrochemistry. Prentice-Hall Int., London; (b) Hamann CH, Hamnett A, Vielstich W (1998) Electrochemistry. Wiley-VCH, Weinheim, New York
2. (a) Bard AJ, Parsons R, Jordan J (eds) (1985) Standard potentials in aqueous solution. Marcel Dekker, New York; (b) Pourbaix M (1966) Atlas of electrochemical equilibria in aqueous solutions. Pergamon Press, Elmsford; (c) Kotrlý S, Šůcha L (1985) Handbook of chemical equilibria in analytical chemistry. Ellis Horwood Ltd., Chichester
3. Galus Z (1994) Fundamentals of electrochemical analysis, 2nd ed. Ellis Horwood, New York, Polish Scientific Publishers PWN, Warsaw, p 278
4. Galus Z (1994) Fundamentals of electrochemical analysis, 2nd ed. Ellis Horwood, New York, Polish Scientific Publishers PWN Warsaw, p 281
5. Heyrovský M, Vavricka S (1972) J Electroanal Chem 36: 203
6. Heyrovský J, Kuta J (1965) Principles of Polarography. Publishing House of the Czechosloval Academy of Sciences, Prague, p 147
7. Bard AJ, Faulkner LR (2001) Electrochemical methods. Fundamentals and applications, 2nd ed, Wiley & Sons, New York, p 196
8. Scholz F, Schröder U, Gulaboski R (2005) Electrochemistry of immobilized particles and droplets. Springer, Berlin
9. Koryta J, Vanýsek P (1981) Electrochemical phenomena at the interface of two immiscible electrolyte solutions. In: Gerischer H, Tobias ChW (eds) Advances in electrochemistry and electrochemical engineering, vol 12. John Wiley, New York, p 113
10. Girault HH, Schiffrin DJ (1989) Electrochemistry of liquid-liquid interfaces. In: Bard AJ (ed) Electroanalytical chemistry, vol 15. Marcel Dekker, New York, p 1
11. Vanýsek P (1996) Liquid-liquid electrochemistry. In: Vanýsek (ed) Modern techniques in electroanalysis. John Wiley, New York, p 346
12. Data base on http://lepa.epfl.ch/

1.2.6.2 Data Derived from Formal Potentials

[illegible]

Chapter I.3
Kinetics of Electrochemical Reactions

György Inzelt

I.3.1 Introduction

It is a matter of common knowledge, not only in science but also in everyday life, that spontaneous processes can take place at different rates and that the velocity of chemical, physical, and biological changes can be influenced by a variation of the conditions (e.g., with an increase or decrease in temperature). Electrochemical reactions involve charged species whose energy depends on the potential of the phase containing these species.

In electrochemical (galvanic or electrolytic) cells a chemical reaction (cell reaction) takes place. The essential step is the transfer of charged species (ions or electrons) across the interface of two adjacent phases. The rate of this process is related to the potential difference between these phases.

This potential difference can conveniently be varied; therefore, at least within certain limits, we can regulate the reaction rate. This reaction occurs spatially separated at the two electrodes and electric current flows through the cell and the outer conductor connecting the two electrodes. If one of the electrodes is at equilibrium (two-electrode arrangement), or a separate reference electrode is used (three-electrode arrangement), we can investigate a given electrode reaction at the so-called working electrode. Therefore, we focus on the events occurring at a single electrode although it should be kept in mind that this electrode cannot be studied alone, as it is part of an electrochemical cell and its potential (electrode potential) is related to another electrode which is kept at an equilibrium potential.

The electrode reaction is an *interfacial reaction* that necessarily involves a charge transfer step. It is well known that the rate of this type of reaction is determined by one of the consecutive steps (i.e., by the most hindered or "slowest" one) and the overall rate is related to the unit area of the interface. The *electrode* (or interfacial) *reaction* involves all the processes (chemical reaction, structural reorganization, adsorption) accompanying the charge transfer step.

G. Inzelt (✉)
Department of Physical Chemistry, Eötvös Loránd University, 1518 Budapest 112, Hungary
e-mail: inzeltgy@chem.elte.hu

F. Scholz (ed.), *Electroanalytical Methods*, 2nd ed.,
DOI 10.1007/978-3-642-02915-8_3,

We also have to define the term "electrode process." *Electrode processes* involve all the changes and processes occurring at the electrode or in its vicinity while current flows through the cell. Electrode processes consist of the electrode reaction and the mass transport processes.

Diffusion, migration, and convection are the three possible *mass transport processes*. Diffusion should always be considered because, as the reagent is consumed or the product is formed at the electrode, concentration gradients between the vicinity of the electrode and the bulk solution arise, which will induce diffusion processes [reactant species move in the direction of the electrode surface and product molecules (ions) leave the interfacial region (interphase)].

The simplest way to determine the reaction rate is to measure the current flowing in the electrical circuit ($I = \mathrm{d}Q/\mathrm{d}t$). Because the current is proportional to the surface area (A) of the electrode, in order to characterize the rate of the reaction the *current density* ($j = I/A$) is used. The relationship between the current density and *reaction rate related to unit surface area* (v) is as follows:

$$j = nFv \tag{I.3.1}$$

It should be mentioned that – in contrast to the usual study of reaction kinetics – it is possible to control the reaction rate (j = const., galvanostatic method).

I.3.2 Relationship Between the Current Density and Potential Under Steady-State Conditions

Let us begin by investigating a simple reversible redox reaction:

$$\mathrm{O} + ne^- \underset{k_{\mathrm{ox}}}{\overset{k_{\mathrm{red}}}{\rightleftarrows}} \mathrm{R}^{n-} \tag{I.3.2}$$

where O and R are the oxidized and reduced forms of a redox couple.

We may write for the rate of reduction, v_{red}, and the rate of oxidation, v_{ox}:

$$v_{\mathrm{red}} = k_{\mathrm{red}}\, c_{\mathrm{O}}\,(0, t) = -j_{\mathrm{c}}/nF \tag{I.3.3}$$

$$v_{\mathrm{ox}} = k_{\mathrm{ox}}\, c_{\mathrm{R}}\,(0, t) = j_{\mathrm{a}}/nF \tag{I.3.4}$$

where j_{c} and j_{a} are the partial cathodic and anodic current densities, respectively, while c_{O} (0, t) and c_{R} (0, t) are the concentrations of the oxidized and reduced forms at the electrode surface (surface concentrations), respectively, at time t.

Under steady-state conditions there is no time dependence. However, if the system is not at equilibrium, current will flow:

$$v = v_{\mathrm{red}} - v_{\mathrm{ox}} = k_{\mathrm{red}}\, c_{\mathrm{O}}\,(x = 0) - k_{\mathrm{ox}}\, c_{\mathrm{R}}\,(x = 0) = j/nF \tag{I.3.5}$$

Since the reaction considered is an interfacial, first-order reaction, the unit of v is mol s^{-1} cm^2, if c is measured in mol cm^{-3}; therefore, the unit of the rate constant is cm s^{-1}.

For any chemical reaction, the rate constant is related to the Gibbs free energy of activation ($\Delta^{\ddagger}G$) by

$$k = \chi\, Z \exp\left(-\Delta^{\ddagger}G/RT\right) \tag{I.3.6}$$

where $\chi \leq 1$ is the transmission coefficient and Z is given by $k_{\mathrm{B}}T\delta/h$ within the framework of the activated complex or absolute rate theory. The so-called reaction length (δ) is comparable to the molecular diameter.

The product of χZ – which is related to the vibrational frequency of the activated complex – determines the upper limit of k, which is less than 10^5 cm s^{-1}.

As already mentioned, the energy of any charged species depends on the potential of the phase.

By introducing the electrochemical potential $\tilde{\mu}_{\mathrm{i}}$ of species i, it can be written as

$$\left(\partial G^{\alpha}/\partial n_{\mathrm{i}}\right)_{p,T} = \tilde{\mu}_{\mathrm{i}}^{\alpha} = \mu_{\mathrm{i}}^{\alpha} + z_{\mathrm{i}}F\Phi^{\alpha} \tag{I.3.7}$$

where G^{α} is the Gibbs free energy of the phase α at constant pressure and temperature, n_{i}, $\mu_{\mathrm{i}}^{\alpha}$, and z_{i} are the number of particles, the chemical potential, and the charge of species i, respectively, F is the Faraday constant, and Φ^{α} is the inner potential of phase α.

In the case of an electrode, a Galvani potential difference exists between the two phases (e.g., between the metal and the electrolyte solution):

$$\Delta_{\alpha}^{\beta}\Phi = \Phi^{\beta} - \Phi^{\alpha} \tag{I.3.8}$$

Variation of the inner potential of a metal can be executed by applying a potential to the metal [in this way the Fermi level of the metal is changed, which is related to the electrochemical potential of electrons ($\tilde{\mu}_{\mathrm{e}}^{\mathrm{M}}$) in the metal]. The electrochemical potential of the electrons in the electrolyte ($\tilde{\mu}_{\mathrm{e}}^{\mathrm{s}}$) can be varied by changing the concentration ratio of the components of the redox couple in the solution.

Taking into account that the change in the free energy of activation is proportional to the variation in the free energy of the reaction (ΔG) – this is a consequence of the properties of the potential energy curves – $\Delta^{\ddagger}G$ of a given charge transfer reaction will increase or decrease by a $\Delta\Phi$ change of the potential as follows:

$$\mathrm{d}\Delta^{\ddagger}G = \alpha \mathrm{d}\Delta G \tag{I.3.9}$$

$$\Delta^{\ddagger}G_{\mathrm{a}} = \Delta^{\ddagger}G_{\mathrm{a}}^{\ominus} - \alpha_{\mathrm{a}} zF\Delta\Phi \tag{I.3.10}$$

$$\Delta^{\ddagger}G_{\mathrm{c}} = \Delta^{\ddagger}G_{\mathrm{c}}^{\ominus} + \alpha_{\mathrm{c}} zF\Delta\Phi \tag{I.3.11}$$

where α_{a} and α_{c} are the anodic and cathodic transfer or symmetry coefficients, respectively. In general, α is called the transfer coefficient (it can be determined from the current–potential function); the name symmetry factor refers to the fact that its value depends on the symmetry of the potential barrier. For a symmetric barrier, $\alpha_a = \alpha_c = 0.5$, but, in general, $0 \leq \alpha \leq 1$and, for a simple reaction

$$\alpha_a + \alpha_c = 1$$

We may collect all the potential independent terms in the standard rate constant (k_s) and introduce the electrode potential (E) instead of $\Delta\Phi$ ($E = \Delta\Phi + B$) because only E can be determined by measurements. B is a constant and characteristic of a given reference electrode.

The potential dependence of rate constants can be written as

$$k_{\text{red}} = k_s \exp\left[-a_c nf\left(E - E_c^{\ominus'}\right)\right] \tag{I.3.12}$$

$$k_{\text{ox}} = k_s \exp\left[a_a nf\left(E - E_c^{\ominus'}\right)\right] \tag{I.3.13}$$

where $f = F/RT$ and $E_c^{\ominus'}$ is the formal potential of the electrode reaction.

I.3.2.1 Equilibrium

At equilibrium, $j = 0, j_a = |j_c| = j_o$ (dynamic equilibrium), $E = E_e$

$$nFk_s c_O(x=0) \exp\left[-\alpha_c nf\left(E_e - E_c^{\ominus'}\right)\right] = nFk_s c_R(x=0) \exp\left[\alpha_a nf\left(E_e - E_c^{\ominus'}\right)\right] \tag{I.3.14}$$

where E_e is the equilibrium electrode potential and j_o is the exchange current density.

At equilibrium there is no difference between the surface and bulk concentrations, i.e.,

$$c_O(x=0) = c_O(x=\infty) = c_O^* \text{ and } c_R(x=0) = c_R(x=\infty) = c_R^*$$

Assuming that $\alpha_a = \alpha_c$, the Nernst equation is obtained:

$$\exp\left[nf\left(E_e - E_c^{\ominus'}\right)\right] = c_O^*/c_R^* \tag{I.3.15}$$

The Nernst equation is valid and can be used only under equilibrium conditions!

The expressions for j_o are as follows:

$$j_o = -nFk_s c_O \exp\left[-\alpha_c nf\left(E_e - E_c^{\ominus'}\right)\right] \tag{I.3.16}$$

$$j_o = nFk_s \left(c_O^*\right)^{(1-\alpha_c)} \left(c_R^*\right)^{\alpha_c} \tag{I.3.17}$$

if

$$c_{\mathrm{O}}^{*} = c_{\mathrm{R}}^{*} = c^{*} \tag{I.3.18}$$

$$j_{\mathrm{o}} = nFk_{\mathrm{s}}c^{*}$$

From these equations a and n can be determined by measuring j_{o} at different values of E_{e} (at different $c_{\mathrm{O}}^{*}/c_{\mathrm{R}}^{*}$ ratios).

I.3.2.2 Rate Controlled by the Charge Transfer Step

In order to overcome the activation barrier and thus enhance the desirable reaction, an overpotential (η) has to be applied:

$$\eta = E - E_{\mathrm{e}} \tag{I.3.19}$$

By using Eqs. (I.3.14), (I.3.16), and (I.3.19) we obtain the following general relationship:

$$j = j_{\mathrm{o}}\left[-\frac{c_{\mathrm{O}}(x=0)}{c_{\mathrm{O}}^{*}}\exp(-\alpha_{\mathrm{c}}nf\eta) + \frac{c_{\mathrm{R}}(x=0)}{c_{\mathrm{R}}^{*}}\exp(\alpha_{\mathrm{a}}nf\eta)\right] \tag{I.3.20}$$

If the solution is intensively stirred and/or j is small so that $c_{\mathrm{i}}\ (x=0) \approx c_{\mathrm{i}}^{*}$, then

$$j = j_{\mathrm{o}}\left[-\exp(-\alpha_{\mathrm{c}}nf\eta) + \exp(\alpha_{\mathrm{a}}nf\eta)\right] \tag{I.3.21}$$

which is the equation of the steady-state polarization curve (voltammogram) when the charge transfer is the rate-determining step. It is called the Erdey-Grúz–Volmer or Butler–Volmer equation.

At low overpotentials (because $\exp x \approx 1 + x$)

$$j = j_{\mathrm{o}}nf\eta \tag{I.3.22}$$

Equation (I.3.22) is similar to Ohm's law; thus we may define the so-called charge transfer resistance (R_{ct}):

$$R_{\mathrm{ct}} = RT/nFj_{\mathrm{o}} \tag{I.3.23}$$

At high overpotentials ($|\eta| > 0.118/n$, V), the reaction will take place in either the cathodic or anodic direction; then

$$j_{\mathrm{c}} = -j_{\mathrm{o}}\exp(-\alpha_{\mathrm{c}}nf\eta)\ \text{if}\ \eta \ll 0 \tag{I.3.24}$$

$$j_{\mathrm{a}} = j_{\mathrm{o}}\exp(\alpha_{\mathrm{c}}nf\eta)\ \text{if}\eta \gg 0 \tag{I.3.25}$$

The logarithmic form of Eqs. (I.3.24) and (I.3.25) is called the Tafel equation:

$$\eta = a + b \log |j| \qquad \text{(I.3.26)}$$

where $a = (2.3\,RT/\alpha nF)\log j_o$ and $b = \pm 2.3\,RT/\alpha nF = \pm 0.0591/\alpha n$, V decade^{-1} at 25°C.

The log j (ln j) vs η plots are called Tafel plots.

I.3.2.3 Effect of Mass Transport on the Kinetics of Electrode Processes

Mass transport to an electrode can be described by the Nernst–Planck equation. For one-dimensional mass transport along the x-axis, it can be written as

$$J_i(x) = -D_i\left[\partial c_i(x)/\partial x\right] - (z_iF/RT)\,D_ic_i\left[\partial \Phi(x)/\partial x\right] + c_iv(x) \qquad \text{(I.3.27)}$$

where J_i (mol cm^{-2} s^{-1}) is the flux of species i, D_i (cm^2 s^{-1}) is the diffusion coefficient, $\Phi(x)$ is the potential at a distance x from the electrode surface, and v (cm s^{-1}) is the velocity of a volume element of the solution moving in the x direction. The first term contains the concentration gradient related to the diffusion, the second one expresses the migration (motion of charged species forced by the potential gradient), and the third term is the convection due to the stirring of the solution.

The particular solution of Eq. (I.3.27) depends on the conditions, and a rigorous mathematical solution is generally not very easy.

In many cases, experiments are designed so that the effects of one or more transport processes are negligible or can be handled in a straightforward way. We will survey only the essential information regarding the role of mass transport processes.

I.3.2.3.1 Diffusion

The flux in the case of planar diffusion can be described by Fick's first law:

$$-J_o(x,t) = D_O\left[\partial c_O(x,t)/\partial x\right] \qquad \text{(I.3.28)}$$

If the concentration of species O at a given location x changes with time, Fick's second law should be considered:

$$\left[\partial c_O(x,t)/\partial x\right]_x = D_O\left[\partial^2 c_O(x,t)/\partial x^2\right]_t \qquad \text{(I.3.29)}$$

For reaction (I.3.2), the following relationship exists between the current density and the diffusional flux of O:

$$j/nF = -J_o(0,t) = D_O\left[\partial c_O(x,t)/\partial x\right]_{x=0} \qquad \text{(I.3.30)}$$

In order to obtain the concentrations of O and R and other possible components at a location x and time t, the partial differential equations should be solved. This is possible if the initial (values at $t = 0$) and boundary conditions (values at certain location x) are known, i.e., if we have some idea (model) concerning the diffusion process.

Initial conditions for a homogeneous solution if only one component (O) is present at the start of the experiment are

$$c_{\mathrm{O}}\,(x{,}0) = c^* \tag{I.3.31}$$

$$c_{\mathrm{R}}\,(x{,}0) = 0 \tag{I.3.32}$$

Boundary conditions depend on the model of the diffusion. The most important cases are as follows.

Semi-infinite diffusion:

$$\lim_{x \to \infty} c_{\mathrm{O}}\,(x{,}t) = c_{\mathrm{O}}^* \tag{I.3.33}$$

$$\lim_{x \to \infty} c_{\mathrm{R}}\,(x{,}t) = 0 \tag{I.3.34}$$

where $x \to \infty$ means a large distance from the electrode, where the concentration does not change during the experiment. This is the case when no stirring is applied, and

$$(x = \infty) \gg \sqrt{2D\tau} \tag{I.3.35}$$

where τ is the duration of the experiment.

Finite diffusion

$$c_{\mathrm{O}}\,(x > \delta{,}t) = c_{\mathrm{O}}^* \tag{I.3.36}$$

This situation exists in well-stirred solutions when, beyond a given distance (δ), called the diffusion layer thickness, the concentration is constant and the reservoir is infinite for the reactants. It should be mentioned that δ is an approximation and the real concentration profile is not linear, as illustrated in Fig. I.3.1.

Limited diffusion:

$$c_{\mathrm{O}}\,(x > L) = 0 \tag{I.3.37}$$

Typical cases are thin-layer cells and electrochemically active species confined to the electrode surface (e.g., chemisorbed H-atoms, polymer films, and microparticles on electrode surfaces).

In this case, the total amount of the electrochemically active species is constrained in a layer of thickness L, i.e., no infinite supply of material exists.

If $L \gg \sqrt{2D\tau}$, the semi-infinite diffusion conditions can be applied (large L or small τ).

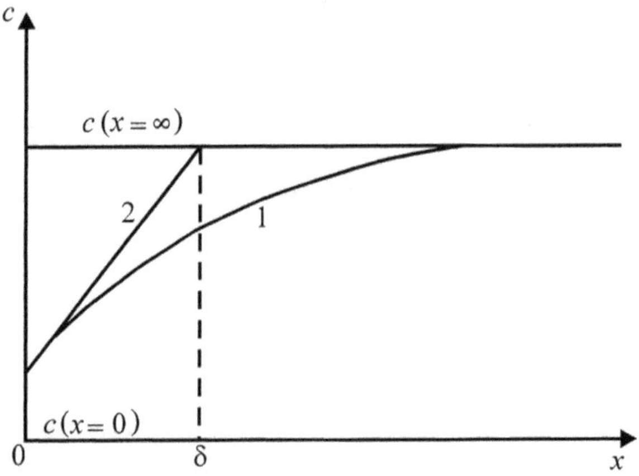

Fig. I.3.1 Variation of the concentration (c) of a species as a function of the distance from the electrode surface (x) in the case of a steady-state electrode reaction. 1 True concentration profile; 2 fictitious profile of Nernst. Thickness of diffusion layer (δ) depends on the rate of stirring, while $c(x = 0)$ depends on the overpotential

Boundary conditions related to the electrode surface:
It follows from the law of conservation of matter that

$$J_{\mathrm{o}}\,(0,t) = -J_{\mathrm{R}}\,(0,t) \tag{I.3.38}$$

There might be other specific conditions, e.g., the ratio $c_{\mathrm{O}}\,(0,t)\,/c_{\mathrm{R}}\,(0,t)$ obeys the Nernst equation at a wide potential interval.

The mathematical methods and the derivation can be found in several electrochemistry books [1–4] and reviews [5]. Herein, we present the most important considerations and formulae for steady-state electrolysis conditions. It is assumed that the solution is well stirred (the concentration gradient at the electrode surface is constant) and that both the reactant and product molecules are soluble. By combining Eqs. (I.3.28), (I.3.29), and (I.3.30) and considering the respective initial and boundary conditions [Eqs. (I.3.31), (I.3.32), and (I.3.36)], we obtain

$$-J_{\mathrm{o}}\,(0,t) = D_{\mathrm{O}}\left[c_{\mathrm{O}}^{*} - c_{\mathrm{O}}\,(0,t)\right]/\delta_{\mathrm{O}} = k_{\mathrm{mO}}\left[c_{\mathrm{O}}^{*} - c_{\mathrm{O}}\,(0,t)\right] \tag{I.3.39}$$

where k_{mO} is the mass transport coefficient.

The term $c_{\mathrm{O}}(0,\ t)$ depends on the electrode potential (overpotential). For a reversible system (see below), the potential dependence of the concentrations can be described by the Nernst equation substituting the respective surface concentrations, but Eq. (I.3.39) and the following expressions are valid regardless of the actual kinetic properties of the system.

If the charge transfer is so facile that every reactant species arriving at the electrode surface immediately reacts, then the concentration of the reacting species at the electrode surface approaches zero, $c_O\,(x=0) = 0$; the current becomes independent of potential and reaches a maximum value which depends only on the actual hydrodynamic conditions.

This maximum current is called the diffusion limiting current, and it follows from Eq. (I.3.39) that in this case the current density (j_L) can be given as

$$j_L = nFk_{mO}c_O^* \tag{I.3.40}$$

Combining Eqs. (I.3.39) and (I.3.40) we arrive at

$$c_O\,(0,t)\,/c_O^* = 1 - (j/j_L) \tag{I.3.41}$$

Substituting Eq. (I.3.41) into Eq. (I.3.20) the equation of the polarization curve is obtained:

$$j = j_o\left(1 - j/j_{L,c}\right)\exp\left(-\alpha_c nf\eta\right) + \left(1 - j/j_{L,a}\right)\exp\left(\alpha_a nf\eta\right) \tag{I.3.42}$$

At high (e.g., cathodic) overpotential

$$j = j_o\left(1 - j/j_{L,c}\right)\exp\left(-\alpha_c nf\eta\right) \tag{I.3.43}$$

or

$$\eta = \frac{RT}{\alpha_c nF}\ln\frac{j_o}{j_{L,c}} + \frac{RT}{\alpha_c nF}\ln\frac{j_{L,c} - j}{j_{L,c}} \tag{I.3.44}$$

For small overpotentials ($\alpha n\eta \ll 1$), using the Taylor expansion we obtain

$$j/j_o = \left(c_O\,(0,t)\,/c_O^*\right) - \left(c_R\,(0,t)\,/c_R^*\right) - (nF/RT)\,\eta \tag{I.3.45}$$

or

$$\eta = -\frac{RT}{nF}j\left(\frac{1}{j_o} + \frac{1}{j_{L,c}} - \frac{1}{j_{L,a}}\right) \tag{I.3.46}$$

$$\eta = -j\left(R_{ct} + R_{d,c} - R_{d,a}\right) \tag{I.3.47}$$

If $R_{ct} \ll R_{d,c} + R_{d,a}$, we speak of concentration polarization, while if $R_{ct} > R_{d,c} + R_{d,a}$, activation polarization prevails. This is meant to express that the overpotential is related either to the diffusion or to the activation processes. The diffusion overpotential can be given as

$$\eta_{diff} = (RT/nF)\sum \nu_i \ln\left[c_i^*/c_i\,(x=0)\right] \tag{I.3.48}$$

The overpotential applied may be related to other processes (e.g., chemical reaction, crystallization, or high resistance of a surface layer); thus there are also reaction, crystallization, and resistance overpotentials.

I.3.2.3.2 Migration

In a bulk electrolyte solution the current is related to the motion of ions under the influence of the electric field (potential gradient); the diffusion plays a minor role, if any, because the concentration gradient is generally small in the bulk phase.

The electrochemically active species may not carry electricity if these species are uncharged (e.g., methanol molecules) or non-reacting ions are present in a large excess (supporting electrolyte), because their transport number (t_i) – which gives the fraction of the total electric current that a given ion carries – is small in the electrolytic solution used.

For ion i the transport number

$$t_i = |z_i|\, u_i c_i \Big/ \sum_{j=1}^{n} |z_j|\, u_j c_j \tag{I.3.49}$$

where z_j, u_j, and c_j are the charge, mobility, and concentration of ion j, respectively.

Near the electrode, the charged reacting species are, in general, transported by both diffusion and migration. The current can be separated into diffusion (I_d) and migration (I_m) components:

$$I = I_d + I_m \tag{I.3.50}$$

The moles of species i oxidized or reduced per unit time is $|I|/nF$. The moles arriving at the electrode or moving away from the electrode in the same period of time by migration is $\pm t_i I/z_i F$, i.e., the migrational flux can be given as

$$I_m/nF = \pm t_i I/z_i F \tag{I.3.51}$$

I and I_d always have the same sign, but I_m may have an opposite one [e.g., in the case of the reduction of anions ($I_c < 0, z_i < 0$) when anions migrate away from the negatively charged electrode surface]; then $I < I_d$.

It is evident that the migration flux of the electroactive ions can be decreased or even eliminated by the addition of an excess of indifferent electrolyte since, in this case, $t_i \to 0$, consequently $I_m \to 0$. This means that the limiting current in the case of the reduction of a cation (e.g., Cu^{2+}) will decrease as the concentration of the inert electrolyte (e.g., KNO_3) is increased.

This practice is frequently used because the mathematical treatment of electrochemical systems becomes simpler due to the elimination of the $\partial \Phi / \partial x$ term in the mass transport equations.

The role of the supporting electrolyte is, however, more complex. It decreases the cell resistance, strongly influences the double-layer structure – at high concentration of supporting electrolyte the charging and faradaic processes can be separated, which allows simplification of the modeling of the cell impedance and the mathematical treatment – and, in analytical applications, may decrease matrix effects.

I.3.2.3.3 Convection

Convection may occur due to density gradients (natural convection). This phenomenon, however, is rare in electroanalytical techniques. A density gradient may arise at high currents, especially in technical electrolysis and in coulometric experiments.

Forced convection may be unintentional, e.g., due to the vibration of a building, but usually stirring is applied to enhance the rate of the mass transport process. Stirring can be achieved by stirring the solution with the help of a separate stirrer, or the electrode itself can rotate (rotating disc electrode), vibrate (vibrating electrodes), or even simply expand its volume (which is a movement of its surface against the solution) as the dropping mercury electrode does. Convection is also essential in all flow-through electrodes.

We now consider forced convection. We have seen that the diffusion layer thickness (δ) is a crucial parameter in the diffusion equations. It is a fitting parameter; in fact, a thickness from the electrode surface within which no hydrodynamic motion of the solution is assumed, i.e., the mass transport occurs by molecular mechanism, mostly by diffusion. The exact solution of the respective convective–diffusion equations is very complicated; therefore, only the essential features are surveyed for two cases: stirring of the solution and rotating disc electrode (RDE).

In the case of a laminar flow, the flow velocity is zero at the plane electrode surface, then continuously increases within a given layer (the Prandtl boundary layer), and eventually reaches the value characteristic of the stirred liquid phase.

For this situation the thickness of the fictitious diffusion layer can be given as follows:

$$\delta_{\mathrm{O}} \sim y^{1/2} \upsilon^{1/6} D_{\mathrm{O}}^{1/3} \left[v\left(y\right) \right]^{-1/2} \tag{I.3.52}$$

where $v(y)$ is the flow velocity at a distance (y) measured parallel to the surface ($y = 0$ is the center of the plane electrode), υ is the kinematic viscosity, and D_{O} is the diffusion coefficient of reactant O (note that δ_{O} depends on D_{O}, i.e., it differs for different species).

According to this relationship, δ varies along the plate; it is highest at the edge of the electrode; consequently, a current distribution should also be considered:

$$j \sim nFD_{\mathrm{O}}^{2/3} y^{-1/2} \left[v\left(y\right) \right]^{1/2} \upsilon^{1/6} \left[c_{\mathrm{O}}\left(x = 0\right) - c_{\mathrm{O}}^{*} \right] \tag{I.3.53}$$

If the electrode is disc shaped, imbedded in a rod of an insulating material and rotating with an angular velocity ω_{r} (the axis of rotation goes through the center of the disc and is perpendicular to the surface), δ is independent of coordinate y. The convective-diffusion equations have been solved for the case of this situation (RDE):

$$\delta_{\mathrm{O}} = 1.61\, D_{\mathrm{O}}^{1/3} \omega_{\mathrm{r}}^{1/2} \upsilon^{1/6} \tag{I.3.54}$$

[we may obtain Eq. (I.3.54) from (I.3.51) by substituting $y = r$ and $v = r\omega_r$] and

$$j = 0.62\, nFD_O^{2/3}\omega_r^{1/2}\upsilon^{1/6}\left[c_O\,(x=0) - c_O^*\right] \tag{I.3.55}$$

When $c_O\,(x=0) = 0$, i.e., $j = j_L$ Eq. (I.3.53) is called the Levich equation. In this case, the Prandtl boundary thickness (x_h) is

$$x_h = 3.6\,(\upsilon/\omega_r)^{1/2} \tag{I.3.56}$$

In aqueous solution δ_O is ~ 0.05 x_h. When ω_r is small, x_h may approach the disc radius. In this case the derived equations are no longer valid. If $r = 0.1$ cm, ω should be larger than $10\,\upsilon/r^2$, i.e., $\omega_r > 10\,\mathrm{s}^{-1}$ is the lower limit. The upper limit is the transition to turbulent flow, which, under the considered conditions, means that ω_r should be less than 200,000 s^{-1}.

The product of an electrode reaction can be detected by applying an additional ring-shaped electrode, which is situated concentrically and insulated from the disc (rotating ring-disc electrode, RRDE). The potential of the ring can be independently controlled using a bipotentiostat.

The rotating disc electrode can be used for kinetic studies. A deviation of a plot j vs $\omega_r^{-1/2}$ from a straight line that intersects the origin suggests a slow kinetic step. In this case $j^{-1/2}$ can be plotted against $\omega_r^{-1/2}$ and is known as the Koutecky–Levich plot. This plot should be linear and can be extrapolated to $\omega_r \to \infty \left(\omega_r^{-1/2} = 0\right)$ to obtain $1/j_K$ where j_K represents the current in the absence of any mass transport effect [see Eqs. (I.3.24) or (I.3.25) and (I.3.18) for charge transfer controlled reactions]. It follows that the kinetic parameters k_s and α can be determined if j_K values are measured at different overpotentials. RDEs and RRDEs are powerful tools to study the kinetics of complex electrode processes (e.g., multistep charge transfer reactions and coupled chemical and charge transfer reactions).

General treatments of hydrodynamic problems in electrochemistry [1, 2] and details of the application of RDEs and RRDEs can be found in the literature [6–10].

I.3.2.4 Reversibility, Quasi-reversibility, and Irreversibility

These concepts are used in several ways. We may speak of *chemical reversibility* when the same reaction (e.g., cell reaction) can take place in both directions. *Thermodynamic reversibility* means that an infinitesimal reversal of a driving force causes the process to reverse its direction. The reaction proceeds through a series of equilibrium states; however, such a path would require an infinite length of time. *Electrochemical reversibility* is a practical concept. In short, it means that the Nernst equation ([Eq. (I.3.15)] can be applied also when $|E| \geq E_e$. Therefore, such a process is called a reversible or Nernstian reaction (reversible or Nernstian system, behavior). This is the case when the activation energy is small; consequently k_s and j_o are high.

If $j_{\mathrm{o}} \gg j$, i.e., $j/j_{\mathrm{o}} \to 0$ by rearranging Eq. (I.3.20) we obtain

$$c_{\mathrm{O}}(0,t)/c_{\mathrm{R}}(0,t) = \left(c_{\mathrm{O}}^{*}/c_{\mathrm{R}}^{*}\right) \exp\left[nf\left(E - E_{\mathrm{e}}\right)\right] \tag{I.3.57}$$

By substituting $(c_{\mathrm{O}}^{*}/c_{\mathrm{R}}^{*})$ from Eq. (I.3.15) into Eq. (I.3.57)

$$c_{\mathrm{O}}(0,t)/c_{\mathrm{R}}(0,t) = \exp\left[nf\left(E - E_{\mathrm{c}}^{\ominus\prime}\right)\right] \tag{I.3.58}$$

or

$$E = E_{\mathrm{c}}^{\ominus\prime} + \frac{RT}{nF} \ln \frac{c_{\mathrm{O}}(0,t)}{c_{\mathrm{R}}(0,t)} \tag{I.3.59}$$

i.e., the Nernst equation expresses the relationship between the surface concentrations and the electrode potential regardless of the current flow. The surface equilibrium is a consequence of the very fast charge transfer kinetics. Equation (I.3.58) contains no kinetic parameter. Kinetic parameters can be determined when the irreversible kinetics prevails and Eqs. (I.3.21–I.3.26) can be used.

The kinetics is called *irreversible* in electrochemistry when the charge transfer step is very sluggish (k_{s} and j_{o} are very small). In this case, the anodic and cathodic reactions are never simultaneously significant. In order to observe any current, the charge transfer reaction has to be strongly activated either in the cathodic or in the anodic direction by application of overpotential. When the electrode process is neither very facile nor very sluggish, we speak of *quasi-reversible* behavior.

It should also be mentioned that the appearance of reversible behavior depends on the relative value of k_{s} and k_{m}, since no equilibrium exists between the surface and bulk concentrations; reactants are continuously transported to the electrode surface by mass transport (diffusion).

Of course, the current can never be higher than the rate of the most hindered (slowest) step, which is either the electron transfer or the mass transport. The problem of reversibility in terms of the relative ratio of k_{s} and k_{m} is illustrated in Fig. I.3.2. Two systems with different electron transfer rates [k_{s} (1) and k_{s} (2)] are considered and the effect of forced convection (the variation of the rotation rate of the electrode, ω_{r}) is presented. The rate constants (k_{red} and k_{ox}) can be altered by the potential applied. The effect of the overpotential on a reversible [j_{o} (2) is very large] and an irreversible [j_{o} (1) is small] system, respectively, is schematically shown in Fig. I.3.3. For the sake of simplicity, the values of the diffusion coefficient of the electroactive species in both systems are equal.

The current–potential relationship for a reversible redox system (note the very high value of the exchange current density) is displayed in Fig. I.3.4. As can be seen in Fig. I.3.4b, there is no Tafel region. On the other hand, in the case of an irreversible system, the Tafel region may spread over some hundreds of millivolts if the stirring rate is high enough (Fig. I.3.5).

In electrochemistry books it is often stated that the charge (electron) transfer is slow. This is not entirely correct because the act of the electron transfer itself is

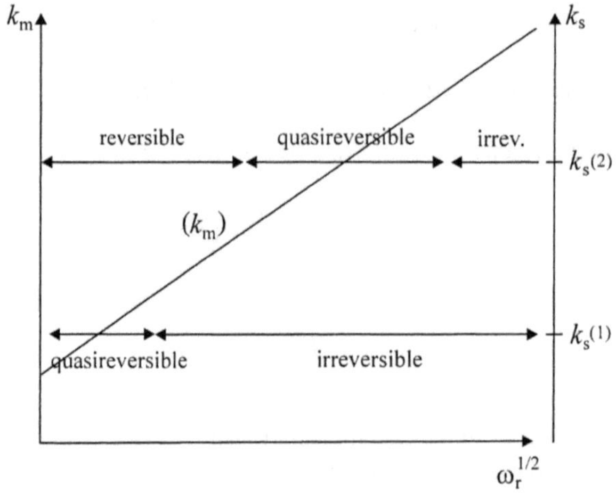

Fig. I.3.2 Illustration of the reversibility problem. The standard rate constants [k_s (1) and $k_s(2)$] are characteristic of the charge transfer rate of the given systems. The diffusion rate constants (k_{mR} or k_{mO}) are varied by the rotation rate of the electrode. If $k_s \gg k_m$ the system is reversible, while in the case of $k_m \gg k_s$ irreversible behavior can be observed. The values of the diffusion coefficients are taken as equal for both systems

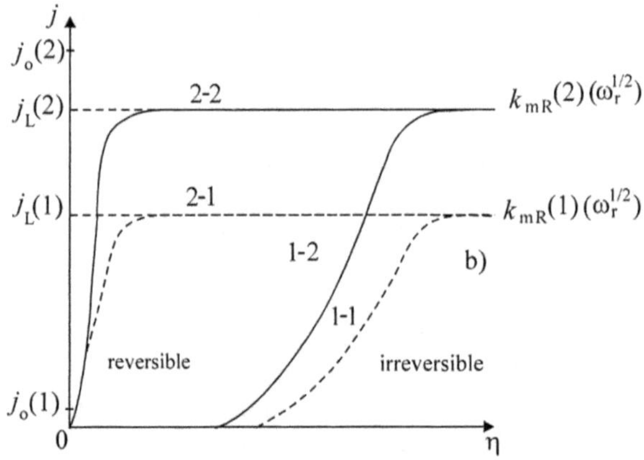

Fig. I.3.3 Steady-state polarization (j–η) curves for reversible [j_o (2)] and irreversible [j_o (1)] systems, respectively, at two different stirring rates. Curves belong to the following parameters: (1–1): j_o (1) and k_{mR} (1), (1–2): j_o (1) and k_{mR} (2), (2–1): j_o (2) and k_{mR} (1), and (2–2): j_o (2) and k_{mR} (2). The same diffusion coefficients for both systems are considered

very fast; it occurs within 10^{-16} s; however, according to the advanced theories (e.g., Marcus theory [11–14]), the reorganization of the structure of the reactants and products and that of their solvation sphere or ligands need more time (10^{-11} – 10^{-14} s).

The exchange current density also depends on the nature of the electrode. For instance, the rate of the hydrogen evolution reaction may change about 8 orders in

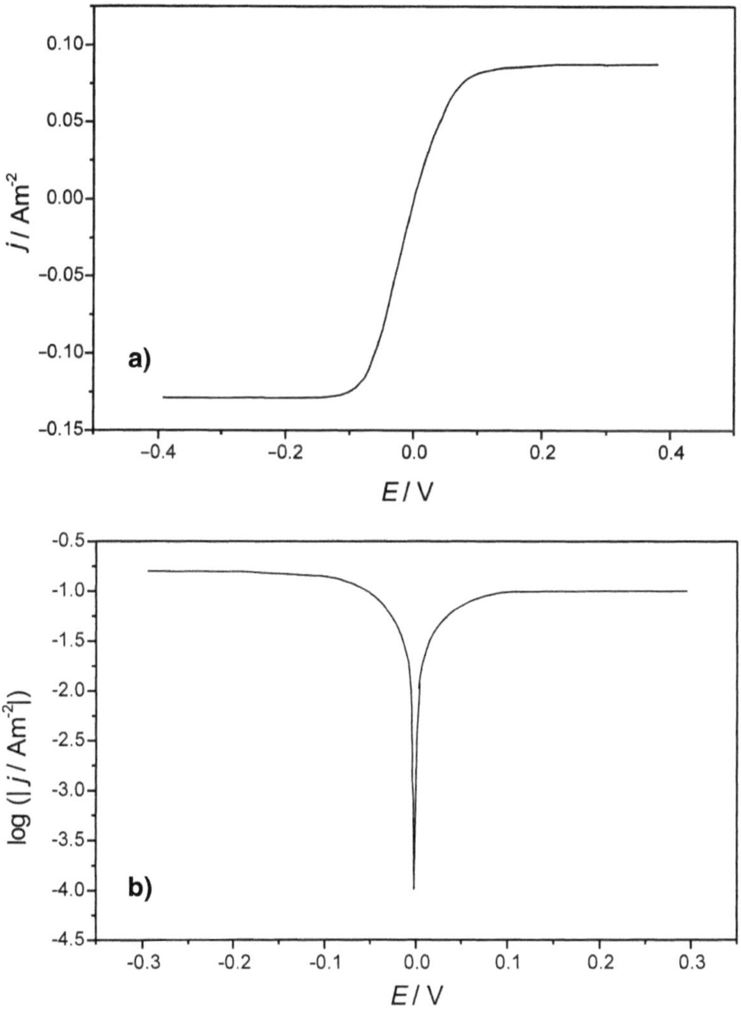

Fig. I.3.4 Reversible system. (**a**) Current–potential curve and (**b**) log $|j| - E$ plot. Characteristic parameters are $E_e = 0.0\,V$, $n = 1$, $D_R = 5 \times 10^{-10} m^2 s^{-1}$, $D_O = 10^{-9} m^2 s^{-1}$, $c_R^* = c_0^* = 10^{-4} mol\,dm^{-3}$, $\omega_r = 50\,s^{-1}$, $\nu = 10^{-6} m^2 s^{-1}$, $\alpha_a = \alpha_c = 0.5$, $T = 298.15\,K$

magnitude, $j_o \sim 10^{-4} Acm^{-2}$ for Pt, while $j_o 10^{-12} A\,cm^{-2}$ for Hg or Pb. This is the very reason that hydrogen evolution needs practically no overpotential at Pt but a rather high one, ca. −2 V, at Hg. Because the electrode material acts as a catalyst, i.e., it specifically enhances the electrode reaction without being consumed, we speak of *electrocatalysis*. There can be several reasons why a metal or other electrode increases the speed of the reaction. In this case the main event is the dissociative chemisorption of hydrogen molecules at Pt (it decreases the energy of activation) that does not occur at Hg.

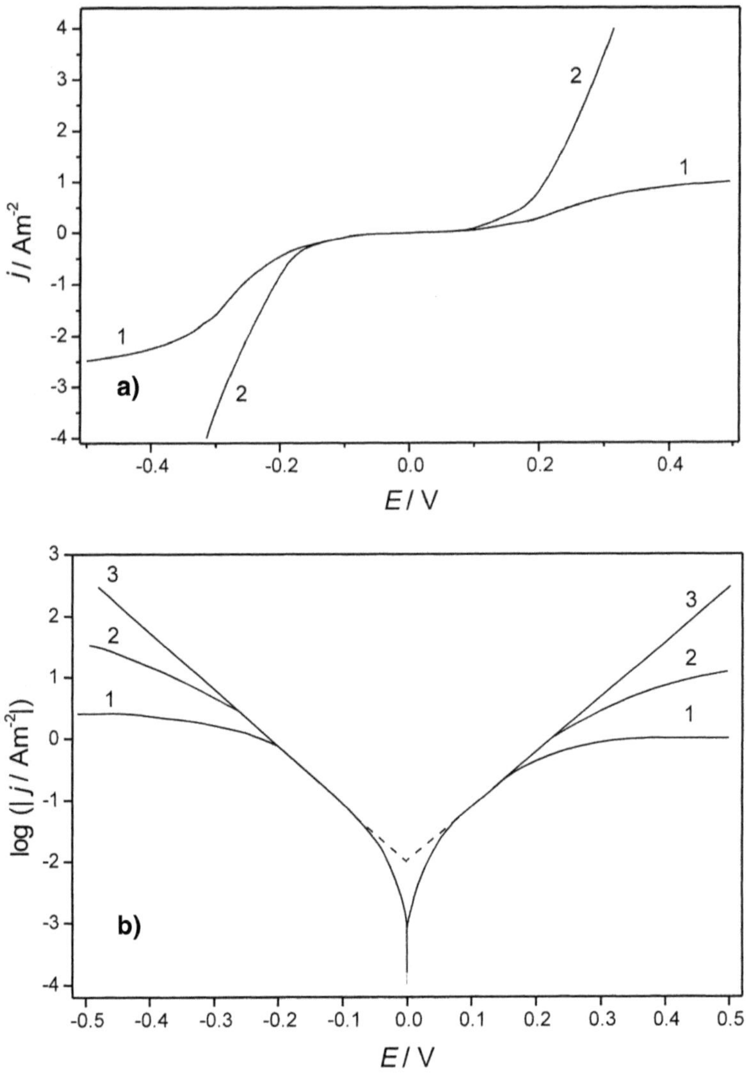

Fig. I.3.5 Quasi-reversible and irreversible behavior. (**a**) Current–potential curves and (**b**) log $|j|$ – E plots (Tafel plots) for a redox system at different angular velocities of a rotating disc electrode: $\omega_r = 50\,\mathrm{s}^{-1}$ (*1*) and $E_e = 0.0\mathrm{V}$, $j_o = 0.01\,\mathrm{Am}^{-2}$, $n = 1$, $D_R = 5 \times 10^{-10}\,\mathrm{m}^2\,\mathrm{s}^{-1}$, $D_O = 10^{-9}\mathrm{m}^2\mathrm{s}^{-1}$, $c_R^* = 10^{-3}\mathrm{mol\,dm}^{-3}$, $c_0^* = 2 \times 10^{-3}\mathrm{mol\,dm}^{-3}$, $\upsilon = 10^{-6}\mathrm{m}^2\mathrm{s}^{-1}$, $\alpha_a\alpha_c = 0.5$, $T = 298.15\,\mathrm{K}$. For the sake of comparison, the pure activation-controlled case (curves *3*) is also displayed

I.3.2.5 Effect of the Double-Layer Structure on the Rate of the Charge Transfer Reaction

Owing to the structure of the double layer, the concentration of ions participating in the charge transfer reaction is different from their bulk concentration. (Only those ions that are close to the electrode surface can participate in the reaction.) The electric potential (Φ), in general, varies with the distance from the electrode surface and differs from the inner potential of the bulk phase (Φ^s). Because the concentration of the reacting ions will depend on the actual potential at the outer Helmholtz plane (Φ_2 at $x = x_2$) and the driving force is not the whole potential drop between the electrode surface (Φ^M) and bulk phase (Φ^s) but only ($\Phi^M - \Phi_2$), the rate equations should be corrected. For instance, for the cathodic current, we obtain

$$j_c = -nFk_s c_O^* \exp\left[-\frac{z_O F(\Phi_2 - \Phi^s)}{RT}\right] \exp\left[-\frac{\alpha_c nF(\Phi^M - \Phi^s)}{RT}\right] \quad \text{(I.3.60)}$$

which is called the Frumkin correction.

There is no such effect for uncharged species ($z_i = 0$) and not in cases of a high concentration of supporting electrolyte. In the latter case the whole potential drop is in the compact Helmholtz layer [there is no diffuse double layer, $(\Phi_2 - \Phi^s) \to 0$].

It follows that in dilute solutions, especially if the same ions build the double layer up and react, this effect might be substantial. If $\Phi_2 - \Phi^s > 0$ (($\sigma^M > 0$, i.e., the electrode has a positive charge), anions will be attracted (their concentration will be higher at the surface) while cations will be repelled. At $\Phi_2 - \Phi^s < 0$, the opposite effect will hold, while at the potential of zero charge (pzc) $(\Phi_2 - \Phi^s) = 0$, $c_O(x_2,t) = c_O^*$, i.e., no increase or decrease of concentration near the electrode surface.

It should also be noted that the actual charge on the metal is of importance; reduction of anions may occur at potentials higher or lower than E_{pzc}. In some cases (e.g., reduction of $S_2O_8^{2-}$), nonlinear Tafel plots have been observed as a consequence of the Frumkin effect [1, 4, 15].

I.3.3 Current–Potential Transients

I.3.3.1 Charging the Double Layer

Even during potentiostatic or galvanostatic experiments over shorter or longer periods of time, no steady state or equilibrium prevails during setting a new potential or current value by a potentiostat. In the absence of a charge transfer reaction (ideally polarizable electrodes), a current flows, which is due to the charging (or discharging) of the double layer. This current is called the capacitive or charging current. This process also occurs simultaneously with the charge transfer (faradaic current).

A two-electrode system can be represented by an equivalent circuit that consists of two capacitances and a resistance in series. Because the capacitance of the reference electrode is usually much higher than that of the electrode under study, it is enough to investigate the effect of different electrical perturbations on an RC circuit with R_s (the ohmic resistance of the solution) and C_d (the double-layer capacitance of the working electrode) in series. (Similarly, the effect of the auxiliary electrode can also be neglected in three-electrode arrangements if Z (working electrode) $\gg Z$ (auxiliary electrode), which is usually the case.)

The relationships between the electrode potential and capacitive current ($I_c = dQ/dt$) in the three most important cases are as follows.

(i) Potential step, assuming that, at $t = 0$, $Q = 0$

$$I_c = (E/R_s)\exp(-t/R_s C_d) \quad \text{(I.3.61)}$$

This means that, after applying a potential step of magnitude E, an exponentially decaying current is obtained with a time constant $\tau = R_s C_d$.
It should be mentioned that the potential E and the maximum current $I_c\, E/R_s$ ($t = 0$) cannot be reached immediately either because the cell also has a time constant (that depends on the cell design) or because the potentiostat does not have enough power.

(ii) Current step
If $I = 0$ ($t = 0$) and the RC circuit is charged by a constant current I_c

$$E = I_c\,(R_s + t/C_d) \quad \text{(I.3.62)}$$

The change of potential due to the ohmic drop is instantaneous since if $I = 0$, then $IR = 0$. (This is the basis of the ohmic drop compensation by interruption techniques.) It is obvious that from the E vs. t function C can be determined. This is the principle of the determination of pseudocapacitance (e.g., chemisorbed hydrogen on platinum) by the method of charging curves. In this case, after the adsorption of hydrogen, a not too high anodic current ($j < j_o$) is applied and E is followed as a function of time.
Since there are 1.31×10^{15} Pt atoms/cm^{-2}, i.e., 2.18×10^{-9} mol cm^2, it follows that a charge of 2.1×10^{-4} C cm^{-2} is needed for the oxidation of the adsorbed hydrogen. From the charge consumed the real surface area of the electrode can easily be determined.

(iii) Potential sweep ($E = E_i + vt$, $v = dE/dt$)

$$I_c = vC_d + \left[(E/R_s) - vC_d\right]\exp(-t/R_s C_d) \quad \text{(I.3.63)}$$

It can be seen that there is a steady-state and a transient component of the current. It is of importance that I_c is proportional to v, and this is so not only in the case of the double-layer capacitance, but also for any pseudocapacitance (electrochemically active surface layer, absorbed atoms, etc.) as well as for thin-layer cells. It follows that during cyclic voltammetric experiments the capacitive current may exceed the faradaic current – which is proportional to $v^{1/2}$ – at high sweep rates that causes problems in the study of fast kinetics by microelectrodes.

I.3.3.2 Faradaic Current

Most of the voltammetric techniques used in electroanalytical chemistry are based on a programmed perturbation of the potential of the working electrode. After the electric perturbation, it is not possible to attain a steady state for a shorter or longer period of time. In the case of widely used transient techniques (cyclic voltammetry and chronoamperometry at conventional-sized electrodes), a steady state can never be reached. There is no stirring of the solution and supporting electrolyte applied; therefore, except in extreme cases, only diffusion has to be considered as a mass transport process. In general, during the experiment, the concentration of the bulk solution does not change; semi-infinite diffusion conditions prevail.

The partial differential equations of Fick's diffusion laws can be solved. The initial and boundary conditions of semi-infinite diffusion have already been surveyed in Sect. I.3.2.3.1. In addition, a flux equation and boundary conditions related to the electrode surface must be taken into account. The latter condition is related to the type of perturbation applied.

For instance, if the potential is stepped (chronoamperometry) into the region of limiting current, i.e., $c(0, t) = 0$, at $t > 0$ the solution of Eq. (I.3.29) is

$$I_{\mathrm{L}}(t) = nFAD^{1/2}c^{*}(\pi t)^{-1/2} \tag{I.3.64}$$

which is called the Cottrell equation.

In contrast to steady-state conditions, the current decreases with time because the concentration gradient decreases:

$$\partial c(x,t)/\partial x = c^{*}(\pi Dt)^{-1/2}\exp\left(-x^{2}/\pi Dt\right) \tag{I.3.65}$$

and

$$\left[\partial c(x,t)/\partial x\right]_{x=0} = c^{*}(\pi Dt)^{-1/2} \tag{I.3.66}$$

It is worth noting that, in all diffusion equations, t is on 0.5 power, and c is at all time proportional to c^{*}. It follows that the diffusion proceeds with $t^{1/2}$, the diffusion layer thickness increases with $t^{1/2}$, each $xt^{-1/2}$ product corresponds to a given concentration, i.e., the concentration will be the same at x_1 and x_2 when $x_1/x_2 = \sqrt{t_1}/\sqrt{t_2}$.

If the charge transfer is very fast, Eqs. (I.3.38) and (I.3.59) can be used as boundary conditions at the electrode surface. Then, the diffusion current will change again with $t^{1/2}$; however, the potential and the diffusion coefficients will appear in the equation

$$I(t) = nFAD^{1/2}c^{*}(\pi t)^{-1/2}\left\{1 + (D_{\mathrm{O}}/D_{\mathrm{R}})^{1/2}\exp\left[nf\left(E - E_{\mathrm{c}}^{\ominus'}\right)\right]\right\}^{-1} \tag{I.3.67}$$

If the potential is stepped to the diffusion limiting current region $E \gg E_c^{\ominus'}$, Eq. (I.3.67) is simplified to Eq. (I.3.64).

Equations (I.3.64) and (I.3.67) are valid only for planar electrodes of infinite size. For spherical electrodes or microelectrodes – where hemispherical diffusion conditions exist – the solution of the diffusion equation leads to the following equation:

$$I(t) = nFADc^* \left[(\pi Dt)^{-1/2} + r_o^{-1} \right] \quad \text{(I.3.68)}$$

where r_o is the radius of the spherical electrode or microelectrode disc.

Equation (I.3.68) turns into the usual Cottrell equation if $r_o \to \infty$.

It can be seen that at short times the spherical correction can be neglected. On the other hand, at large t values, a steady-state current will flow. The smaller the electrode radius, the faster the steady state is achieved. The steady state can easily be reached at microelectrodes; however, at electrodes of ordinary size ($r_o > 1$ mm), steady-state current is seldom observed due to the effect of natural convection.

If the special conditions (e.g., nernstian behavior or diffusion limiting current) are not valid, the solution of the differential equations are more difficult. In electroanalytical chemistry, techniques are used where the current–potential relationship is relatively simple; consequently, the evaluation of the data is straightforward. This means that the current is directly proportional to the concentration (quantitative analysis) or the characteristic potential (peak potential, half-wave potential, etc.) values can easily and unambiguously be determined (qualitative analysis). In most cases, the experimental conditions can be varied in such a way that the desirable situation is realized.

The majority of electrode processes take place via a number of consecutive (and/or simultaneous), respectively, competitive steps. Even such apparently simple processes as the electrodeposition of univalent ions consist of at least two steps, viz. neutralization and incorporation into the crystal lattice. If the electrode reaction involves the transfer of more than one electron it usually occurs in two steps. Complications may arise from preceding and subsequent chemical reactions, adsorption and desorption, etc. The rate will be determined by the step with the smallest rate constant (rate-determining, hindered or "slowest" step).

If the charge transfer reaction is the rate-determining step, the equations introduced earlier will describe the current–potential relationship; however, the charge number of the electrode reaction will differ from the n value determined, because the latter is characteristic of the rate-determining step.

Although on the basis of current–potential relationships important conclusions can be drawn regarding the mechanism of the electrode processes – especially if the experimental parameters are varied over a wide range – the use of combined electrochemical and nonelectrochemical methods is inevitable to elucidate the mechanism of the complex electrode processes. As we will see later in this volume, a great variety of advanced electrochemical and in situ probes are available, which give different types of information and therefore provide a better insight into the nature of the chemical events that occur during electrochemical reactions. The solid theoretical foundations and the relative simplicity of the final formulae and techniques make electroanalysis an attractive and powerful tool to obtain fast and reliable information on chemical systems.

References

1. Bard AJ, Faulkner LR (2001) Electrochemical methods, fundamentals and applications, 2nd edn. John Wiley, New York
2. Oldham HB, Myland JC (1994) Fundamentals of electrochemical science. Academic Press, San Diego
3. Rieger PH (1987) Electrochemistry. Prentice Hall, Oxford
4. Gileadi E (1993) Electrode kinetics. VCH, New York
5. Bard AJ (ed) Electroanalytical chemistry, a series of advances. Marcel Dekker, New York (22 volumes until 2004)
6. Levich VG (1962) Physicochemical hydrodynamics. Prentice Hall, Englewood Cliffs, NJ
7. Albery WJ, Hitchman ML (1971) Ring-disc electrodes. Clarendon Press, Oxford
8. Newman JS (1973) Electrochemical systems. Prentice Hall, Englewood Cliffs, NJ
9. Pleskov Yu V, Filinovskii V Yu (1976) The rotating disc electrode. Consultants Bureau, New York
10. Brett CMA, Oliveira Brett AM (1993) Electrochemistry. Oxford University Press, Oxford
11. Miller CJ (1995) Heterogeneous electron transfer kinetics at metallic electrodes. In: Rubinstein I (ed) Physical electrochemistry. Marcel Dekker, New York, pp 27–79
12. Marcus RA (1997) J Electroanal Chem 438: 251
13. Schmickler W (1996) Interfacial electrochemistry. Oxford University Press, Oxford
14. Hush NS (1999) J Electroanal Chem 460: 5
15. Damaskin BB, Petrii OA (1983) Introduction to electrochemical kinetics. Vysshaya Skola, Moscow

Part II
Electroanalytical Techniques

Chapter II.1
Cyclic Voltammetry

Frank Marken, Andreas Neudeck, and Alan M. Bond

II.1.1 Introduction

Although one of the more complex electrochemical techniques [1], cyclic voltammetry is very frequently used because it offers a wealth of experimental information and insights into both the kinetic and thermodynamic details of many chemical systems [2]. Excellent review articles [3] and textbooks partially [4] or entirely [2, 5] dedicated to the fundamental aspects and applications of cyclic voltammetry have appeared. Because of significant advances in the theoretical understanding of the technique today, even complex chemical systems such as electrodes modified with film or particulate deposits may be studied quantitatively by cyclic voltammetry. In early electrochemical work, measurements were usually undertaken under equilibrium conditions (potentiometry) [6] where extremely accurate measurements of thermodynamic properties are possible. However, it was soon realised that the time dependence of signals can provide useful kinetic data [7]. Many early voltammetric studies were conducted on solid electrodes made from metals such as gold or platinum. However, the complexity of the chemical processes at the interface between solid metals and aqueous electrolytes inhibited the rapid development of novel transient methods.

A very important development in voltammetry was the analysis of electrochemical processes at mercury electrodes (polarography [8]), based on the pioneering and Nobel Prize winning work of Heyrovský [9]. Mercury may be regarded as an ideal electrode material. At room temperature, the surface is clean, microscopically featureless, and may be continuously renewed, e.g. by dropping from a fine capillary or mechanical removal of drops. Furthermore, potential scans to very negative potentials are possible due to the high hydrogen evolution overpotential on mercury. This feature allowed the alloy (amalgam) formation of a wide range of metals with mercury to be studied and exploited in electroanalysis [10]. Subsequently, it was quickly realised that interesting information was available, not only by applying a

F. Marken (✉)
Department of Chemistry, University of Bath, BA2 7AY, UK
e-mail: F.Marken@bath.ac.uk

F. Scholz (ed.), *Electroanalytical Methods*, 2nd ed.,
DOI 10.1007/978-3-642-02915-8_4, © Springer-Verlag Berlin Heidelberg 2010

constant potential, but also by scanning the potential at a static mercury droplet. The latter methodology led to the fundamentals of linear (potential) sweep voltammetry (LSV) being developed [11]. A key publication by Nicholson and Shain [12] finally allowed simple criteria for the analysis of cyclic voltammograms (triangular waveform) to be applied without full analysis of the shape of the voltammogram. Via application of the theory, data quantifying the rate of the heterogeneous electron transfer or homogeneous chemical processes accompanying the charge transfer step could be determined based simply on the measurement of peak potential and peak current data as a function of scan rate. Subsequently, the Nicholson and Shain approach was extended to embrace the analysis of numerous reaction types, e.g. follow-up versus preceding reactions or concerted versus non-concerted processes, and this methodology remains an important tool in the analysis of voltammetric data. However, commercial software packages [13] for the analysis and fitting of *all* data points in experimental cyclic voltammograms are now available and have broadened the range of possibilities considerably.

To illustrate the uses, benefits, and pitfalls of cyclic voltammetry, an example of a system with complex chemical processes being disentangled step by step based on a systematic simulation procedure may be given [14] (Fig. II.1.1). Prenzler

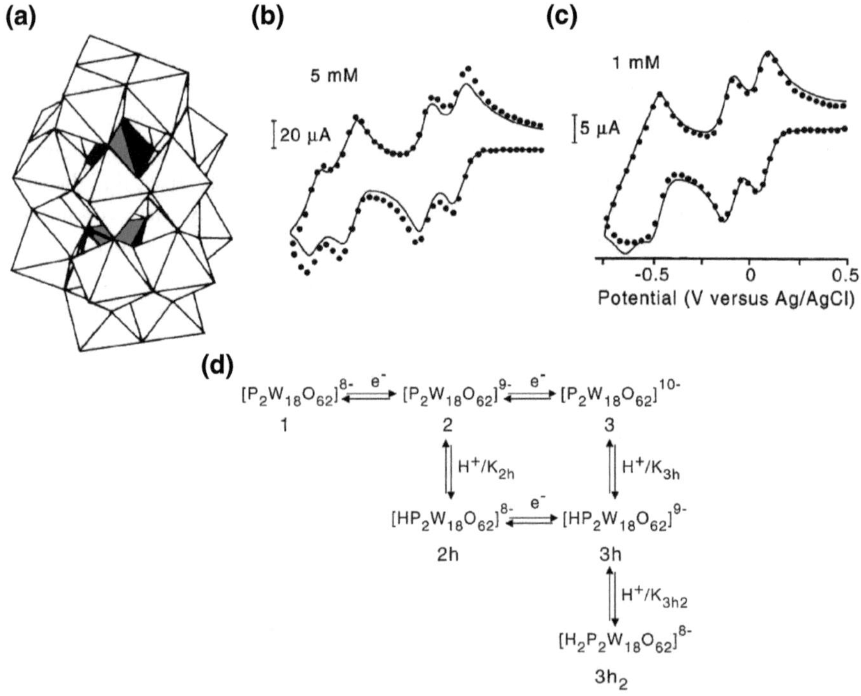

Fig. II.1.1 (**a**) Drawing of a $P_2W_{18}O_{62}^{6-}$ polyoxotungstate cluster anion with Dawson structure. (**b, c**) Cyclic voltammograms of solutions of $K_6[P_2W_{18}O_{62}]_{14}H_2O$ (**b**) 5 mM and (**c**) 1 mM in aqueous 0.5 M NaCl (scan rate 0.1 V s^{-1}). The *dotted line* indicates the results of numerical simulation studies [14]. (**d**) Reaction scheme proposed for the multi-step reduction/protonation reaction of $P_2W_{18}O_{62}^{6-}$ with respect to $[P_2W_{18}O_{62}]^{8-/9-/10-}$

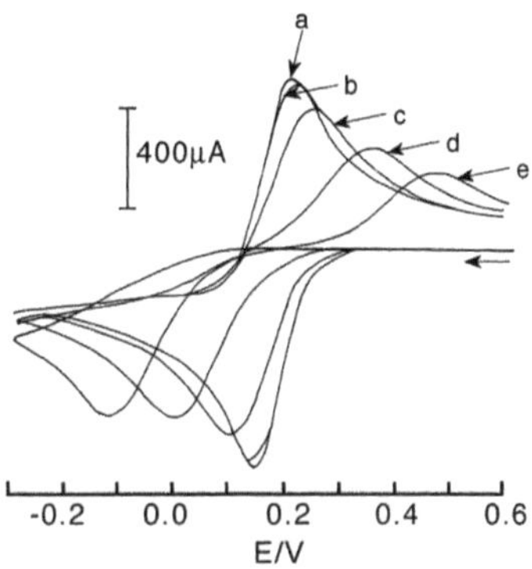

Fig. II.1.2 Simulated cyclic voltammograms (DigiSim) for a reduction process affected by heterogeneous and homogeneous kinetics (see text) [16]

et al. studied the scan rate, pH, and concentration dependence of the reduction of a solution of polyoxoanion, $[P_2W_{18}O_{62}]^{6-}$ (see Fig. II.1.1a), in aqueous media by cyclic voltammetry (Fig. II.1.1b, c) and determined the equilibrium constants for the protonation processes (see Fig. II.1.1d). Additional disproportionation and cross-redox reactions associated with the redox system described in Fig. II.1.1d (Eqs. II.1.1 and II.1.2) are difficult to monitor directly by cyclic voltammetry but still may have subtle effects on cyclic voltammetric data. These so-called thermodynamically superfluous reactions [15] can be derived from the voltammetric data because the equilibrium constant data can be calculated from the formal potentials and protonation equilibrium constants. For the resolution of this type of complex reaction scheme, data obtained over a wide range of conditions and at different concentrations are required.

$$2 + 2h \rightleftarrows 1 + 3h \tag{II.1.1}$$

$$2h + 2h \rightleftarrows 1 + 3h_2 \tag{II.1.2}$$

Unfortunately, in some situations, the interpretation of cyclic voltammetric data can be inconclusive, especially when based solely on the analysis of an inappropriately small data set. Figure II.1.2 shows simulated cyclic voltammograms [16] at a scan rate of 1.0 V s^{-1} for a reaction scheme (Eqs. II.1.3 and II.1.4) involving an electron transfer process with a charge transfer coefficient $\alpha = 0.5$ and standard rate constant k_s = (a) 10^2 m s^{-1}, (b) 3.16×10^{-2} m s^{-1}, (c) 3.16×10^{-3} m s^{-1}, (d) 3.16×10^{-4} m s^{-1}, and (e) 3.16×10^{-5} m s^{-1}, followed by a rapid equilibrium process (Eqs. II.1.3 and II.1.4).

$$A + e^- \rightleftarrows A^- \tag{II.1.3}$$

$$A^- \rightleftarrows B \tag{II.1.4}$$

In these equations, A is reduced in a one-electron process to the product A^-, which in a fast equilibrium process forms B. An equilibrium constant $K_4 = 1000$ and rate constant $k_4 = 10^{10}\,s^{-1}$ for the forward direction of the chemical reaction step were also included in these simulations and it can be seen that the shape and peak-to-peak separation change characteristically with the k_s value. However, essentially the same set of cyclic voltammograms (only offset in potential) may be simulated for the case of a slow electron transfer (Eq. II.1.3) without the chemical follow-up process (Eq. II.1.4), and with $\alpha = 0.5$ and $k_{s,\,app} =$ (a) 3.16 m s^{-1}, (b) 1.0×10^{-3} m s^{-1}, (c) 1.0×10^{-4} m s^{-1}, (d) 1.0×10^{-5} m s^{-1}, and (e) 1.0×10^{-6} m s^{-1}. That is, if the chemical step (Eq. II.1.4) is sufficiently fast in both the forward and backward direction, so that it is in equilibrium on the voltammetric timescale, then the apparent standard rate constant $k_{s,\,app}$ can be written as $k_s \times (K_4)^{-\alpha}$. In this expression α denotes the transfer coefficient ($\alpha \approx 0.5$) that is used to describe the potential dependence of the rate of heterogeneous electron transfer (*vide infra*). Thus, voltammetric data sometimes may be interpreted with more than one mechanistic scheme. In this case, only further experimental evidence can resolve the ambiguity, e.g. voltammetric data obtained over a sufficiently wide range of scan rates or access to independent spectroelectrochemical or other data defining the thermodynamics or kinetics of processes. Finally, it may be noted that the mechanistic cases of a chemical equilibration reaction (Eq. II.1.4) occurring in the solution phase (homogeneous) or on the electrode surface (heterogeneous) may both give rise to essentially the same voltammetric features. The experimental dissection of this and other types of mechanistic ambiguity is aided by use of spectroelectrochemical techniques.

II.1.1.1 Shape of Cyclic Voltammograms

The characteristic shape of 'reversible' voltammetric current responses is governed by mass transport/diffusion processes in the solution phase. In order to understand the origin of this shape, it may be helpful to think of the observed current response at each potential as being 'composed' of two 'simpler' current responses based on (i) the conventional 'transient' or potential step technique (chronoamperometry), in which the decay of the current at a given potential is monitored as a function of time, and (ii) the conventional 'steady-state' technique (polarography, hydrodynamic or micro electrode methods) in which the current is independent of time. Reinmuth [17] depicted this concept of cyclic voltammetry in a three-dimensional plot, as shown in Fig. II.1.3. The current trace in the diagonal cross section (Fig. II.1.3b) shows features similar to those of a voltammogram. Since it combines features of both transient and 'steady-state' techniques in a single experiment, cyclic voltammetry creates the need for a more complex analysis of experiments. However, this enables the required information to be more readily derived compared to what can be discerned from separate application of chronoamperometric and steady-state techniques.

Figure II.1.4 compares the characteristic features of (a) a steady-state process, (b) a potential step experiment, and (c) a cyclic voltammogram. The steady-state experiment is independent of time and gives a sigmoidally shaped response. Most important is the extent to which the concentration profile penetrates into the solution phase. For a steady-state process, there is no time dependence and the diffusion layer thickness, δ, remains constant. In a chronoamperometric or potential step

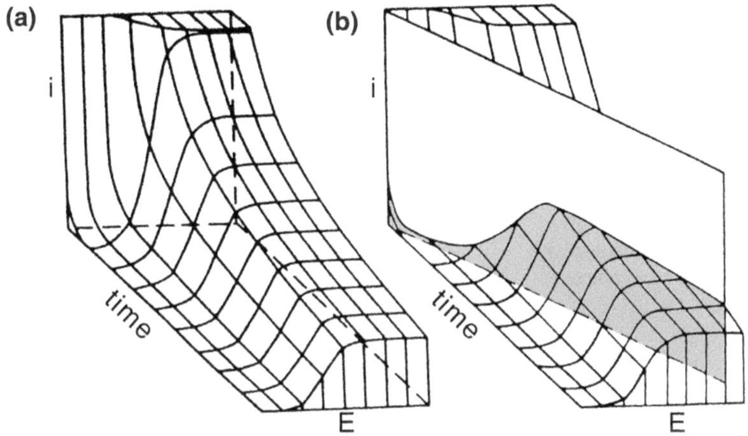

Fig. II.1.3a, b Schematic drawing of a 3D plot with characteristic current–potential–time relationships for chronoamperometric and steady-state responses. The trace following the intersecting plane shows approximately the peak characteristics of a linear sweep voltammogram

experiment, the diffusion layer thickness continuously moves into the solution phase. During the initial course of a cyclic voltammetric experiment, the diffusion layer also moves into the solution phase. However, this is of course followed by a

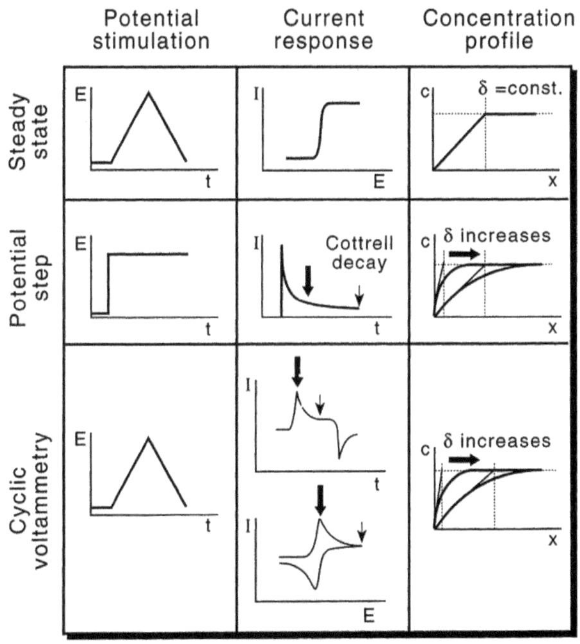

Fig. II.1.4 Plots of the potential stimulation applied in steady-state, chronoamperometric, and cyclic voltammetric experiments together with the resulting current response and the concentration profile in the vicinity of the electrode surface

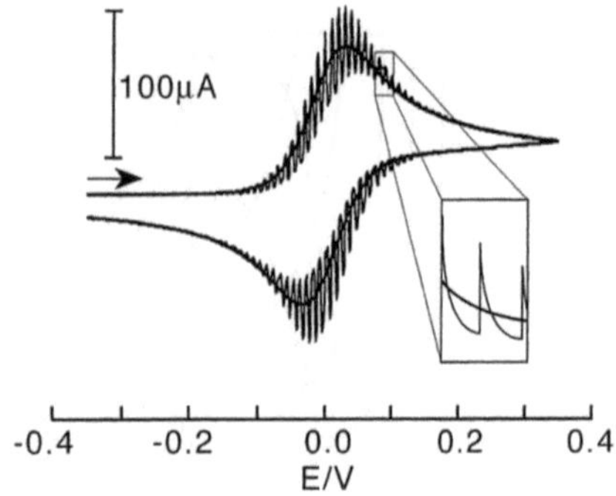

Fig. II.1.5 Comparison of the current responses characteristic for a conventional analogue cyclic voltammogram and for a cyclic staircase voltammogram. Sampling and averaging is used in staircase voltammetry to give conventional voltammograms for sufficiently small potential steps

second change in concentration generated after the reversal of the scan direction. The diffusion layer thickness for an electrochemically reversible process at the time when the peak occurs is a useful benchmark and given approximately by $\delta_{\mathrm{peak}} = const\sqrt{\frac{RTD}{nFv}}$ with $const = 0.446^{-1}$, T, the temperature in K, $R = 8.134\,\mathrm{J\,K^{-1}\,mol^{-1}}$ the gas constant, D, the diffusion coefficient in $\mathrm{m^2\,s^{-1}}$, n, the number of electrons transferred per molecule diffusing to the electrode, $F = 96485\,\mathrm{C\,mol^{-1}}$ the Faraday constant, and v the scan rate in $\mathrm{V\,s^{-1}}$ (*vide infra*).

Although the theory and practice of cyclic voltammetry, as initially developed, was based strictly on the use of a linear potential ramp excitation waveform, there are numerous other voltammetric techniques in which the derivative of the cyclic voltammetric response is analysed [18], or in which sequences of steps (pulse and staircase techniques [19]) or more complex waveforms (see also cyclic AC voltammetry [20] and Fourier transform voltammetry [21]) are employed. The close relationship between the different applied potential techniques in electrochemistry can be understood by considering cyclic staircase voltammetry (Fig. II.1.5). Most commercial computerised instrumentation relies on digital electronics, and, instead of applying an analogue potential ramp, a staircase potential is applied. Measurements are taken in the last part of the step interval. The current response from the staircase voltammogram can be seen to oscillate above and below the current expected for a conventional cyclic voltammogram. By choosing sufficiently small interval times and potential steps, current responses essentially identical to those obtained by analogue instrumental forms of cyclic voltammetry are obtained. However, the distortion of voltammetric signals caused by the use of large steps has to be considered and, indeed, may even be usefully applied for particular purposes. If staircase voltammetry is applied uncritically, considerable errors, e.g. in the determination of the charge under a peak by integration, are possible.

Applications of cyclic voltammetry are extensive and include the analysis of solids [22] as well as solutions, media with and without added supporting electrolyte, emulsions and suspensions [23], frozen solutions [24], polymers [25], membrane and liquid/liquid systems [26], and biological systems such as enzymes [27] or cultures of bacteria [28].

Fundamental limitations of cyclic voltammetry, such as the low resolution with respect to structural information, need to be addressed by coupling of the technique with structurally richer spectroscopic or other methods of detection. For example, simultaneous application of cyclic voltammetry and various types of spectroscopy [29], diffraction techniques [30], quartz crystal microbalance [31] and plasmon resonance [32] experiments provide the detailed structural information of chemical changes that accompany an electron transfer process. Thus, a warning needs to be provided that proposals of complex mechanistic schemes based on cyclic voltammetry *only* should be treated with caution. Credibility requires that such mechanisms be verified by independent measurements based on in situ or ex situ spectroelectrochemical or other techniques combined with voltammetric measurements.

II.1.2 Basic Principles

The most common experimental configuration for recording cyclic voltammograms consists of an electrochemical cell (Fig. II.1.6a) that has three electrodes, i.e., counter or auxiliary electrode (C), reference electrode (R), and working electrode (W), all immersed in a liquid and connected to a potentiostat. The potentiostat [33] (a simplified schematic is shown in Fig. II.1.6b) allows the potential difference between the reference and working electrode to be controlled with minimal interference from *IR* (ohmic) drop. In this configuration, the current flowing through the reference electrode also can be minimised thereby avoiding polarisation of the reference electrode and hence keeping the applied potential distribution between the working and reference electrode stable (see also Sect. III.4.5).

Positioning the reference electrode (or Luggin probe [34]) close to the working electrode further helps to minimise the *IR* drop between the reference and working

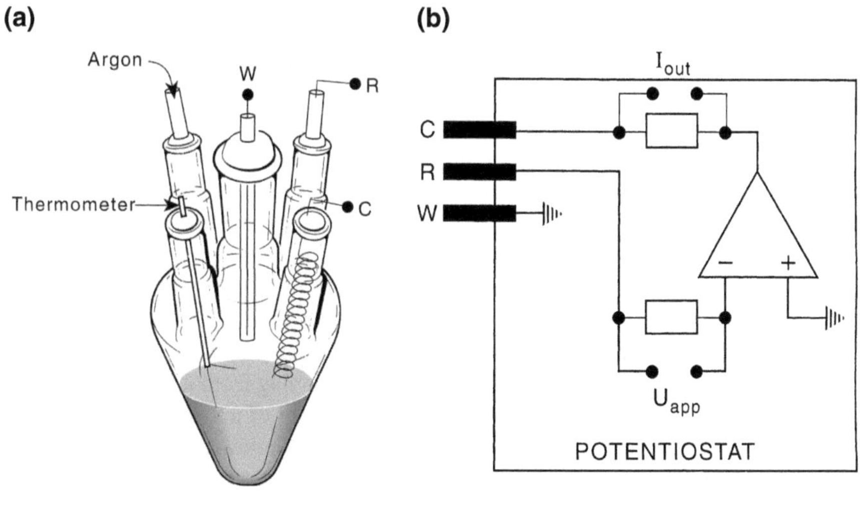

Fig. II.1.6 (**a**) Cell for cyclic voltammetric experiments with three-electrode configuration. (**b**) Schematic drawing of the simplified circuit diagram for a three-electrode potentiostat system

electrode due to the resistivity of the solution phase. Instrumental methods, based on positive feedback or other circuitry, for compensation of uncompensated resistance are also available [4]. The electrodes commonly used for cyclic voltammetric experiments are:

(i) *Reference electrode*: commonly used are aqueous Ag/AgCl or calomel half cells [34], which can be obtained commercially or easily prepared in the laboratory. Sometimes, when a suitable conventional reference electrode is not available (e.g. for some dry organic solvents or room temperature ionic liquids) or when they introduce problems with salt leakage or junction potentials, pseudo-reference electrodes, such as a simple silver or platinum wire, are used in conjunction with an internal potential reference provided by ferrocene [35]. Experimentally, ferrocene is added into the cell at the end of a series of measurements, and the reversible voltammetric response for the $Fc^{+/0}$ couple is taken as reference point on the potential scale ($E_{mid} = 0.0$V, where $E_{mid} = {}^{1}/_{2}\left(E_{p,c} + E_{p,a}\right)$), and it is assumed that $E_{mid} \approx E_{c}^{\ominus'}$ for the case of equal diffusion coefficients for Fc^{+} and Fc^{0}) (see also Chap. III.2).
(ii) *Counter electrode*: non-reactive high surface area electrode, commonly platinum gauze.
(iii) *Working electrode*: most commonly used are inlaid disc electrodes (Pt, Au, graphite, glassy carbon, etc.) of well-defined area. Other geometries may be employed beneficially in appropriate circumstances (dropping or hanging mercury hemisphere, cylinder, band, arrays, or grid electrodes) with the appropriate modification to the data analysis. (See also Chap. III.1)

For chemical systems requiring a very dry and oxygen-free environment in organic solvents such as acetonitrile, tetrahydrofuran, or dichloromethane, it may be necessary to work under more demanding conditions than required when an electrochemical cell is operated standing on a laboratory bench top. For example, reduction processes at potentials negative of ca. −2.0 V vs Ag/AgCl are often very sensitive to traces of water. In these situations, glove-box techniques [36] have been used or alternatively sophisticated closed glassware systems (Kiesele cell [37]) connected to a Schlenk line apparatus [38]. An alternative and versatile experimental system, which allows voltammetric experiments to be undertaken under dry conditions and in inert atmosphere, is shown in Fig. II.1.7. With this kind of apparatus, the electrochemical cell may be dried initially under vacuum by heating with a hot air gun. Next the supporting electrolyte, e.g. NBu_4PF_6, is placed into the cell under argon and is converted to the molten state by heating under vacuum. Dry solvent is allowed into the cell under an inert and dry atmosphere by pumping from a reservoir through a stainless steel cannula or freshly dried from a column of activated alumina [39]. Significant improvements of the accessible potential window are also possible by directly applying freeze–pump–thaw cycles [40].

The liquid phase in an electrochemical experiment typically consists of a solvent containing the dissolved material to be studied and a supporting electrolyte salt to

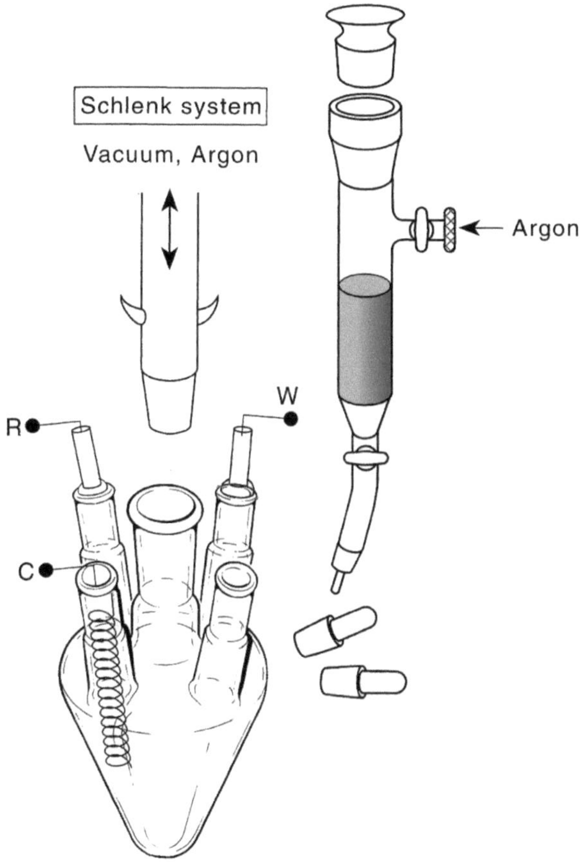

Fig. II.1.7 Schematic drawing of the three-electrode cell connected to a vacuum system

achieve the required conductivity and hence minimise the IR_u potential drop. With sufficient supporting electrolyte, the electrical double layer (see also Chap.I.1) at the working electrode occupies a distance of about 1 nm from the electrode surface (Fig. II.1.8). Note that the length scale in Fig. II.1.8 is not linear. This layer has been shown to consist of a compact or 'inner Helmholtz' layer and the diffuse (*not* diffusion layer) or 'Gouy–Chapman' layer [41]. The extent to which the diffuse layer extends into the solution phase depends on the concentration of the electrolyte and the double layer may in some cases affect the kinetics of electrochemical processes. Experiments with low concentrations or no added supporting electrolyte can be desirable [42] but, since the double layer becomes more diffuse, they require careful data analysis. Furthermore, the IR drop is extended into the diffusion layer [43] (see also Sect. III.4.5).

Under most experimental conditions, the size of the diffusion layer (see above) is several orders of magnitude larger than that of the diffuse layer (see Fig. II.1.8).

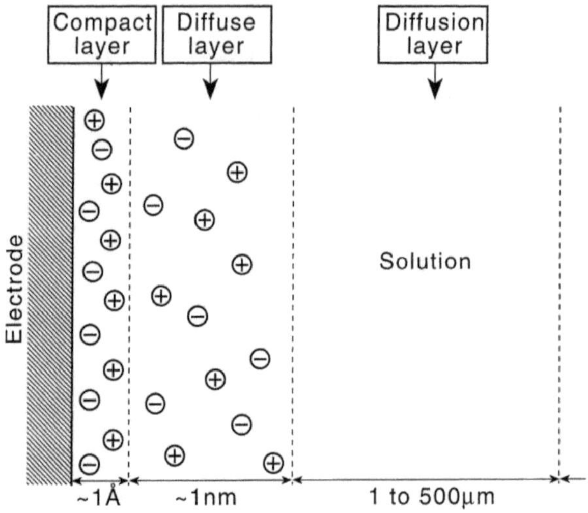

Fig. II.1.8 Schematic representation of the composition of the solution phase in the vicinity of the electrode surface (not to scale)

Initiated by a change in the electrode potential, a concentration perturbation travels away from the electrode surface into the solution phase and the time, t, required for a distance $\delta_{\text{diffusion}}$ to be affected, may be estimated via the use of Eq. (II.1.5) [44] where D is the diffusion coefficient.

$$\delta_{\text{diffusion}} = \sqrt{\frac{4Dt}{\pi}} \tag{II.1.5}$$

With very fast scan rate cyclic voltammetry, an upper limit of the scan rate where standard theory prevails is given by the condition that the *diffusion* layer becomes equal in size to the *diffuse* layer (see Fig. II.1.8). It has been estimated that this limit occurs at a scan rate of 1–2 × 10^6 V s^{-1} [45]. In the other extreme, at very slow scan rates, natural convection is known to affect the shape of experimental cyclic voltammograms.

A series of conventional cyclic voltammograms obtained for a model system, the oxidation of 3.3 mM ferrocene in acetonitrile containing 0.1 M NBu_4PF_6 as supporting electrolyte at a 0.4-mm diameter platinum electrode, is shown in Fig. II.1.9. The current scale has been normalised by dividing by the square through the square root of scan rate in order to compare data obtained at different scan rates. Data obtained from these cyclic voltammograms are summarised in Table II.1.1.

The analysis and methodology for the extraction of characteristic parameters obtained from cyclic voltammograms is shown in Fig. II.1.9b. A zero current line for the forward scan data has to be chosen (dashed line) as baseline for the determination of the anodic peak current. For the reverse sweep data the extended forward scan (dashed line with Cottrell decay) is folded backwards (additionally accounting for capacitive current components) to serve as the baseline for the determination of the

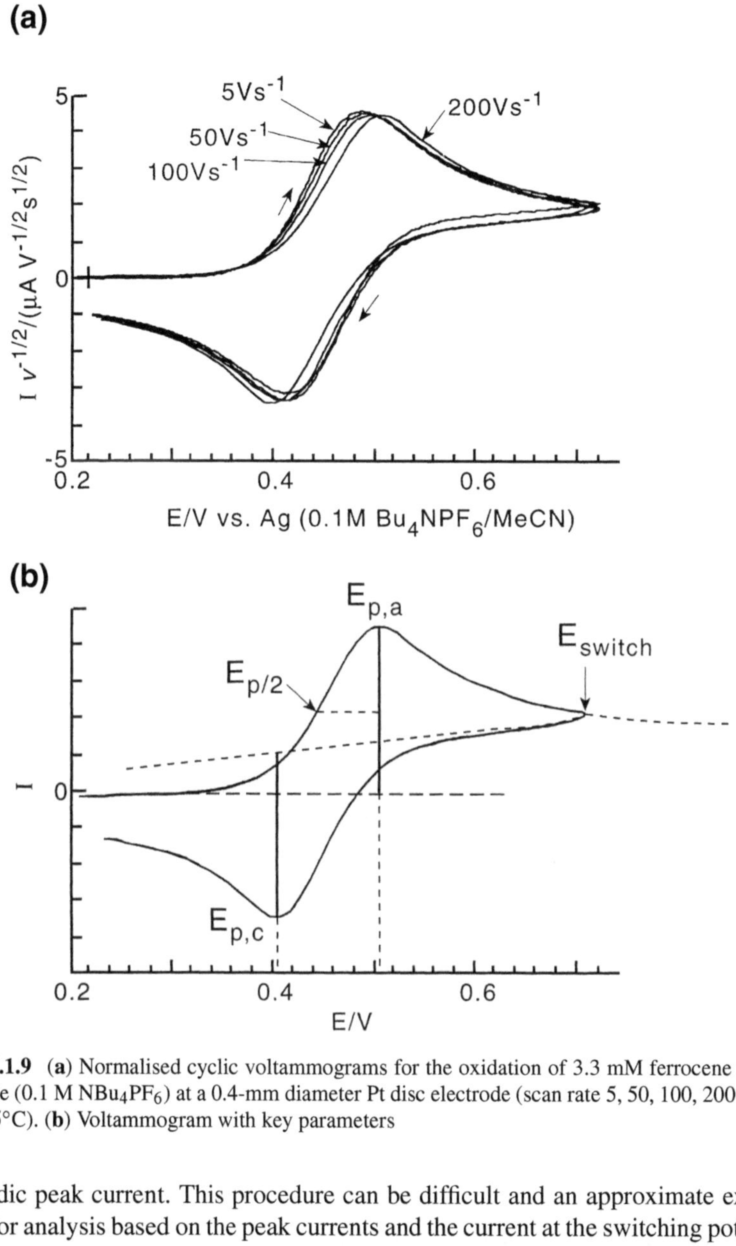

Fig. II.1.9 (**a**) Normalised cyclic voltammograms for the oxidation of 3.3 mM ferrocene in acetonitrile (0.1 M NBu_4PF_6) at a 0.4-mm diameter Pt disc electrode (scan rate 5, 50, 100, 200 V s^{-1}, $T = 25°C$). (**b**) Voltammogram with key parameters

cathodic peak current. This procedure can be difficult and an approximate expression for analysis based on the peak currents and the current at the switching potential has been proposed as an alternative [46]. If the blank current before the anodic peak starts cannot be neglected, this current has to be extrapolated into the range where the peak occurs, or, if possible, has to be subtracted from the sample voltammogram. Also, when the sample solution does not contain only the reduced form (as supposed in Fig. II.1.9), but the oxidised form as well (or only), the evaluation of the peak current needs more care (see Fig. II.1.15). The peak current data, $I_{p,a}$ and $I_{p,c}$,

Table II.1.1 Cyclic voltammetric data for the oxidation of 3.3 mM ferrocene in 0.1 M NBu_4PF_6 in acetonitrile at a 0.4-mm diameter Pt disc electrode at 25°C

Scan rate/V s^{-1}	E_{mid}/V vs. Ag[(a)]	$\Delta E_p/10^{-3}$ V [(b)]	$I_p/10^{-6}$ A [(c)]
5	0.45	59	10.3
50	0.45	71	32.5
100	0.45	85	46.2
200	0.45	104	65.2

[a] Measured from the value $1/2\left(E_{p,c} + E_{p,a}\right)$; versus a silver pseudo-reference
[b] $\Delta E_p = E_{p,a} - E_{p,c}$.
[c] Peak current for the oxidation component of the process.

and peak potentials, $E_{p,a}$ and $E_{p,c}$, for the anodic and cathodic processes, respectively, are then determined and the midpoint potential, E_{mid}, is calculated from $1/2\left(E_{p,c} + E_{p,a}\right)$ (see also Sect. I.2.4). For a fully reversible voltammetric response E_{mid} often coincides with the half-wave potential obtained based on the half-peak potential, the potential at which the current reaches half of the peak current, $E_{p/2,a}$ or $E_{p/2,c}$ (see Fig. II.1.9b). The midpoint potential E_{mid} (often also called half-wave potential $E_{1/2}$) is linked to the thermodynamically defined formal potential $E_c^{\ominus'}$ via $E_{mid} = E_c^{\ominus'} + \frac{RT}{2nF} \ln\left(\frac{D_{Red}}{D_{Ox}}\right)$ (See also Sect. I.1.2.3). Since D_{Red} and D_{Ox} (D_{Red} and D_{Ox} denote the diffusion coefficients for the reduced and the oxidised forms, respectively) are normally close to equal, $E_{mid} = E_{1/2} \approx E_c^{\ominus'}$. The term $E_{1/2}$ has its origin in the polarography literature, where it represents the potential at half the limiting current from a sigmoidally shaped polarogram [8]. This term is retained in cyclic voltammetry and will appear in a number of relationships given later in the text noting that it can be equated to $E_c^{\ominus'}$ in most practical situations. Another important parameter for the analysis of voltammetric responses, i.e., the peak current ratio, $I_{p,a}/I_{p,c}$, can be determined. This ratio should be 1.0 for a reversible voltammetric response.

It can be seen that cyclic voltammograms at low scan rate have peak-to-peak separations close to the value theoretically expected for a reversible process of $\Delta E_p = 2.218 \times RT/F = 57\,\text{mV}$ at 298 K [47] and the peak current increases with the square root of the scan rate. Under these conditions, the process is diffusion controlled and termed electrochemically 'reversible' or 'Nernstian' within the timescale applicable to the experiment under consideration. Hence, as with all reversible systems operating under thermodynamic rather than kinetic control, no information concerning the rate of electron transfer at the electrode surface or the mechanism of the process can be obtained from data obtained at slow scan rate. The increase of ΔE_p at faster scan rate may be indicative of the introduction of kinetic control on the shorter timescale now being applied (hence the rate constant could be calculated) or it may arise because of a small amount of uncompensated resistance. Considerable care is required to distinguish between these two possible origins of enhancement of ΔE_p. For example, repetition of the experiments in Table II.1.1 at a different concentration of ferrocene would exhibit a systematic dependence of ΔE_p with concentration if uncompensated resistance is present. Alternatively, the value of the rate constant for heterogeneous electron transfer (*vide infra*) should be shown to be independent of concentration. Ideally, all electrochemical kinetic studies should be undertaken over a range of concentrations!

The nature of the physical processes responsible for the shape of an electrochemically 'reversible' voltammogram is based on only two laws: (i) Fick's law of diffusion for the case of a planar electrode (Eq. II.1.6), and (ii) Nernst's law (Eq. II.1.7).

$$\frac{d[A]}{dt} = D\frac{d^2[A]}{dx^2} \tag{II.1.6}$$

$$\frac{[A]_{x=0}}{[B]_{x=0}} = e^{\frac{nF}{RT}(E(t)-E_c^{\ominus'})} \tag{II.1.7}$$

Nernst's law (Eq. II.1.7, which is simply the Nernst equation written in the exponential form) defines the surface concentrations of the oxidised, $[A]_{x=0}$, and the reduced form, $[B]_{x=0}$, of the redox reagents for a reduction process $A + ne^- \rightleftarrows B$ as a function of $E(t)$ and $E_c^{\ominus'}$, the applied and the formal reversible potential [4], respectively, where t is time, n is the number of electrons transferred per molecule of A reacting at the electrode surface, F, the Faraday constant, R, the constant for an ideal gas, and T, the absolute temperature. Fick's second law of diffusion (Eq. II.1.6) governs the mass transport process towards the electrode where D is the diffusion coefficient. The parameter x denotes the distance from the electrode surface.

For a reversible diffusion-controlled process, the concentration profiles for [A] and [B] may be elegantly expressed as a function of current and time (Eqs. II.1.8 and II.1.9) via a mathematical procedure called semi-integration $\left(d^{-1/2}y/dx^{-1/2}\right)$ or convolution [48]. Semi-integration is an option built into the control software of many modern computer-controlled instruments. With the semi-integral technique, the concentrations $[A]_{x=0}$ and $[B]_{x=0}$ can be expressed as a function of the current, $I(t)$, and since $E(t) = E_{\text{initial}} - vt$ (v is the scan rate), the current response at each potential for a reversible cyclic voltammogram may be calculated (Eq. II.10) from Nernst's law (Eq. II.1.7) and Eqs. (II.1.8) and (II.1.9).

$$[A]_{x=0} = [A]_{\text{bulk}} + \frac{1}{nFA\sqrt{D}}\frac{d^{-1/2}I(t)}{dt^{-1/2}} \tag{II.1.8}$$

$$[B]_{x=0} = \frac{1}{nFA\sqrt{D}}\frac{d^{-1/2}I(t)}{dt^{-1/2}} \tag{II.1.9}$$

$$I(t) = -nFA\,[A]_{\text{bulk}}\sqrt{D}\frac{d^{1/2}}{dt^{1/2}}\left(\frac{1}{1+e^{\frac{nF}{RT}\left(E_{\text{initial}}-vt-E_c^{\ominus'}\right)}}\right) \tag{II.1.10}$$

The relationship given in Eq. (II.1.10) can be used directly to derive the theoretical shape of a reversible cyclic voltammogram by applying semi-differentiation to the expression in brackets [49]. Additionally, the concentrations of $[\mathrm{A}]_{x=0}$ and $[\mathrm{B}]_{x=0}$ at the electrode surface at any time t during the course of the cyclic voltammetric experiment may be calculated by semi-integration of the experimental voltammogram (Eqs. II.1.8 and II.1.9).

Many years ago, a theoretical expression for the peak current for a reversible cyclic voltammogram was derived as a function of the scan rate to give the Randles–Ševčik expression [50] (Eq. II.1.11). According to this relationship, the dependence of the peak current, I_p, on scan rate, v, follows a characteristic square-root law, which provides a tell-tale sign of the presence of a diffusion-controlled process.

$$I_\mathrm{p} = -0.446\, nFA\, [\mathrm{A}]_\mathrm{bulk} \sqrt{\frac{nFvD}{RT}} \tag{II.1.11}$$

The negative sign is used here in order to conform to a reduction process. However, usually, this equation is stated without sign. Note the $n^{3/2}$ dependence of the peak current, which implies a non-linear increase for the case of n simultaneously transferred electrons (*vide infra*).

The diagnostic criteria and characteristics of the cyclic voltammetric response for a reversible heterogeneous electron transfer process, denoted E_rev, are summarised below together with expressions for the determination of $E_\mathrm{c}^{\ominus'}$ and n. The determination of n as well as of several other important parameters is dependent on the knowledge of the diffusion coefficient. In some cases the diffusion coefficient may be estimated based on values for similar compounds in the same solvent or based on empirical equations [51].

– **$\mathbf{E_{rev}}$ Diagnostics:**

$|I_\mathrm{p,a}/I_\mathrm{p,c}| = 1.0$ and $\Delta E_\mathrm{p} = 2.218\, RT/nF = 57/n$ mV at 298 K independent of the scan rate v. I_p is proportional to $v^{1/2}$ (Randles–Ševčik).

$|E_\mathrm{p} - E_\mathrm{p/2}| = 2\,|E_\mathrm{p} - E_{1/2}| = 2.218\,(RT/nF) \approx \ln(9)\,(RT/nF)$

$E_\mathrm{c}^{\ominus'} = E_\mathrm{mid} - (RT/2nF)\ln(D_\mathrm{Red}/D_\mathrm{Ox})$ where D_Ox and D_Red denote the diffusion coefficients for the oxidised and reduced forms, respectively)

n: (with D known or estimated) calculate n from the peak currents (Randles–Ševčik) or, better, from the steady-state limiting current (most reliable), or from the convoluted voltammogram.

Reversible cyclic voltammograms are not always governed by diffusion-controlled processes. For example, the cases of a redox reagent adsorbed onto an electrode surface or confined to a thin layer of solution adjacent to the electrode surface are also of considerable importance. In fact, the same theory may be applied to both adsorbed layers [52] and processes that occur in thin layers [53] (thinner than the diffusion layer). In both these cases, the current for the reversible process can be derived by substitution of the expression $I(t) = nFA\frac{\mathrm{d}[\mathrm{A}]_{x=0}}{\mathrm{d}t}$ into the Nernst equation (Eq. II.1.7) and noting that $E(t) = E_\mathrm{initial} - vt$, $[\mathrm{B}]_{x=0} = [\mathrm{A}]_\mathrm{bulk} - [\mathrm{A}]_{x=0}$, and V is the volume of the thin layer (Eq. II.1.12a)

$$I(t) = \frac{-n^2F^2vV\,[\mathrm{A}]_{\mathrm{bulk}}\,\mathrm{e}^{\frac{nF}{RT}\left(E_{\mathrm{initial}}-vt-E_{\mathrm{c}}^{\ominus\prime}\right)}}{RT\left\{1+\mathrm{e}^{\frac{nF}{RT}\left(E_{\mathrm{initial}}-vt-E_{\mathrm{c}}^{\ominus\prime}\right)}\right\}^2} \tag{II.1.12a}$$

Strongly adsorbed materials are described by the surface coverage, Γ_0 (in mol m^{-2}), which may be related to the concentration in the thin layer case by $A\Gamma_0 = V\,[\mathrm{A}]_{\mathrm{bulk}}$ (Eq. II.1.12b)

$$I(t) = \frac{-n^2F^2vA\Gamma_0\,[\mathrm{A}]_{\mathrm{bulk}}\,\mathrm{e}^{\frac{nF}{RT}\left(E_{\mathrm{initial}}-vt-E_{\mathrm{c}}^{\ominus\prime}\right)}}{RT\left\{1+\mathrm{e}^{\frac{nF}{RT}\left(E_{\mathrm{initial}}-vt-E_{\mathrm{c}}^{\ominus\prime}\right)}\right\}^2} \tag{II.1.12b}$$

From these expressions it can be deduced that, in contrast to cyclic voltammetric responses for solution systems with semi-infinite planar diffusion, for redox processes for confined or thin layer systems close to the electrode or adsorbed molecules at the electrode surface, the peak current expression becomes *linear* with respect to the scan rate (Eqs. II.1.13a and II.1.13b).

$$I_{\mathrm{p}} = -\frac{n^2F^2}{4RT}vV\,[\mathrm{A}]_{\mathrm{bulk}} \tag{II.1.13a}$$

$$I_{\mathrm{p}} = -\frac{n^2F^2}{4RT}vA\Gamma_0 \tag{II.1.13b}$$

The width at half height, ΔE_{hh}, of the voltammetric peak shown in Fig. II.1.10 (for both the adsorption and the thin layer case) can be determined as $\Delta E_{\mathrm{hh}} = 3.53\,(RT)\,/\,(nF) = 90.6\,\mathrm{mV}$ for a one-electron process at 25°C [54]. For the case of strongly adsorbed systems, any deviation from the ideal or 'Langmurian' case of no interaction between individual redox centres on the electrode surface manifests itself as a change of ΔE_{hh}. Interpretations based on regular solution theory models have been suggested to account for non-ideal behaviour [55]. Of course, departure from reversibility also leads to changes in wave shape as a function of scan rate [56]. Further mechanistic details for the case of surface-confined reactions have been discussed [57].

II.1.3 Effects Due to Capacitance and Resistance

Cyclic voltammetric data, even when using a three-electrode potentiostat form of instrumentation, can be affected by the presence of uncompensated resistance in the solution phase between the reference and the working electrode and of the working electrode itself, and by the capacity of the working electrode. A 'simple' electrochemical cell may be described by an equivalent circuit, in which electronic

Fig. II.1.10 Theoretical cyclic voltammogrammetric trace for a reversible system adsorbed on an electrode surface

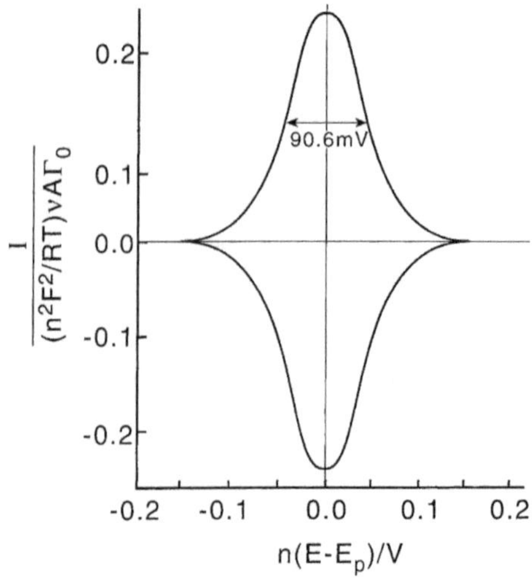

components are used to describe the behaviour of the cell [58]. In Fig. II.1.11 the uncompensated solution resistance and the double-layer capacity at the working electrode are represented by R_u and C_w, respectively. More complex, but more realistic, equivalent circuits have been proposed [59] (see also Chap. II.5).

The separation between the reference and the working electrode as well as the cell geometry has a considerable effect on the IR_u drop present between reference and working electrode. It is quite clear that the configuration shown in Fig. II.1.11b creates a higher IR_u drop between the working and reference electrode (reference located in the 'current path') compared to the configuration shown in Fig. II.1.11c. This effect becomes dramatic in flow systems in which reference, working, and counter electrodes are connected via tubing in a flow system filled with electrolyte solution [60]. The configuration shown in Fig. II.1.11c minimises the effect of R_u and is therefore preferable as long as instability of the potentiostat-cell system resulting from a higher resistance between counter and reference electrode can be avoided.

In Fig. II.1.12, cyclic voltammograms incorporating both IR_u drop and capacitance effects are shown. Effects for the ideal case of a potential independent working electrode capacitance give rise to an additional non-Faradaic current (Fig. II.1.12b) that has the effect of adding a current, $I_{capacitance} = C_W \times v$, to both the forward and backward Faradaic current responses. The capacitance, C_W, is composed of several components, e.g. double layer, diffuse layer, and stray capacitance, with the latter becoming relatively more important for small electrodes [61]. On the other hand, the presence of uncompensated resistance causes a deviation of the applied potential from the ideal value by the term $R_u \times I$, where R_u denotes the uncompensated resistance and I the current. In Fig. II.1.12, the shift of the peak potential, and indeed the entire curve due to the resistance, can clearly be seen. If the value of R_u is known (or can be estimated from the shape of the electrochemically reversible

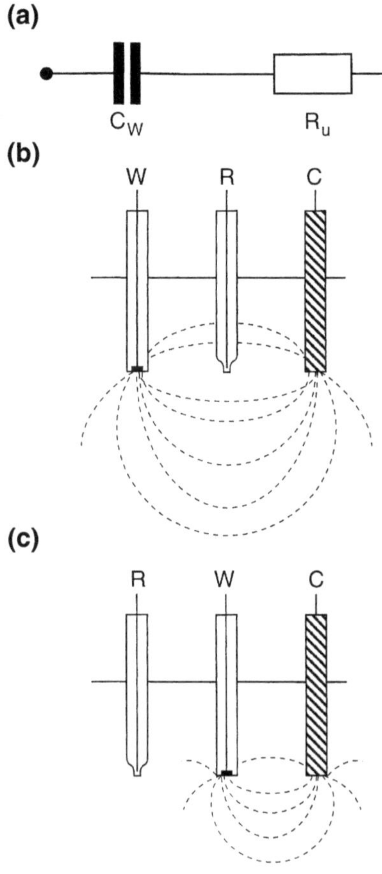

Fig. II.1.11 (**a**) RC representation of the simplest equivalent circuit for an electrochemical cell. (**b, c**) Electrode configuration with *dashed lines* indicating the flow of current accompanied by potential gradients through the solution phase

voltammetric response as an 'internal standard' redox reagent), the voltammograms may be corrected approximately by adjusting the potential axis. However, when both capacitance and resistance effects are accounted for simultaneously via this method, effectively, the IR_u drop contribution modifies the scan rate, which can lead

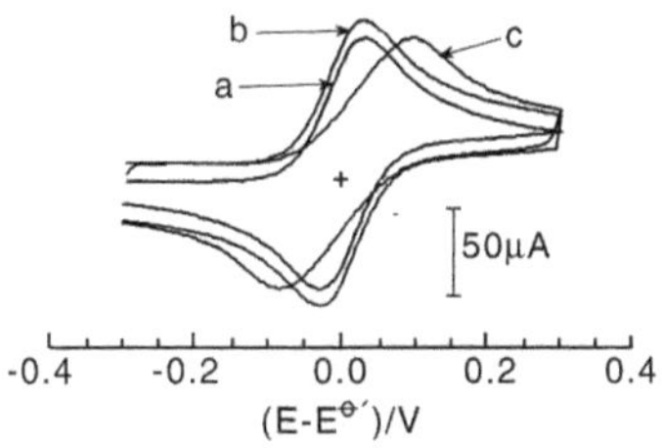

Fig. II.1.12 Simulated cyclic voltammograms (DigiSim) indicating the effect of interfacial capacity and solution resistance: *a* reversible cyclic voltammogram ($D = 10^{-9}\,m^2 s^{-1}$, $c = 1$ mM, $A = 10^{-6} m^2$, scan rate 10 V s^{-1}), *b* with the effect of capacity, $C_W = 10^{-7}$ A s $V^{-1} = 10^{-7}$ Farad, and *c* with additional uncompensated resistance, $R_u = 600\ \Omega$

to difficulties in the comparison of theory and experiment. More accurate correction terms have been introduced [62] (Eq. II.1.14).

$$I_F = I_{total} - I_{residual} - I_{capacitance} + R_u\, C_W\, dI_{total}/dt \qquad \text{(II.1.14)}$$

With this method, the Faradaic current I_F can be obtained from the total current after subtraction of residual and capacitive current components (the 'background' current) and addition of a term reflecting the distortion of the potential scale caused by the IR_u drop. Another important aspect of solution resistance is revealed by examination of the schematic drawing in Fig. II.1.11b, c. Different points on the surface of the working electrode have a different IR_u drop relative to the reference electrode and therefore a potential gradient exists across the working electrode. With this configuration, and when the electrode potential is scanned, the electrochemical process will commence initially on the side closer to the reference electrode and then spread over the entire electrode surface. Accounting for this form of distortion of the cyclic voltammogram is very difficult.

II.1.4 Electrode Geometry, Size, and Convection Effects

Most of the initial practical and theoretical work in cyclic voltammetry was based on the use of macroscopic-sized inlaid disc electrodes. For this type of electrode, planar diffusion dominates mass transport to the electrode surface (see Fig. II.1.13a). However, reducing the radius of the disc electrode to produce a micro disc electrode leads to a situation in which the diffusion layer thickness is of the same dimension as the electrode diameter, and hence the diffusion layer becomes non-planar. This non-linear or radial effect is often referred to as the 'edge effect' or 'edge diffusion'.

In Fig. II.1.13, it can be seen that reducing the size of the electrode increases the mass flux and hence the current density. For the case when radial diffusion becomes dominant, a sigmoidally shaped steady-state voltammetric response is established, which is independent of the scan rate. The transition between planar and radial (or spherical) diffusion is dependent on the scan rate and the size of the disc electrode (Fig. II.1.13b). Horizontal lines in this figure indicate regimes where steady-state conditions corresponding to dominance of radial diffusion prevail.

In more general terms, the geometry of the diffusion field may be divided into convex (e.g. spherical electrodes), planar, and concave (e.g. tubular electrodes, or finite boundary cells) cases. The functional dependence of the peak current on the scan rate (Randles–Ševčik), $I_p = F(v^x)$, may change continuously with $x \to 0$ for convex (convergent) geometries and $x \to 1$ for concave (divergent) geometries. Table II.1.2 contains a summary of Randles–Ševčik expressions for the scan rate dependence of the peak current, I_p, for various electrode geometries. The general Randles–Ševčik expression for a reduction process is given in Eq. (II.1.15).

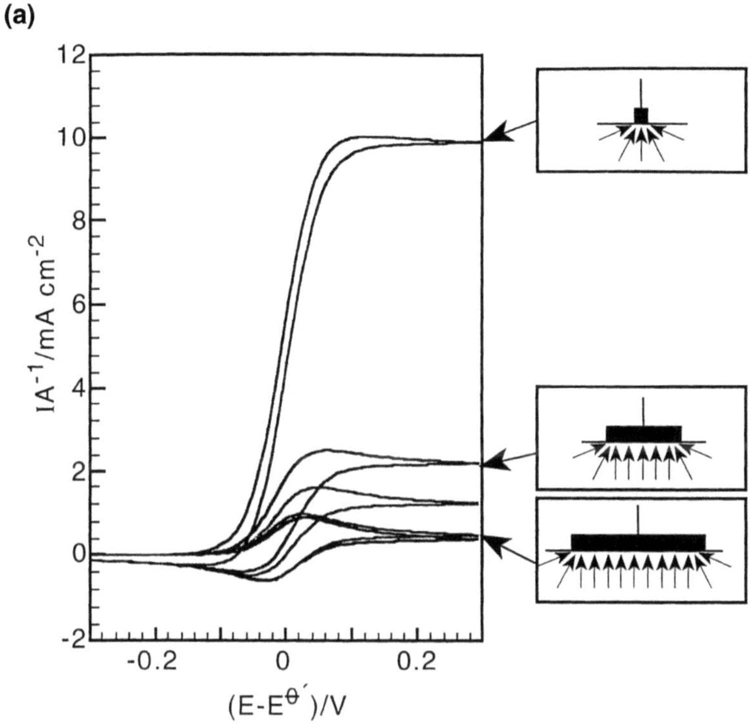

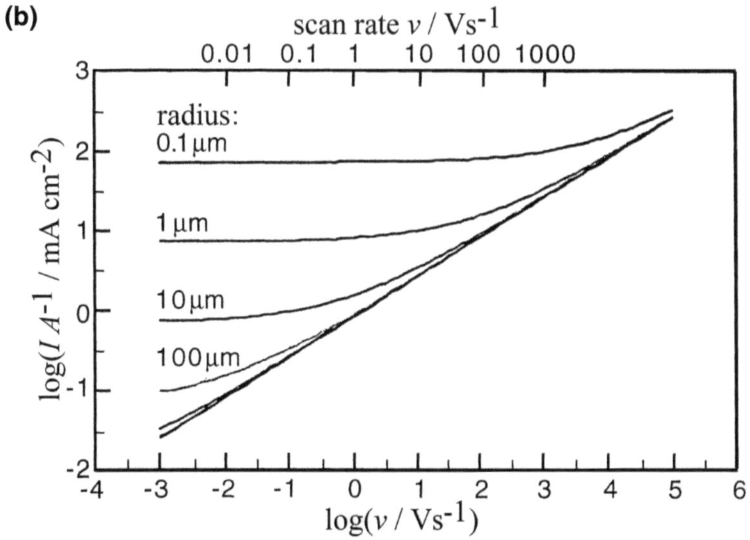

Fig. II.1.13 (**a**) Effect of the electrode geometry on the shape of reversible cyclic voltammograms and schematic drawings of the corresponding transition from planar semi-infinite to spherical diffusion. (**b**) Plot of the peak or limiting current vs. scan rate for different electrode radii assuming $D = 10^{-9}\ \mathrm{m^2\ s^{-1}}$

Table II.1.2 Randles–Ševčik expressions (see Eq. II.1.15) for reversible electron transfer and commonly used electrode geometries [11, 63–65]

Geometry	$\Psi_{\text{peak}}(p)$
In stagnant solution Planar disk electrode(r = radius)	0.446
Spherical or hemispherical electrode (r = radius)	$0.446 + 0.752p^{-1}$
Small disk electrode (r = radius)	$0.446 + \left(0.840 + 0.433 \times \mathrm{e}^{-0.66p} - 0.166\,\mathrm{e}^{-11/p}\right)p^{-1}$ $\approx 0.446 + 4/\pi p^{-1}$
Cylinder or hemicylinder (r = radius)	$0.446 + 0.344p^{-0.852}$
Band electrode ($2r$ = width)	$0.446 + 0.614\left(1 + 43.6p^2\right)^{-1} + 1.323p^{0.892}$ $\approx 0.446 + 3.131p^{-0.892}$
In hydrodynamic systems Planar diffusion to a uniformly accessible electrode, e.g. for rotating disk electrodes (hypothetical Nernst model with δ = diffusion layer thickness)	$0.446 + \left[\delta\sqrt{\frac{nFv}{RTD}}\right]^{-1}$

$$I_{\text{p}} = -\psi_{\text{peak}}(p)\sqrt{\frac{n^3F^3vD}{RT}}A\,[\text{A}]_{\text{bulk}} \tag{II.1.15}$$

with $p = r\sqrt{\frac{nFv}{RTD}}$.

In these expressions, p is a dimensionless parameter describing the scan rate for a given characteristic length, r. In particular, for $p = 1$, a transition occurs from one type of diffusion to another. For the case of a disc electrode (see Fig. II.1.13a), the characteristic length r can be identified as the radius of the disc. For sufficiently fast scan rates or $p \gg 1$, the case of planar diffusion applies (see Eq. II.1.11) and $\psi_{\text{peak}}(p) = 0.446$ (see Table II.1.2). On the other hand, for very slow scan rates or $p \ll 1$, spherical diffusion towards the disc electrode has to be considered dominant and therefore $\psi_{\text{peak}}(p) \approx 0.446 + 4/\pi p^{-1}$. In the steady-state limit $\psi_{\text{steady state}}(p) = p^{-1}\lim\limits_{p\to 0}\left[p\psi_{\text{peak}}(p)\right] = p^{-1}4/\pi$, and Eq. II.1.15 becomes $I_{\text{lim}} = -4\,\text{n}FDr\,[\text{A}]_{\text{bulk}}$, the expression for the limiting current for a reduction process at a microdisc electrode with radius r.

It is important to note that, in all experiments in which a transition of diffusion geometries is observed and I_{p} is a function of p (or δ), the voltammetric measurement may be used to determine simultaneously both the concentration and the diffusion coefficient of the redox active compound. For example, one measurement in the steady-state limit and the other in the planar diffusion limit can be used. Alternatively, data obtained over a range of different scan rates can be analysed. For the case of a disc electrode, a plot of $I_{\text{p}} \times p^{1/2}$ versus p, the dimensionless scan rate, is predicted to result in a parabola. After fitting the experimental data to the

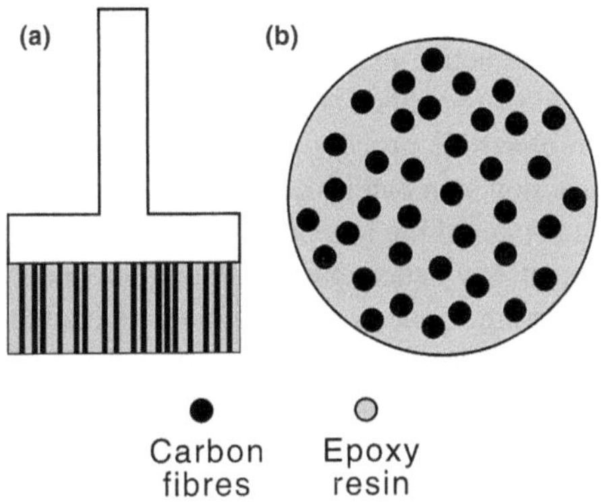

Fig. II.1.14 (a, b) Schematic drawing of the geometry of random assemblies of microelectrodes (RAM) [69]

parabola, the minimum value of the graph can be used for the determination of the diffusion coefficient, the concentration, $[A]_{bulk}$, or the characteristic length, r, of the electrode.

The use of microelectrodes under fast-scan-rate cyclic voltammetric conditions is now common. The benefits of undertaking cyclic voltammetric experiments at microelectrodes are numerous (e.g. fast timescale, less problem with IR_u drop, etc.) and have been summarised in several review articles [66]. The use of intermediate sized or 'mini-electrodes' has not been very widespread because the quantitative interpretation of data obtained in a diffusion regime that lies in the interesting transition region between planar and convex is complex. However, recently, simulation software [16] has become available for the case of mini-disc electrodes, for both the hemispherical geometry [16a] and the true mini-disc geometry [16b]. The advantages of 'mini-electrodes' are similar to those of microelectrodes. However, they are more robust and much easier to polish and problems with ill-defined geometry can be avoided. A major advantage of mini-electrodes is the possibility to cover a range of different diffusion regimes in the transition from radial to planar diffusion and therefore to achieve the enhanced 'kinetic resolution' required for the investigation of more complex mechanistic problems.

Instead of employing a single electrode, an array of electrodes [67] or an interdigitated electrode [68] may be used to study electrochemical systems. Similar to advantages achieved by variations in electrode geometry, the use of several 'communicating' electrodes poised at the same or different potentials opens up new possibilities for the study of the properties or the kinetics of chemical systems. An interesting development is the 'random assemblies of microelectrodes' (RAM) (see Fig. II.1.14), which promises the experimental timescale of microelectrodes but with considerably improved current-to-noise levels [69].

Although of considerable importance in many practical applications of cyclic voltammetry, theoretical models based on the 'simple' Nernstian or electrochemically reversible electrode process are usually not sufficient to explain all aspects of experimentally obtained data. By changing the time domain of the voltammetric experiment (scan rate), it is possible to modify the relative influence of the kinetics of homogeneous or heterogeneous processes associated with an electrode process. Therefore, the mechanism and rate constant of the chemical reaction step can be quantified by variation of the scan rate. A distinction between the effect of the kinetics of the heterogeneous electron transfer and that of the kinetics of any other chemical/potential independent reaction steps is often made by referring to electrochemically irreversible (slow electron transfer) and chemically irreversible (fast chemical follow-up reaction step) voltammetric responses. In the following sections the effects of heterogeneous and homogeneous kinetics on cyclic voltammograms will be discussed.

II.1.5 Determination of Redox State and Number of Transferred Electrons

An important initial problem to consider in the study of a redox system with unknown redox state and redox characteristics is the potential range in which no electrochemical process occurs. With this range of zero current known, it is then possible to progress to further voltammetric study of the system.

II.1.5.1 Chronoamperometric Test Procedure

The potential range in which the initial redox state of a compound is stable can be explored by monitoring the 'switching-on' current with time (chronoamperometry). By using this simple technique, the potential range in which no Faradaic and only capacitive currents are detected can be clearly distinguished from the higher Faradaic currents observed when oxidation or reduction processes occur. Unless a well-considered reason exists, cyclic voltammetric experiments should always be started in this 'zero current' potential region!

II.1.5.2 Method Based on Cyclic Voltammetry

Cyclic voltammetry can be used directly to establish the initial redox state of a compound if data analysis is applied in a careful manner [70]. In Fig. II.1.15, simulated and experimental cyclic voltammograms are shown as a function of the ratio of $Fe(CN)_6^{4-}$ and $Fe(CN)_6^{3-}$ present in the solution phase. It can clearly be seen that the current at the switching potential, $i_{\lambda,a}$ or $i_{\lambda,c}$, is affected by the mole fraction m_{red}. Employing multi-cycle voltammograms at slow scan rates is recommended. Quantitative analysis of mixed redox systems with this method may be based on the plot shown in Fig. II.1.15d.

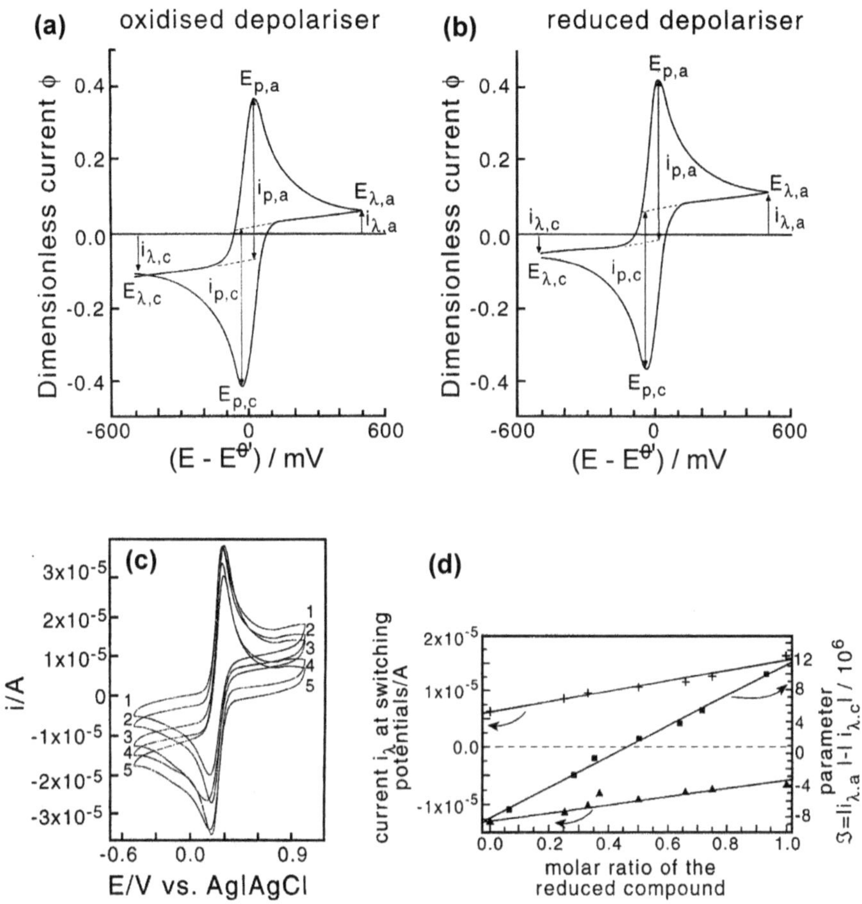

Fig. II.1.15 (**a, b**) Simulated cyclic voltammograms (2nd cycle) of an electrochemically reversible system with the depolariser in the bulk of the solution being in its reduced and oxidised forms, respectively. (**c**) Experimental cyclic voltammograms of a Pt electrode in 0.5 M KNO_3 solutions with defined additions of $K_3Fe(CN)_6$ and $K_4Fe(CN)_6$ (scan rate 20 mV s^{-1}, 5th cycle shown, 3 mm diameter inlaid disc electrode). *1* 125 μM $K_4Fe(CN)_6$; *2* 93.75 μM $K_4Fe(CN)_6$/31.25 μM $K_3Fe(CN)_6$; *3* 62.5 μM $K_4Fe(CN)_6$/62.5 μM $K_3Fe(CN)_6$; *4* 31.25 μM $K_4Fe(CN)_6$/93.75 μM $K_3Fe(CN)_6$; and *5* 125 μM $K_3Fe(CN)_6$. (**d**) Dependence of parameter $\Im = |i_{\lambda,a}| - |i_{\lambda,c}|$ and the currents at the anodic and cathodic switching potentials, $i_{\lambda,a}$ and $i_{\lambda,c}$, respectively, on the molar ratio $m_{red} = c_{red}/(c_{red} + c_{ox})$ of the reduced depolariser

II.1.5.3 Methods Based on Steady-State Techniques

Very informative and reliable data are available from complementary steady-state voltammetric techniques, such as rotating disc voltammetry or slow scan rate micro-electrode voltammetry. Preliminary data may be obtained even by simply stirring the solution phase during the course of a cyclic voltammetric experiment. In Fig. II.1.16

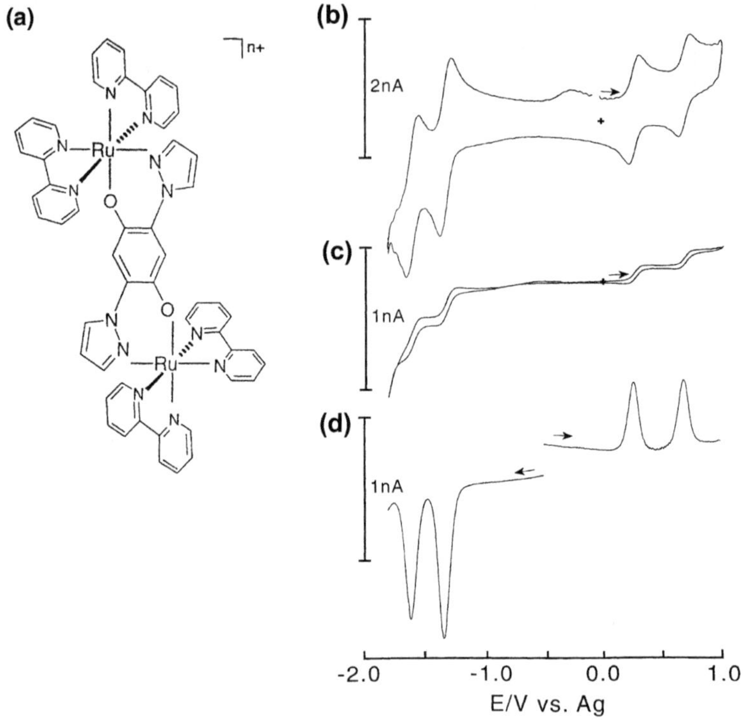

Fig. II.1.16 (**a**) Structure of the binuclear ruthenium complex $[\{Ru(bipy)_2\}_2(\mu\text{-}L)](PF_6)_2$ (bipy = 2,2′-bipyridyl, L = 1,4-dihydroxy-2,5-bis(pyrazol-1-yl)benzene dianion) and (**b**–**d**) voltammograms for the oxidation and reduction of $[\{Ru(bipy)_2\}_2(\mu\text{-}L)](PF_6)_2$ in dimethylsulphoxide (0.1 M NBu_4PF_6) at a 25 μm diameter platinum disc electrode employing (**b**) a scan rate of 2 V s^{-1}, (**c**) a scan rate of 20 mV s^{-1}, and (**d**) square-wave detection with frequency 50 Hz, amplitude 25 mV, step potential 5 mV

typical cyclic voltammograms [71] characterising the redox behaviour of a binuclear metal complex, $[\{Ru(bipy)_2\}_2(\mu\text{-}L)](PF_6)_2$ (bipy = 2,2′-bipyridyl, L = 1,4-dihydroxy-2,5-bis(pyrazol-1-yl)benzene dianion, see Fig. II.1.16a), are shown. The cyclic voltammogram obtained at a 25 μm diameter Pt microdisc electrode with a scan rate of 2 V s^{-1} (Fig. II.1.16b) indicates the presence of four redox processes in the potential range from +1 V to –1.8 V vs Ag. However, the assignment of the zero current region is not immediately obvious. At a lower scan rate of 20 mV s^{-1}, the voltammogram shows the characteristics of a quasi steady-state experiment (Fig. II.1.16c) and the zero current region can be unequivocally identified. Two oxidation processes at 0.25 V and at 0.65 V vs. Ag and two reduction processes at −1.30 V and at −1.60 V vs. Ag are detected. The mass transport limited current plateau detected at the microdisc electrode, which is given by $I_{lim} = 4nFDrc$, also allows the number of transferred electrons, n, to be determined for the case that D is known or can be estimated [51]. Alternatively, the cyclic voltammetric data

(Fig. II.1.16b) can be analysed based on the change of the peak current with scan rate (plot of peak current versus $v^{1/2}$) with the help of the corresponding Randles–Ševčik equation. However, the latter method is much less reliable due to the sensitivity of the peak current towards kinetic effects and adsorption phenomena.

Finally, other types of voltammetric experiments may be employed beneficially for the characterisation of the redox properties of redox active compounds. Figure II.1.16d shows square wave voltammograms [72] for the oxidation and reduction of the binuclear ruthenium complex. Well-defined peak responses indicate the presence of a reversible redox process. In this situation, the peak position corresponds closely to the reversible potential for the process and the peak height is related to the number of transferred electrons. Square-wave voltammetry may be employed to enhance reversible redox processes and to discriminate against irreversible and background processes (see also Chap. II.3).

II.1.6 Heterogeneous Kinetics

The rate of the electron transfer between the redox reagent and the electrode surface is finite and may be limited by the rate of electron exchange at the electrode–redox active reagent interface. The commonly observed process $A + n\,e^- \rightleftarrows B$ is usually described in terms of the Butler–Volmer equation (Eq. II.1.16) [73].

$$I = -nFAk_s\left\{[A]_{x=0}e^{-\alpha\frac{nF}{RT}\eta} - [B]_{x=0}e^{\alpha\frac{nF}{RT}\eta}\right\} \tag{II.1.16}$$

In this equation, which may be regarded as the electrochemical equivalent of the well-known Arrhenius expression with two exponential terms representing anodic (oxidation) and cathodic (reduction) currents [74], the current I observed at the electrode, when both A and B are soluble in solution, is related to the electrode area A, the standard rate constant k_s (in m s^{-1}), the surface concentrations $[A]_{x=0}$ and $[B]_{x=0}$, the transfer coefficient α, and the overpotential $\eta = E - E_e$. Further, n denotes the number of electrons transferred per reacting molecule, F, Faraday's constant, R, the gas constant, and T, denotes the absolute temperature. Under equilibrium conditions currents for the anodic and for the cathodic processes cancel each other out and no net current is observed. Under these conditions, and for the general case of $[A]_{bulk} \neq [B]_{bulk}$, the 'exchange current' is given by $I_0 = nFAk_s\,[A]_{bulk}^{1-\alpha}\,[B]_{bulk}^{\alpha}$. The standard rate constant for the heterogeneous electron transfer, k_s, has been found to span a wide range [75] from 10^{-10} m s^{-1} for $[Co(NH_3)_6]^{3+/2+}$ to $>10^{-2}$ m s^{-1} for anthracene$^{0/-}$ or ferrocene$^{+/0}$. The fundamental basis of the Butler–Volmer equation can be explained by considering the energy profile along the reaction coordinate for the heterogeneous electron transfer process (Fig. II.1.17).

It can be seen that a change in the overpotential, η, causes a change in the driving force for the heterogeneous electron transfer, $\Delta G = nF\eta$ and simultaneously a

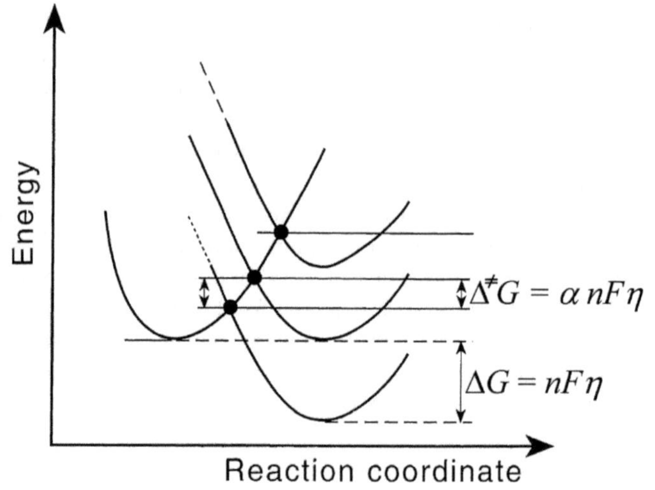

Fig. II.1.17 Schematic drawing of the energy profile along the reaction coordinate for a heterogeneous electron transfer with the electrode poised at three different potentials

change in the kinetic barrier for the reaction $\Delta^{\neq}G = \alpha nF\eta$. In a first approximation, simple lines of identical slope replacing the parabolic profiles may be considered. For this case, straightforward geometric considerations result in a transfer coefficient $\alpha = 0.5$ for the case of a symmetric energy profile. This case corresponds to the most frequently observed behaviour of electrochemical systems. Transition states with $\alpha \approx 0.5$ are observed when the widths of the two parabolas are very different, a phenomenon encountered, for example, in concerted electron transfer processes [76]. If the full parabolic energy profile is considered (Fig. II.1.17), a related, but more complex, expression is obtained in the context of Marcus theory [77].

Cyclic voltammograms for the case of a transition from reversible to irreversible characteristics are shown in Fig. II.1.18. It can be seen that the peak current for the current response in the forward direction decreases as the peak is shifted to more positive potentials. The Randles–Ševčik expression for the irreversible electrochemical response is different to the reversible case by a factor 0.351/0.446 (see Fig. II.1.18a) or $0.496(\alpha)^{0.5}/0.446$ taking account of the change in wave shape (Eq. II.1.17).

$$I_{\mathrm{p}} = -0.496\sqrt{\alpha n'} nFA[\mathrm{A}]_{\mathrm{bulk}}\sqrt{\frac{F\nu D}{RT}} \qquad \text{(II.1.17)}$$

In this expression, n' denotes the number of electrons transferred in the rate-determining reaction step, whereas n denotes the overall numbers of transferred electrons per molecule diffusing to the electrode surface. It can be seen in Fig. II.1.18 that a transition from electrochemically reversible to irreversible characteristics occurs even though a peak remains after reversal of the scan direction

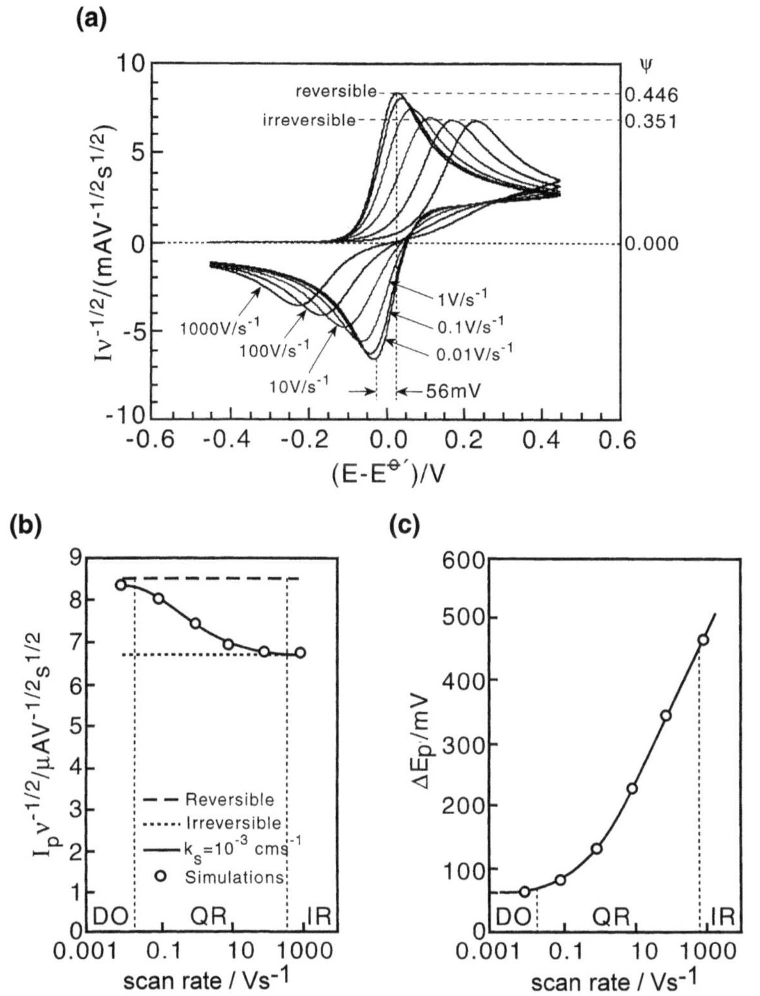

Fig. II.1.18 (**a**) Simulated cyclic voltammograms for a quasi-reversible charge transfer with $k_s = 10^{-4}$ m s^{-1}, $E_c^{\ominus'} = 0.0$ V, $\alpha = 0.5$, $D = 10^{-9}$ m^2 s^{-1}, $A = 10^{-6}$ m^2, $c = 1$ mM, and $T = 298$ K. (**b**) Plot of the normalised peak current vs. scan rate showing the transition from reversible ('diffusion only' DO) to quasi-reversible (QR), and finally to irreversible (IR), charge transfer kinetics. (**c**) Plot of the peak-to-peak separation, ΔE, as a function of scan rate

remains. The term 'quasi-reversible' is sometimes used to describe the intermediate situation between reversible and irreversible. The peak-to-peak separation can be seen to change in the irreversible limit by $\ln(10) \times RT/\alpha F = 118$ mV (if $\alpha = 0.5$) at 298 K for a decade change of scan rate. Alternatively, in the absence of a peak during the reverse scan, the shift of the peak potential E_p is given by $|dE_p/d\log v| = \ln(10) \times RT/2\alpha F$ for processes with electrochemically irreversible kinetics. For the quasi-reversible voltammetric responses obtained under planar semi-infinite diffusion conditions, the transition regions between diffusion only (DO), quasi-reversible (QR), and irreversible (IR) characteristics are plotted in Fig. II.1.18b, c. Assuming $\alpha = 0.5$ and with a known diffusion coefficient D, the experimentally determined peak-to-peak separation, ΔE_p, may be used directly for the determination of the standard rate constant for heterogeneous electron transfer, k_s, based on the approximate expression given in Eq. (II.1.18).

$$\log_{10}\left(k_s\sqrt{\frac{RT}{nFvD}}\right) = 0.294\left(\frac{nF}{RT}\Delta E_p - 2.218\right)^{-1} \\ -0.0803 - 0.108\left(\frac{nF}{RT}\Delta E_p - 2.218\right) \tag{II.1.18}$$

This useful expression allows the standard rate constant, k_s, to be determined directly from the peak-to-peak separation, ΔE_p, and as a function of temperature and scan rate, v.

The wave shapes observed for electrochemically irreversible or quasi-reversible voltammograms are governed by the Fick's law of diffusion (Eq. II.1.6) and the Butler–Volmer expression (Eq. II.1.16). By rewriting the Butler–Volmer equation for the case of a reduction A + n e$^-$ $\rightarrow$ B (Eq. II.1.19), it can be shown that, for the limit of extremely fast electron transfer kinetics, $k_s \rightarrow \infty$, the Nernst law (Eq. II.1.7) is obtained as anticipated.

$$\frac{I}{nFAk_s}e^{\alpha\frac{nF}{RT}\eta} = -\left\{[A]_{x=0} - [B]_{x=0}\, e^{\frac{nF\eta}{RT}}\right\} \tag{II.1.19}$$

Therefore, the expression for the case of a reversible cyclic voltammogram derived above (Eq. II.1.10) may also be regarded as a special case of a more general expression, *including* the effects of heterogeneous electron transfer. Use of Eqs. (II.1.8), (II.1.9), (II.1.16) and $E = E_{\text{initial}} - vt$ gives Eq. (II.1.20):

$$I(t) = -\left\{nFAk_s[A]_{\text{bulk}} + \frac{k_s\, d^{-1/2}}{\sqrt{D}\, dt^{-1/2}}I(t)\right\} e^{-\alpha\frac{nF}{RT}\left(E_{\text{initial}} - vt - E_c^{\ominus'}\right)} \\ -\left\{\frac{k_s\, d^{-1/2}}{\sqrt{D}\, dt^{-1/2}}I(t)\right\} e^{-\alpha\frac{nF}{RT}\left(E_{\text{initial}} - vt - E_c^{\ominus'}\right)} \tag{II.1.20}$$

As required, this expression also converges to the reversible case (Eq. II.1.10) under conditions where diffusion rather than the electron transfer rate dominates the overall rate of the process. This can be demonstrated by rearranging Eq. (II.1.20) into an expression similar to Eq. (II.1.19) and deriving the limiting case of $k_s \rightarrow \infty$. Because of its general importance, Eq. (II.1.20) in a slightly more general form has been termed 'the universal equation of transient voltammetry' [78].

Unfortunately, because Eq. (II.1.20) contains the current terms in the form of $I(t)$ and $\left(\mathrm{d}^{-1/2}/\mathrm{d}t^{-1/2}\right) I(t)$, it is not easily solved analytically.

In practice, analytical approaches based, e.g. on Eq. II.1.20 are of limited use and advanced numerical simulation methods [16] and fitting of experimental to simulation data for a range of experimental data sets is the most reliable procedure to confirm the Butler–Volmer kinetics and to accurately obtain kinetic parameters such as k_s and α from cyclic voltammetric data. Some criteria for the case of irreversible electron transfer (symbol $\mathrm{E_{irrev}}$) are listed below.

– **$\mathrm{E_{irrev}}$ Diagnostics:**

For reduction processes $\left|I_{p,a}/I_{p,c}\right| < 1.0$ and for oxidation $\left|I_{p,c}/I_{p,a} < 1.0\right|$.
Peak shift $\left|\Delta E_p\right| = \ln(10) \times (RT)/(2\alpha nF) = 30/\alpha n\,\mathrm{mV}$ at 298 K per decade change in scan rate v. I_p is proportional to $v^{1/2}$ (Randles–Ševčik)

$\left|E_p - E_{p/2}\right| = (1.857\,RT)/(\alpha nF) = 47.7/\alpha n\,\mathrm{mV}$ at 298 K (useful for α evaluation)

$E_c^{\ominus'} = E_{\mathrm{mid}} - (RT/nF)\ln(D_{\mathrm{red}}/D_{\mathrm{ox}})^{1/2}$ (under quasi-reversible conditions, D_{red} and D_{ox} denote the diffusion coefficients for the reduced and oxidised forms, respectively)

n, n': usually 1, but otherwise (with D known or estimated) may be calculated from peak current (use the appropriate Randles–Ševčik equation, Eq. II.1.17) or, better, from the steady-state limiting current, or from the convoluted voltammogram.

k_s: use Eq. (II.1.18) or numerical simulation for quasi-reversible systems or, if $E_c^{\ominus'}$ is known, the following equation for irreversible systems

$$I_p = 0.227nFA\ [\mathrm{A}]_{\mathrm{bulk}}\, k_s \exp\left[-\left(\alpha n' F/RT\right)\left(E_p - E_c^{\ominus'}\right)\right]$$

Alternatively, Tafel plot analysis [1] may be applied to the foot of the voltammetric response.

Finally, it should be noted that non-Butler–Volmer behaviour may be observed in the analysis of cyclic voltammetric data. For example, particularly in the presence of low concentrations of supporting electrolyte, electron transfer kinetics of charged species may be significantly modified due to the double layer or 'Frumkin' effects [79]. Under these conditions, (i) the potential experienced by the reactant at the point of closest approach to the electrode can be different from the applied potential, and (ii) an additional energy barrier for the approach of charged reactants to the electrode may exist. Corrections to account for 'Frumkin' effects have been proposed. Deviations from Butler–Volmer behaviour may also be interpreted in terms of the Marcus theory [80]. A further interesting case of non-Butler–Volmer voltammetric characteristics is observed with semiconducting electrode materials [81].

Commonly, only one electron is transferred between electrode and reactant. However, multi-electron transfer processes are of considerable interest due to their importance in catalytic processes such as those associated with the reduction of

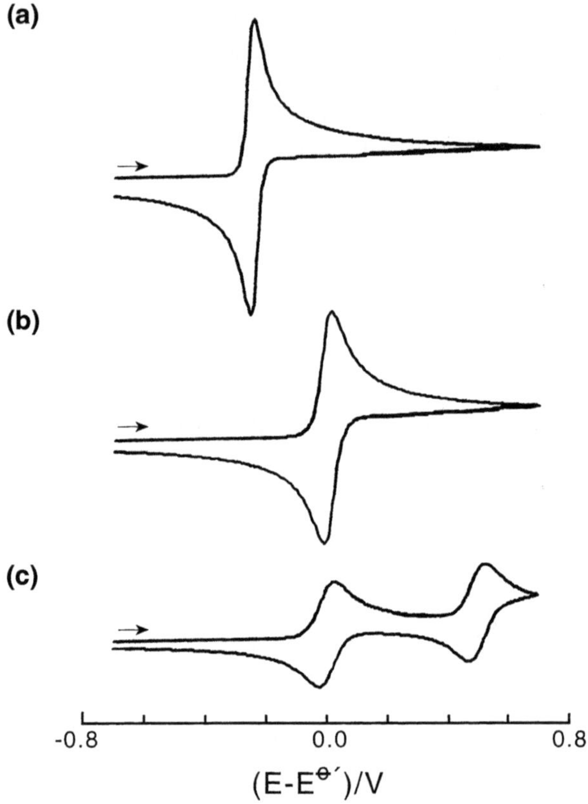

Fig. II.1.19 Simulated cyclic voltammograms (DigiSim) for a reversible oxidation process with consecutive transfer of two electrons ($E_{rev}E_{rev}$ mechanism) with (**a**) $E^{\ominus\prime}_{c,1} > E^{\ominus\prime}_{c,2}$, (**b**) $E^{\ominus\prime}_{c,1} = E^{\ominus\prime}_{c,2}$, and (**c**) $E^{\ominus\prime}_{c,1} < E^{\ominus\prime}_{c,2}$

dioxygen. As would be expected, the nature of the voltammetric responses for the reversible oxidation or reduction of multi-electron mediator systems provides a diagnostic tool to reveal details concerning the standard potential, the extent of interaction between redox centres, and the rate of electron transfer [82]. In Fig. II.1.19, the voltammetric responses for a two-electron oxidation and for the three cases (a) $E^{\ominus\prime}_{c,1} > E^{\ominus\prime}_{c,2}$, (b) $E^{\ominus\prime}_{c,1} = E^{\ominus\prime}_{c,2}$, and (c) $E^{\ominus\prime}_{c,1} < E^{\ominus\prime}_{c,2}$ show that both the peak current and the peak potential strongly depend on the type of system. For two consecutive reversible one-electron transfer processes (case c), two well-separated electrochemically reversible waves each with a value of $\Delta E_p = 57$ mV (at 298 K) are present. If both electron transfer processes occur with identical standard potentials (case b), the peak current, I_p, is a factor 2.41 higher and ΔE_p is 42 mV [83]. Related cases in which two redox centres in a single molecule do not interact have been considered on a statistical basis [84]. Finally, for case (a), in which $E^{\ominus\prime}_{c,1} > E^{\ominus\prime}_{c,2}$, the peak current is enhanced by a factor of $2^{3/2} = 2.82$ compared to the one-electron case, and the predicted peak separation is $\Delta E_{\mathrm{p}} = 2.218 \times RT/nF = 56/2\,\mathrm{mV}$ at 298 K. This is the case treated in the Randles–Ševčik expression for the peak current for a reversible cyclic voltammogram (Eq. II.1.11) predicting peak current enhancements of $n^{3/2}$. The midpoint potential for this type of process is located in good approximation half way between the individual formal potentials, $E_{\mathrm{mid}} = 1/2\left(E^{\ominus\prime}_{\mathrm{c},1} + E^{\ominus\prime}_{\mathrm{c},2}\right)$ [85]. For case (a), the simultaneous transfer of two electrons, additional complications may arise if the first [86] or the second electron transfer proceeds with slow kinetics. Systems showing this latter kind of kinetically controlled behaviour are well known and include, for example, diaminodurene and diaminoanthracene derivatives [87].

II.1.7 Homogeneous Kinetics

Homogeneous chemical steps are ubiquitously present in liquid-phase chemical processes and hence in voltammetry Thus, rapid pre-equilibration processes frequently accompany the electron transfer step, but they may be difficult to detect. The key to their experimental observation is the use of a sufficiently wide time domain with respect to the electrochemical (or spectroelectrochemical) response. Voltammetric experiments designed to cover a certain time domain can only detect homogeneous rate constants with a designated range of values. Conversely, experiments only accessing this given time domain *cannot* be used to obtain rate constants outside this designated range.

The loss of the current peak on the reverse scan of a cyclic voltammogram is the most obvious indicator of a homogeneous chemical reaction step consuming the product generated from the interfacial redox step involving reversible electron transfer kinetics. That is, the reverse current peak is reduced in height if a significant amount of the product from the electron transfer step reacts in the diffusion layer irreversibly to give a new product (electroinactive in the potential range of interest) during the potential scan. Different diagnostic criteria apply to other mechanisms when, for example, a rate-determining chemical reaction may precede the electron transfer step. Figure II.1.20 summarises the use of conventional cyclic voltammetry at macrodisc electrodes (*vide supra*) (which is limited by natural convection on the slow scan rate side), fast scan rate cyclic voltammetry at microelectrodes (*vide supra*), and bulk electrolysis techniques (e.g. in thin layer cells [88]) to address problems that can encompass a huge range of accessible rates. The upper limit for quantifying extremely fast homogeneous reactions by conventional voltammetric techniques is restricted not only by instrumental problems, but also by the time it takes for the double layer at the electrode–solution interface to become charged. At the other extreme, reactions that are so fast that they become part of the reaction coordinate (see Fig. II.1.17) for the interfacial electron transfer step itself (concerted reactions [76]) can be detected even at low scan rate because coupling of

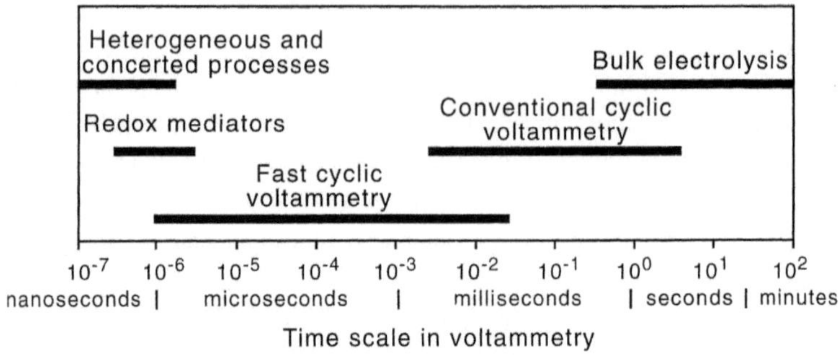

Fig. II.1.20 Schematic representation of timescales accessible with voltammetric techniques

the homogeneous and heterogeneous steps means that the current on the reverse scan and the peak shape in the forward scan change characteristically.

A systematic description of all possible combinations of homogeneous chemical processes coupled to electron transfer at an electrode surface is impossible because an infinite range of theoretically possible reaction schemes can be constructed. Unfortunately, a consistent form of nomenclature for defining the possible web of reaction pathways has not yet been invented. However, the IUPAC nomenclature [89] is of assistance with respect to simple reaction schemes. In this article, the commonly employed descriptors for electron transfer (E) and chemical (C) sequences of reaction steps, e.g. ECEC, will be used for a sequence of reactions involving electron transfer–chemical process–electron transfer–chemical process. Reaction schemes involving branching of a reaction pathway will be considered later.

II.1.7.1 The EC Process

Initially, the simple case of a irreversible first-order chemical reaction step (C_{irrev}) following a reversible heterogeneous charge transfer process (E_{rev}) is considered. The reaction scheme for this type of process is given in Eq. II.1.21 and simulated cyclic voltammograms for the $E_{rev}C_{irrev}$ reaction sequence are shown in Fig. II.1.21a.

$$A + e^- \rightleftarrows B \qquad\qquad B \rightarrow C \tag{II.1.21}$$

In the case given, a chemical reaction B $\rightarrow$ C with $k = 200\ \mathrm{s}^{-1}$ is considered and the effect of varying the scan rate is shown. With a chemical rate constant of $k = 200\ \mathrm{s}^{-1}$, the reaction layer thickness [90] can be estimated from $\delta_{\text{reaction}} = \sqrt{\frac{D}{k}}$ to extend ca. 2.2 μm from the electrode surface into the solution phase. Further, the characteristic diffusion layer thickness at the time that the peak current in the cyclic voltammogram is encountered is given approximately by $\delta_{\text{peak}} = \sqrt{\frac{DRT}{vnF}}$. Equating

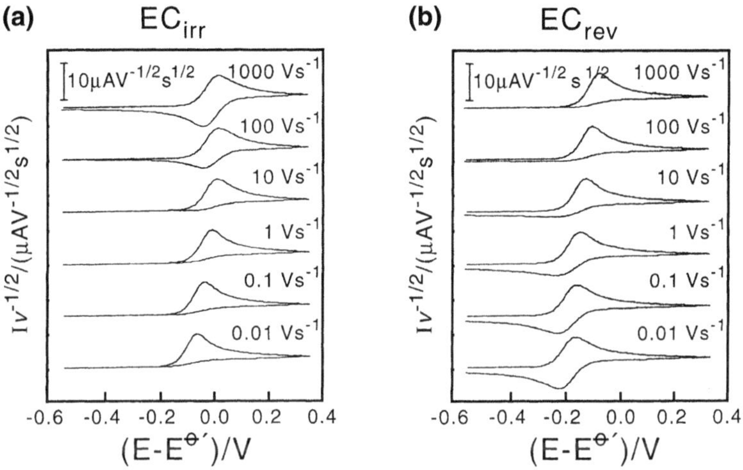

Fig. II.1.21 (**a**) Simulated cyclic voltammograms (DigiSim) for an EC_{irrev} mechanism ($E^{\ominus\prime}_{c,A^+/A} = 0.0\,V$, $\alpha = 0.5$, $k_s = 10^2$ m s^{-1} $k_{chemical} = 200\,s^{-1}$, $D = 10^{-9}\,m^2s^{-1}$, $A = 10^{-6}\,m^2$, $[A] = 1$ mM and $T = 298$ K). (**b**) Simulated cyclic voltammograms (DigiSim) for an EC_{rev} mechanism ($E^{\ominus\prime}_{c,A^+/A} = 0.0\,V$, $\alpha = 0.5$, $k_s = 10^2\,m\,s^{-1}$, $k_{f,chemical} = 10^7\,s^{-1}$, $k_{b,chemical} = 10^4\,s^{-1}$, $D = 10^{-9}\,m^2\,s^{-1}$, $A = 10^{-6}\,m^2$, [A] = 1 mM, and $T = 298$ K)

the reaction and the diffusion layer expressions allows the approximate scan rate at which a match between diffusion and kinetic control occurs to be calculated as $v_{transition} = 5\,V\,s^{-1}$. Figure II.1.21a shows that, at scan rates higher than 5 V s^{-1}, the voltammograms start to become chemically reversible or diffusion controlled and, at very high scan rates, the chemical reaction becomes too slow to significantly affect the response. Vice versa, at lower scan rates, the current peak on the reverse scan is absent and a shift of the peak potential of 30/*n* mV (*n* denotes the number of electrons transferred per reacting molecule) per decade change in scan rate [91] towards more negative potentials occurs. The ratio of forward-to-backward peak current can be used for the determination of the rate constant [92] or, better, quantitative fitting via comparison with theory based on digital simulation should be undertaken. Examples of this EC_{irrev}-type electrode process often involve fast bond-breaking processes, e.g. as observed during reduction of many haloorganics [93]. Pseudo-first-order and -second-order chemical reaction steps are identified based on the characteristic concentration dependence of the voltammetric characteristics.

– **$E_{rev}C_{irrev}$ Diagnostics:**

$|I_{p,a}/I_{p,c}| < 1.0$ and shift peak potential $|\Delta E_p| = 30/n\,mV$ at 298 K per decade change in scan rate v. Transition to E_{rev}- and E_{irrev}-type processes for very slow and extremely fast chemical reaction steps, respectively. I_p follows Randles-Ševčik characteristics $(\propto v^{1/2})$, although with $\psi_{peak}(p) = 0.4956$.

$\delta_{reaction}$: given approximately by $(D/k_h)^{0.5}$ (k_h denotes a first-order rate constant).

$E_{1/2} = E_p + (RT/nF)\,(0.780 - 0.5\ln\{kRT/nFv\})$ (for reduction) or $E_p - (RT/nF)\,(0.780 - 0.5\ln\{kRT/nFv\})$ (for oxidation)
n: (with D known or estimated) calculate from peak current, steady-state limiting current, or from the convoluted voltammogram.
k_h: a good estimate for the case $|I_{p,forward}/I_{p,backward}| = 2.0$ is $k_h \approx RT/2nFv$, alternatively, use a working curve [12, 44] for $I_{p,a}/I_{p,c}$, or numerical simulation.

The simple use of an irreversible chemical reaction step to represent a chemical process is physically unrealistic, because the law of 'microscopic reversibility' or 'detailed balance' [94] is violated. More realistic is the use of an $E_{rev}C_{rev}$ reaction scheme (Eq. II.1.22, Fig. II.1.21b). Even for the relatively simple $E_{rev}C_{rev}$ reaction scheme, interesting additional consequences arise when the possibility of reversibility of the chemical step is considered. In Fig. II.1.21b, cyclic voltammograms for the case of a reversible electron transfer process coupled to a chemical process with $k_f = 10^7\ s^{-1}$ and $k_b = 10^4\ s^{-1}$ are shown. At a scan rate of 10 mV s^{-1}, a well-defined electro-chemically and chemically reversible voltammetric wave is found with a shift in the reversible half-wave (or half-peak) potential $E_{p/2}$ from $E_c^{\ominus\prime}$ being evident due to the presence of the fast equilibrium step. The shift is $\Delta E_{p/2} = RT/F\ln(K) = -177$ mV at 298 K in the example considered. At faster scan rates, the voltammetric response departs from chemical reversibility since equilibrium can no longer be maintained. The reason for this is associated with the back reaction rate of $k_b = 10^4\ s^{-1}$, or, correspondingly, the reaction layer, $d_{reaction} = D/k_b = 0.32\ \mu m$. At sufficiently fast scan rates, the product B is irreversibly converted to C because the rate of reaction back to B is now too slow to maintain equilibrium and, hence, on the voltammetric timescale, the process becomes chemically irreversible rather than chemically reversible. This behaviour is in contrast to that observed for the $E_{rev}C_{irrev}$ mechanism, which may be regarded as the fast scan rate-limiting case of the $E_{rev}C_{rev}$ mechanistic scheme.

$$A + e^- \rightleftarrows B \qquad B \rightleftarrows C \tag{II.1.22}$$

– $E_{rev}C_{rev}$ Diagnostics:

$|I_{p,a}/I_{p,c}| < 1.0$, with increasing scan rate loss of reversibility and transition to the $E_{rev}C_{irrev}$ case. In the high scan rate limit $\Delta E_p = 30/n$ mV at 298 K per decade change in scan rate v. In the low scan rate limit features consistent with a simple E_{rev} case and with a shift in half-wave potential $\Delta E_{1/2} = RT/nF\ln(K)$.
$\delta_{reaction}$: given approximately by $(D/[k_f + k_b])^{0.5}$.
n: (with D known or estimated) calculate from the peak current, steady-state limiting current, or from the convoluted voltammogram.
k_f, k_b: use numerical simulation and fitting for several scan rates and concentrations.

Further increasing the scan rate in the case of the initial $E_{rev}C_{rev}$ mechanism yields cyclic voltammograms with identical characteristics to those shown in Fig. II.1.21 a for the $E_{rev}C_{irrev}$ mechanism. Indeed, the operational rather than the absolute definition of the terms reversible and irreversible is revealed in this example as clearly an $E_{rev}C_{rev}$ process as defined at slow scan rate becomes an $E_{rev}C_{irrev}$ or E_{rev} (or even E_{irrev}) process as the voltammetric timescale becomes progressively decreased. There is abundant experimental evidence [95] to testify to the importance of the $E_{rev}C_{rev}$ mechanistic chemical process. A related and recently extensively studied mechanism has been denoted $E_{rev}C_{dim}$ [96] (Eq. II.1.23) (or more correctly $E_{rev}C_{dim,irrev}$).

$$A + e^- \rightleftarrows B \quad 2B \rightarrow C \tag{II.1.23}$$

– **$E_{rev}C_{dim,irrev}$ Diagnostics:**

$|I_{p,a}/I_{p,c}| < 1$ (for a reduction process). The process is strongly concentration dependent and, in contrast to the $E_{rev}C_{irrev}$ scheme, the shift in peak potential is $|\Delta E_p| = \ln(10) \times RT/3nF = 20/n$ mV at 298 K per decade change in scan rate v. The same shift in peak potential is observed with a tenfold change in $[A]_{bulk}$.

$E_{1/2} = E_p + (RT/nF)\,(0.92 - 0.333 \ln\{2k_{dim}\,[A]_{bulk}/3\,(RT/nFv)\})$ (for reduction).

$\delta_{reaction}$: given approximately by $(D/k_{dim}\,[A]_{bulk})^{0.5}$ (this expression overestimates the reaction layer thickness due to the second-order nature of the reaction).

n: (with D known or estimated) calculate from peak current under pure diffusion control (fast scan rate), or, better, from steady-state limiting current data (not affected by the chemical step), or from the convoluted voltammogram.

k_{dim}: use numerical simulation methods and compare with experimental data obtained at different concentrations and under conditions of mixed kinetic/diffusion control (both anodic and cathodic peaks visible).

In this type of mechanism a second-order dimerisation process coupled to the heterogeneous electron transfer process gives rise to voltammetric characteristics related to those described in Fig. II.1.21. The second-order nature of the dimerisation step is clearly detected from the change in appearance of voltammograms (or the apparent rate of the chemical reaction step) with the concentration of the reactant.

II.1.7.2 The EC′ Process

A special and very important case of EC-type processes is denoted as the ‘catalytic’ or EC′ (or $E_{rev}C'_{rev}$) electrode reaction. In this reaction sequence (see Eq. II.1.24), the heterogeneous electron transfer produces the reactive intermediate B, which upon reaction with C regenerates the starting material A. The redox system A/B may therefore be regarded as redox mediator or catalyst, and numerous applications of this scheme in electro-organic chemistry are known [97]. Furthermore, a redox mediator technique based on the EC′ process has been proposed by Savéant et al. [98] allowing the voltammetric timescale for the study of very fast EC_{irrev} processes to be pushed to the extreme.

$$A + e^- \rightleftarrows B \qquad\qquad B + C \rightarrow A + D \tag{II.1.24}$$

This type of reaction scheme is readily identified in steady-state or cyclic voltammetric [99] investigations based on the effect of the substrate concentration on the current [100]. The reaction layer thickness for the EC′ process is given approximately by $\delta_{reaction} = (D/k_h\,[C]_{bulk})^{0.5}$, where $[C]_{bulk}$ denotes the concentration of C (Eq. II.1.24) and k_h is a second-order rate constant. The ratio of diffusion and reaction layer thickness yields an important kinetic parameter $\lambda = \delta_{peak}/\delta_{reaction} = \left(\left[k_h\,[C]_{bulk}\,RT\right]/[v\,nF]\right)^{0.5}$, which describes the competition between diffusion and chemical reaction in the diffusion layer. With $\lambda < 0.1$, diffusion is fast compared to the chemical reaction (e.g. fast scan rate) and the limiting case of an E_{rev} process is detected. On the other hand, for $\lambda > 1.0$, the process becomes independent from diffusion and instead of a peak-shaped voltammogram, a sigmoidally shaped one with a kinetically controlled limiting current, $I_{lim} = nFA\,[A]_{bulk}\,(Dk_h\,[C]_{bulk})^{0.5}$, is observed. For detailed analysis in the transition region and taking account of the effect of unequal diffusion coefficients, numerical simulation and fitting methods are required.

– **$E_{rev}C_{irrev}$ Diagnostics:**

$|I_{p,a}/I_{p,c}| < 1.0$ (for reduction processes) and a change from peak-shaped voltammograms ($\lambda < 0.1$) to sigmoidal voltammograms ($\lambda > 1.0$) is observed ($\lambda = k_h\,[C]_{bulk}\,(RT/vnF)$).

$\delta_{reaction}$: given by $(D/k_h\,[C]_{bulk})^{0.5}$ where $[C]_{bulk}$ is the substrate concentration.

n: (with D known or estimated) calculate from peak current in the $\lambda < 0.1$ limit, from the steady-state limiting current measured as a function of concentration, or from the convoluted voltammogram in the $\lambda < 0.1$ limit.

k_h: use numerical simulation methods or steady-state experiments (e.g. microdisc electrode or rotating disc voltammetry) and the expression

$$I_{lim} = -nFA\,[A]_{bulk}\left(\frac{\delta}{D} + \frac{1}{\sqrt{Dk_h[C]_{bulk}}}\right)^{-1}.$$

II.1.7.3 The CE Process

Fast pre-equilibration processes may also accompany a reversible electrode reaction. Reactions involving preceding reaction steps are commonly observed in processes in which deprotonation, dehydration, or breaking of a metal–ligand bond occurs [101]. The mechanistic scheme for the CE case is given in Eq. II.1.25, and cyclic voltammograms for this reaction sequence are shown in Fig. II.1.22.

$$A \rightleftarrows B \qquad B \rightleftarrows C^+ + e^- \tag{II.1.25}$$

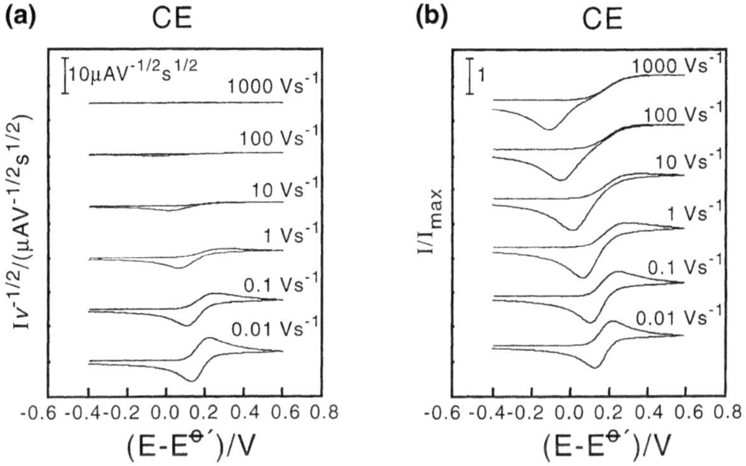

Fig. II.1.22 Simulated cyclic voltammograms (DigiSim) for a CE mechanism with (**a**) normalisation $I/\sqrt{v}$ and (**b**) normalisation I/I_{peak}, ($K = 0.01$, $k_f = 2000\,\text{s}^{-1}$, $E^{\ominus\prime}_{\text{c, A}^+/\text{A}} = 0.0$ V, $\alpha = 0.5$, $k_s = 10^{-3}$ m s^{-1}, $D = 10^{-9}\,\text{m}^2\,\text{s}^{-1}$, $A = 10^{-6}\,\text{m}^2$, $c = 1$ mM, and $T = 298$ K)

For the simulation of the cyclic voltammograms, an equilibrium constant $K = 0.01$ has been selected so that only a small amount of the intermediate B is present in the bulk solution. Thus, with fast scan rates, when little time for conversion of A to B is available, the voltammetric response associated with reduction of the small concentration of B is essentially 'hidden' in the background. Only normalisation using the peak current shows the presence of the small voltammetric signal. The 'step-shaped' characteristic of the oxidation response at high scan rates can be attributed to the potential independent rate constant $k_f = 2000$ s^{-1} limiting the rate. With sufficiently slow scan rate, the cyclic voltammogram becomes reversible, as expected for the case of a very fast pre-equilibrium. In this case, only the shift of the midpoint potential, $E_{\text{mid}} - E^{\ominus\prime}_{\text{c, A}^+/\text{A}} = RT/F \ln(K) = 0.195$ V at 298 K (or spectroelectrochemical data), indicate the existence of a reversible chemical reaction step. The cyclic voltammetric response for a CE-type process can be strongly affected by thermal activation.

The approximate reaction layer thickness for the $\text{C}_{\text{rev}}\text{E}_{\text{rev}}$ process may be written as $\delta_{\text{reaction}} = (D/[k_f + k_b])^{0.5}$ and, combined with the appropriate diffusion layer thickness (see Table II.1.2), the kinetic parameter becomes $\lambda = \delta_{\text{peak}}/\delta_{\text{reaction}} = ([(k_f + k_b)RT]/[vnF])^{0.5}$. Both the equilibrium constant, $K = k_f/k_b$, and the parameter λ affect the appearance of the cyclic voltammogram and distinct 'kinetic zones' can be identified. For $K < 10^{-2}$ and $\lambda < 0.4$ (fast scan rate), diffusion governs the voltammetric response and the limiting case E_{rev} with a simple reversible response (at $E^{\ominus\prime}_{\text{c}}$ in Fig. II.1.22b, not shown) and a peak current according to the equilibrium concentration is anticipated. On the other hand, for $K < 10^{-2}$ and

$\lambda > 10^4$, the kinetics of the chemical reaction step is sufficiently fast (and the scan rate slow) for an E_{rev}-type response to be detected at more positive potential (see Fig. II.1.22). In between these limiting cases a region with purely kinetic control can be identified where a sigmoidal voltammogram with $I_{lim} = nFA\,[A]_{bulk}\,(DKk_f)^{0.5}$ is detected. In the other remaining regions, the observed peak current, $I_{p,CE}$, is empirically related [12] to the peak current $I_{p,E}$ for an E_{rev}-type process by $I_{p,CE} = I_{p,E}\left(1.02 + 0.471\,K^{-1}\lambda^{-1}\right)^{-1}$. Kinetic analysis of the CE-type reaction is often based on steady-state techniques such as rotating disc voltammetry [102].

– **$C_{rev}E_{rev}$ Diagnostics:**

$\left|I_{p,a}/I_{p,c}\right| < 1.0$ (for oxidation processes, see Fig. II.1.22) and two limiting cases, both E_{rev}, for very fast or very slow scan rate are possible. For purely kinetic control, a shift in half-wave potential of $\left|\Delta E_{p/2}\right| = \ln(10) \times RT/2nF = 30/n\,\text{mV}$ at 298 K per decade change in scan rate v can be observed. Under kinetic control a sigmoidal voltammogram with $I_{lim} = nFA\,[A]_{bulk}\,(DKk_f)^{0.5}$ is detected.

$\delta_{reaction}$: given approximately by $(D/\,[k_f + k_b])^{0.5}$.

n: (with D known or estimated) calculate from peak current in the diffusion controlled limit (Randles–Ševčik), from the diffusion controlled steady-state limiting current, or from the convoluted voltammogram obtained under diffusion control.

k_f k_b: use numerical simulation methods or steady-state experiments (e.g. microdisc electrode or rotating disc voltammetry) and the expression [102]

$$I_{lim} = -nFA\,[A]_{bulk}\left(\frac{\delta}{D} + \frac{1}{\sqrt{DKk_f}}\right)^{-1}$$

e.g. plot $1/I_{lim}$ vs. the diffusion layer thickness δ and analyse the intercept.

II.1.7.4 The $\vec{E}C\vec{E}$ Process

When a second electron transfer process exists, the reaction scheme for the overall electrode process becomes considerably more complicated. An example of an $\vec{E}C\vec{E}$ reaction scheme, the $E_{rev}C_{irrev}E_{rev}$ reaction for two consecutive one-electron oxidation processes coupled via a chemical reaction step, is given in Eq. (II.1.26), and voltammetric responses simulated for this type of process are shown in Fig. II.1.23a.

$$\begin{array}{ll} A \rightleftarrows A^+ + e^- & A^+ \rightarrow B \\ B \rightleftarrows B^+ + e^- & \end{array} \qquad \text{(II.1.26)}$$

Overall, three cases for this family of reaction sequences may be considered with (i) the simplest case $E^{\ominus'}_{c,\,A^+/A} < E^{\ominus'}_{c,\,B^+/B}$ directly related to the $E_{rev}C_{irrev}$

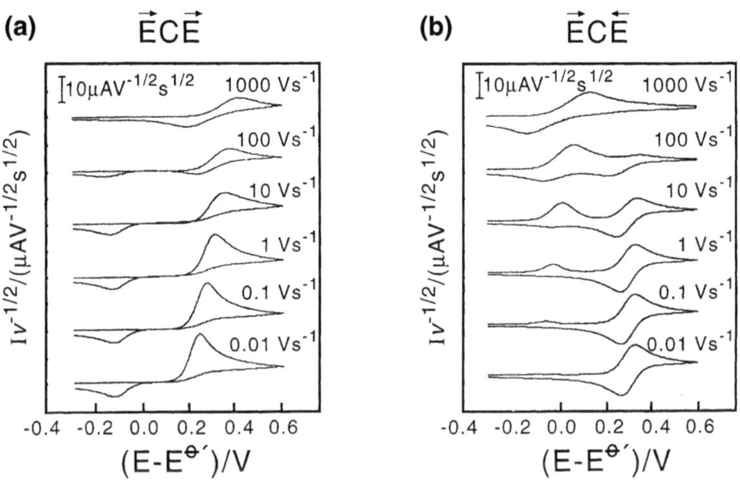

Fig. II.1.23 **a** Simulated cyclic voltammograms (DigiSim) for an ECE mechanism ($E^{\ominus\prime}_{c,\,A^+/A} = 0.3$ V, $\alpha = 0.5$, $k_s = 0.1\ \text{cm s}^{-1}$, $k_f = 200\ \text{s}^{-1}$, $E^{\ominus\prime}_{c,\,B^+/B} = -0.1$ V, $\alpha = 0.5$ $k_s = 10^{-3}\ \text{m s}^{-1}$, $D = 10^{-9}\ \text{m}^2\,\text{s}^{-1}$, $A = 10^{-6}\ \text{m}^2$, [A] = 1 mM, and $T = 298$ K). **b** Simulated cyclic voltammograms (DigiSim) for an $\vec{E}\,C\vec{E}$ mechanism ($E^{\ominus\prime}_{c,\,A^+/A} = -0.1$ V, $\alpha = 0.5$, $k_s = 10^{-3}\ \text{m s}^{-1}$, $k_f = 200\ \text{s}^{-1}$, $E^{\ominus\prime}_{c,\,B^+/B} = 0.3$ V, $\alpha = 0.5$, $k_s = 10^{-3}\ \text{m s}^{-1}$, $k_{\text{disproportionation}} = 300\ \text{dm}^3\,\text{mol}^{-1}\,\text{s}^{-1}$, $D = 10^{-9}\ \text{m}^2\,\text{s}^{-1}$, $A = 10^{-6}\ \text{m}^2$, [A] = 1 mM, $R_u = 100\ \Omega$, and $T = 298$ K)

process (*vide supra*), (ii) $E^{\ominus\prime}_{c,\,A^+/A} > E^{\ominus\prime}_{c,\,B^+/B}$, where the transfer (forward) of the first electron is followed by a chemical reaction step and immediately by the transfer (forward) of the second electron (net two electron transfer), and (iii) $E^{\ominus\prime}_{c,\,A^+/A} > E^{\ominus\prime}_{c,\,B^+/B}$, where the transfer (forward) of the first electron is followed by a chemical reaction step and immediately by the transfer (backward) of the electron back (net zero electron transfer, *vide infra*). The latter two cases are considered in Fig. II.1.23. For the mechanistic scheme II.1.26, a reversible voltammetric response for the A^+/A redox couple at $E^{\ominus\prime}_{c,\,A^+/A} = 0.3$V can be observed at sufficiently fast scan rates. Reducing the scan rate allows the chemical reaction step with $k = 200\ \text{s}^{-1}$ to compete with mass transport in the diffusion layer and the product B is detected on the reverse scan in the form of a B^+/B reduction process with $E^{\ominus\prime}_{c,\,B^+/B} = -0.1$ V. The normalisation of the voltammograms with the square root of the scan rate allows the peak currents for the oxidation response to be compared as a function of the scan rate. It can be seen that the peak current increases considerably from the reversible one-electron process observed at high scan rate to the chemically irreversible two-electron process at slow scan rate. There are many chemical systems that can be described by the reaction scheme given in Eq. II.1.26 [103]. Based on the discussion of the EC reaction it can be anticipated that the $E_{rev}C_{rev}E_{rev}$ reaction scheme (see also square schemes, next section) is also important. Kinetic analysis of steady-state voltammetric data, e.g. obtained by rotating disc voltammetry, can be analysed based on an approximate expression for the limiting current [104].

– $\vec{E}\,C\,\vec{E}$ Diagnostics:

$|I_{p,a}/I_{p,c}| < 1.0$ (for a reduction process). A second voltammetric response occurs in the same potential range.

$\delta_{reaction}$: given approximately by $(D/k_h)^{0.5}$.

n: (with D known or estimated) calculate from peak current under pure diffusion control, or better from the diffusion-controlled steady-state limiting current.

k_h: use numerical simulation methods over a wide range of experimental conditions or steady-state voltammetry (e.g. microdisc or rotating disc techniques) and the expression [104]

$$I_{lim} = I_{lim}\,(n = 1) = \left[2 - \frac{\tanh\,(\delta_{steady\ state}/\delta_{reaction})}{(\delta_{steady\ state}/\delta_{reaction})}\right].$$

The presence of several reactants in different redox states opens up the possibility of further reaction steps consistent with disproportionation reactions. For the case of $E^{\ominus\prime}_{c,\,B^+/B} < E^{\ominus\prime}_{c,\,A^+/A}$, the disproportionation step introduced in Eq. II.1.27 is important and cannot be neglected. The mechanism for the case where the first monomolecular chemical step is rate determining, sometimes denoted DISP1, has been extensively treated in the literature [105].

$$\begin{array}{ll} A \rightleftarrows A^+ + e^- & A \rightarrow B \\ B \rightleftarrows B^+ + e^- & \\ A + B^+ \rightleftarrows A^+ + B & \end{array} \qquad (II.1.27)$$

Alternatively, disproportionation may occur via a bimolecular process involving two A^+ molecules, to give the so-called DISP2 process described in Eq. (II.1.28) [106]. The latter, more complex, mechanistic schemes can be difficult to prove and quantify unequivocally and cyclic voltammetric techniques can give only limited insight depending on the type of electrode geometry and experimental timescale used.

$$\begin{array}{ll} A \rightleftarrows A^+ e^- & 2A^+ \rightarrow B + A \\ B \rightleftarrows B^+ + e^- & \end{array} \qquad (II.1.28)$$

It is also interesting to consider cases in which the second electron transfer occurs at a potential lower than that required for the first electron transfer. Although it may be thought that this condition would be unusual, it is in fact encountered in many electrochemical systems and is of fundamental importance in catalysis.

II.1.7.5 The $\overrightarrow{E}C\overleftarrow{E}$ Process

The $\overrightarrow{E}C\overleftarrow{E}$ mechanism occurs when the initial electron transfer at the electrode–solution interface is followed by a rapid chemical reaction step yielding a product of *higher* 'redox energy' compared to the starting material. That is, if an initial oxidation step occurs, this is followed by a chemical reaction and then reduction of the product so that the net current may be very small. Consequently, if the chemical step is very fast, A may be converted to B without any discernible current. This type of process, which has also been termed electron transfer catalysis (ETC), was reported first by Feldberg [107] and found subsequently for many other chemical systems [108]. The reaction scheme, Eq. (II.1.29), is identical to that of the conventional ECE process ($\overrightarrow{E}C\overleftarrow{E}$, except that, in this case, $E^{\ominus\prime}_{c,\,A^+/A} < E^{\ominus\prime}_{c,\,B^+/B}$ and the cross-redox reaction step is of key importance.

$$\begin{array}{ll} A \rightleftarrows A^+ + e^- & A^+ \rightarrow B^+ \\ B^+ + e^- \rightleftarrows B & \\ A + B^+ \rightarrow A^+ + B & \end{array} \qquad \text{(II.1.29)}$$

Cyclic voltammograms simulated based on the $\overrightarrow{E}C\overleftarrow{E}$ reaction scheme with $E^{\ominus\prime}_{c,\,A^+/A} = -1$ V and $E^{\ominus\prime}_{c,\,A^+/A} = +0.3$ V are shown in Fig. II.1.23b. It can be seen that at the slow scan rate of 0.01 V s^{-1}, a simple reversible voltammetric response at $E^{\ominus\prime}_c = 0.3$ V is detected. Thus, all the A has been converted to B on the timescale prior to reaching the potential of the B^+/B couple. Only a very small pre-wave at $E = -0.1$ V observed at a scan rate of 0.1 V s^{-1} indicates a complication. However, the small amount of charge flowing in this pre-wave is sufficient to catalyse the conversion of most of A in the vicinity of the electrode to B. The fact that the voltammetric response observed at 0.3 V has to be attributed to the B^+/B redox couple rather than to the A^+/A redox couple becomes apparent only at high scan rates. At sufficiently high scan rates, the irreversible chemical reaction step with $k_f = 200$ s^{-1} is too slow to initiate the isomerisation process and the oxidation of A, present in the bulk solution, can be detected. In contrast to the difficulty encountered in cyclic voltammetry in detecting the $\overrightarrow{E}C\overleftarrow{E}$ mechanism at slow scan rates, spectroelectrochemical measurements can provide clear direct experimental evidence for this type of process [109].

II.1.7.6 Square Schemes and More Complex Reaction Schemes

For the case of two reversible equilibria connecting both electrochemical reaction steps (Eq. 1.30), can be drawn with four reversible processes connecting them. This kind of reaction scheme, the so-called square scheme, is commonly observed for redox-state-dependent isomerisation [110], ligand exchange [111], and protonation processes [112] and can be seen to be based on an extended version of the ECE scheme. Under limiting conditions, when chemical steps are in equilibrium on the voltammetric timescale (with a sufficiently small potential gap between $E^{\ominus\prime}_{c, A^+/A}$ and $E^{\ominus\prime}_{c, B^+/B}$, and appropriately slow scan rate), a simple reversible voltammetric response can be observed even when the square scheme is operative. In other circumstances, two scan-rate-dependent processes may be observed.

$$\begin{array}{ccc} A & \rightleftarrows & A^+ + e^- \\ \uparrow\downarrow & & \uparrow\downarrow \\ B & \rightleftarrows & B^+ + e^- \end{array} \qquad \text{(II.1.30)}$$

A complete analysis of the square scheme is complex since disproportionation and/or other second-order 'cross-redox reactions' have to be taken into consideration. However, the limiting cases of the square scheme are much more tractable. An interesting aspect of the square reaction scheme is that, in principle, it applies to all one-electron processes with reaction steps $A^+ \rightleftarrows B^+$ and $A \rightleftarrows B$ coupled to the heterogeneous charge transfer. For example, the redox-induced hapticity change, which accompanies the reduction of $Ru(\eta - C_6Me_6)_2^+$, has been proposed [113] to be responsible for the apparently slow rate of electron transfer. That is, the limiting case of an apparent overall E_{irrev} process is observed for what in reality is a square scheme mechanism.

For processes involving consecutive protonation and redox steps, even more complex reaction schemes such as fences, meshes, or ladders can be constructed. An example for the analysis of this kind of more complex reaction scheme by a stepwise dissection based on numerical simulation is given in the introduction (Fig. II.1.1). Other complex reaction schemes of practical importance commonly encountered involve association reactions such as precipitation or polymerisation. These processes may also be described by a series of consecutive reaction steps. However, as a further complication in these classes of reactions, the properties of the surface of the electrode may change during the course of the process. Extensive studies aimed at elucidating very complex reaction pathways have been conducted for the electrodeposition of conducting polymers and, as might be expected, fast scan rate cyclic voltammetry has made an important contribution to an understanding of the mechanism [114].

II.1.8 Multi-phase Systems

As noted above, cyclic voltammetry is a powerful tool for the investigation of processes combining solution-phase reactions and heterogeneous charge transfer at the electrode surface. However, this technique can also be applied to systems with additional phase boundaries. For example, multi-phase processes in thin films covering an electrode surface (Fig. II.1.24a), particulate solids, bacteria, or micro-droplets attached to the electrode surface (Fig. II.1.24b), or micro-emulsion systems

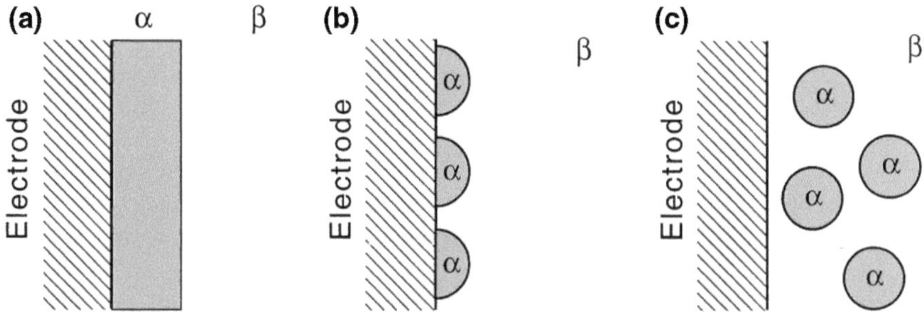

Fig. II.1.24 Schematic representation of three types of electrochemical systems with two phases, α and β, in contact with the electrode surface. (**a**) Film deposit; (**b**) island deposit with three-phase boundary regions; and (**c**) emulsion system

(Fig. II.1.24c), can be studied. An authoritative book dedicated to the voltammetry of microparticles and microdroplets is now available [115].

Processes that may need to be considered in these multi-phase problems are (i) partitioning of solutes, (ii) electro-dissolution of one phase into the other, (iii) the electrochemical conversion of a solute in phase to another redox level accompanied by ion exchange, and (iv) processes involving material adsorbed at the interface between the two phases. In view of the inherent complexity, data analysis based on numerical simulation is almost invariably necessary for detailed quantitative information to be gained.

An extensively studied multi-phase problem is the redox chemistry of a film deposit (Fig. II.1.24a). Well-known characteristic voltammetric responses may be observed for a thin film deposit with a thickness much less than the diffusion layer (see Fig. II.1.10) or for a thick film deposit with diffusion only within the film. The transport of charges in the film deposit may be possible via diffusion or 'hopping' of electrons [116] but transport of ions is also required to achieve charge neutrality.

Interest in liquid–liquid interfaces or liquid–liquid redox systems has been fuelled by their importance in biology. The complexity in electrochemical measurements on such systems led to the development of cyclic voltammetry employing a four-electrode potentiostat, although use of a film [117] or deposit of droplets [118] on the electrode surface has been suggested as a method for undertaking cyclic voltammetric studies of reactions at liquid–liquid interfaces with only a three-electrode configuration. Reviews on the reactivity of the triple-phase boundary formed when droplets are in contact with the electrode surface and immersed in aqueous media have appeared [119].

Figure II.1.25 shows data from a microdroplet liquid–liquid voltammetry experiment in which lactate anions A^- are transferred from the aqueous phase into the organic microdroplet phase (here composed of the organic solvent 4-(3-phenylpropyl)-pyridine or PPP containing a Mn(III/II) redox system and naphthyl-2-boronic acid B [120]). Schematically, the oxidation of each metal complex Mn(II)TPP (with TPP = tetraphenylporphyrinato) is generating a positive charge within the microdroplet and this is coupled to the reversible transfer of the anion A^-

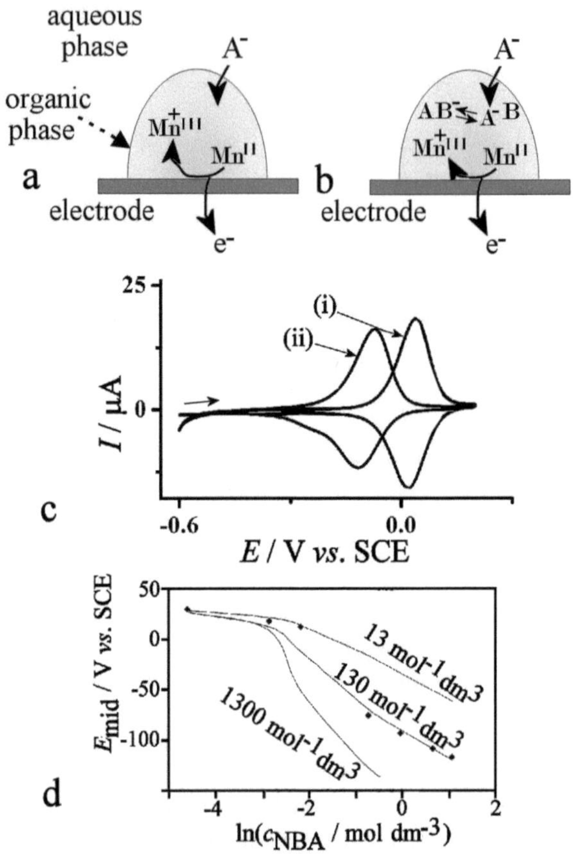

Fig. II.1.25 (**a**) Schematic representation of the transfer of anion A^- from the aqueous into the organic phase upon oxidation of Mn(II)TPP to Mn(III)TPP$^+$. (**b**) In the presence of the boronic acid B as a facilitator the transfer of the anion A^- leads to the formation of the complex AB in the organic phase. (**c**) Cyclic voltammograms [120] (scan rate 10 mVs^{-1}) for the oxidation and re-reduction of 75 mM Mn(II)TPP dissolved in PPP (4-(3-phenylpropyl)-pyridine, 75 nL) and immobilised in the form of microdroplets onto a 4.9-mm diameter graphite electrode immersed in aqueous 0.1 M sodium lactate pH = 7.34. The presence of (i) 0 and (ii) 973 mM naphthyl-2-boronic acid is shown to cause a negative shift in the voltammetric response. (**d**) Plot of the midpoint potential versus the natural logarithm of the naphthyl-2-boronic acid concentration in the microdroplets. Lines indicate calculated data [120] for reversible lactate–boronic acid complex formation for three equilibrium constants

(the EI mechanism, where 'E' indicates electron transfer and 'I' denotes ion transfer, see Fig. II.1.25a). It is possible to facilitate the anion transfer, for example, for lactate in the presence of naphthyl-2-boronic acid. The characteristic shift in the voltammetric response to more negative potentials is clear evidence for the reversible complex formation between lactate and boronic acid (the EIC_{rev} mechanism, see Fig. II.1.25b). Analysis of the midpoint potentials allows the binding

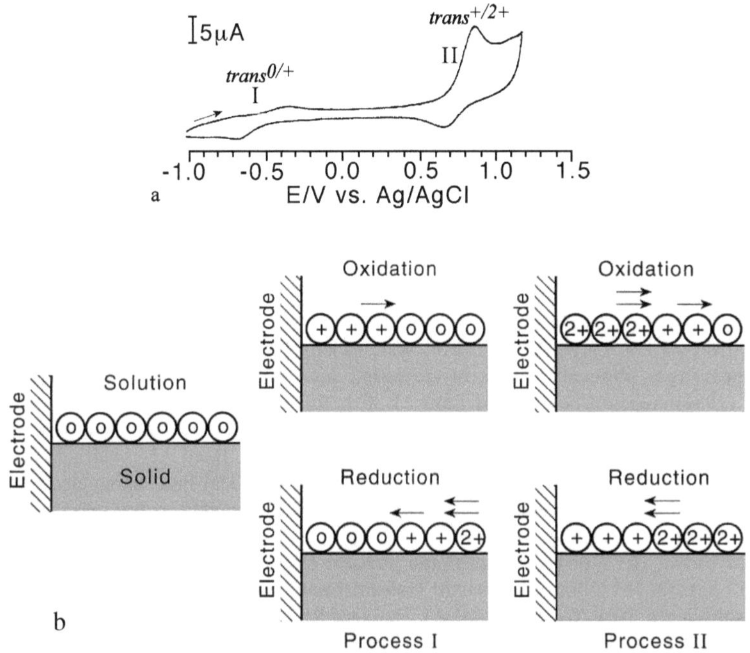

Fig. II.1.26 (**a**) Cyclic voltammogram [123] obtained in aqueous (0.1 M $NaClO_4$) media at 20°C for solid *trans*-$[Cr(CO)_2(dpe)_2]^+$ mechanically attached to a polished basal plane pyrolytic graphite electrode (scan rate = 50 mV s^{-1}). (**b**) Schematic representation of redox processes at the interface between solid, electrode surface, and liquid electrolyte system

constant between the transferring anion and the immobilised facilitator to be determined [120].

Emulsion processes are of considerable importance and cyclic voltammetry allows these to be investigated. Texter et al. [23, 121] and Rusling et al. [122] voltammetrically studied processes in optically clear, stabilised microemulsions, in which the droplet size was smaller than 100 nm. Processes monitored were shown to be consistent with a CE-type reaction scheme in which reactant diffused from the organic emulsion droplet towards the electrode surface.

Finally, voltammetric experiments with micro-crystalline particulate deposits present on the electrode surface provide information on solid-state processes and reactions at the solid–solvent electrolyte interface [22]. Figure II.1.26 shows the voltammetric response observed for the oxidation and reduction of an organometallic complex, *trans*-$[Cr(CO)_2(dpe)_2]^+$(dpe = $Ph_2PCH_2CH_2PPh_2$), together with a schematic representation of the processes which have been proposed to be responsible for the electrochemical inter-conversion of the solid from *trans*-$[Cr(CO)_2(dpe)_2]^0$ to *trans*-$[Cr(CO)_2(dpe)_2]^+$ and to *trans*-$[Cr(CO)_2(dpe)_2]^{2+}$ [123].

The interpretation of cyclic voltammograms in particular for multi-phase systems is often complex and experiments in conjunction with complementary spectroelectrochemical evidence (see also Chap. II.6) can give considerable additional insight into the mechanism of electrochemically driven solid-state reactions.

References

1. Oldham KB, Myland JC (1994) Fundamentals of electrochemical science. Academic Press, London, p 424
2. (a) Compton RG, Banks CE (2007) Understanding voltammetry. World Scientific, Singapore; (b) Noel M, Vasu KI (1990) Cyclic voltammetry and the frontiers of electrochemistry. Aspect Publications, London
3. (a) Brown ER, Large RF (1971) In: Weissenberger A, Rossiter BW (eds) Techniques of chemistry – physical methods of chemistry. John Wiley, New York, p 423; (b) Heinze J (1984) Angew Chem Int Ed Engl 23: 831; (c) Heinze J (1993) Angew Chem Int Ed Engl 32: 1268; (d) Geiger WE (1986) In: Zuckermann JJ (ed) Inorganic reactions and methods, vol 15. VCH, Weinheim, p 110; (e) Eklund JC, Bond AM, Alden JA, Compton RG (1999) Adv Phys Org Chem 32: 1; (f) Bond AM (2002) Broadening electrochemical horizons. Oxford University Press, Oxford; (g) for further information see, for example, http://electrochem.cwru.edu/estir/
4. (a) Girault HH (2004) Analytical and physical electrochemistry. EPFL Press, Lausanne; (b) Savéant JM (2006) Elements of molecular and biomolecular electrochemistry. Wiley-Interscience, New York; (c) Bard AJ, Faulkner RF (1980) Electrochemical methods. John Wiley, New York; (d) Rieger PH (1994) Electrochemistry, 2nd edn. Chapman and Hall, New York; (e) Sawyer DT, Sobkowiak A, Roberts JL Jr (1995) Electrochemistry for chemists, 2nd edn. John Wiley, New York; (f) Brett CMA, Brett AMO (1993) Electrochemistry: principles, methods, and applications. Oxford Univ Press, Oxford; (g) Astruc D (1995) Electron transfer and radical processes in transition-metal chemistry. VCH, Weinheim; (h) Albery J (1975) Electrode kinetics. Clarendon Press, Oxford; (i) Southampton electrochemistry group (1985) Instrumental methods in electrochemistry. Ellis Horwood, Chichester
5. Gosser DK (1993) Cyclic voltammetry: simulation and analysis of reaction mechanisms. VCH, New York
6. For a good historical account, see Adams RN (1969) Electrochemistry at solid electrodes. Marcel Dekker, New York
7. For a review, see Delahay P (1954) New instrumental methods in electrochemistry. John Wiley, New York, Chap. 3
8. Bond AM (1980) Modern polarographic methods in analytical chemistry. Marcel Dekker, New York
9. Heyrovský J, Kůta J (1966) Principles of polarography. Academic Press, New York
10. Wang J (2006) Analytical electrochemistry, 3rd edn. Wiley-VCH, Hoboken; (b) Henze G, Neeb R (1986) Elektrochemische analytik. Springer, Berlin Heidelberg New York; (c) Vydra F, Stulik K, Julakova E (1976) Electrochemical stripping analysis. John Wiley, New York
11. (a) Ševčik A (1948) Coll Czech Chem Commun 13: 349; (b) Randles JEB (1948) Trans Faraday Soc 44: 327
12. Nicholson RS, Shain I (1964) Anal Chem 36: 706
13. For example, the following software packages can be recommended: (a) DigiSim, BAS, 2701 Kent Avenue, West Lafayette, IN 47906, USA; (b) DigiElch from Elchsoft under http://www.elchsoft.com
14. Prenzler PD, Boskovic C, Bond AM, Wedd AG (1999) Anal Chem 71: 3650
15. Luo WF, Feldberg SW, Rudolph M (1994) J Electroanal Chem 368: 109
16. (a) Rudolph M, Reddy DP, Feldberg SW (1994) Anal Chem 66: 589A; (b) See DigiElch from Elchsoft under http://www.elchsoft.com
17. Reinmuth WH (1960) Anal Chem 32: 1509

18. Parker VD (1986) In: Fry AJ, Britton WE (eds) Topics in organic electrochemistry. Plenum Press, New York, p 35
19. Seralathan M, Osteryoung RA, Osteryoung JG (1987) J Electroanal Chem 222: 69
20. Bond AM (1980) Modern polarographic methods in analytical chemistry. Marcel Dekker, New York, p 341
21. Schiewe J, Hazi J, VicenteBeckett VA, Bond AM (1998) J Electroanal Chem 451: 129
22. (a) Scholz F, Meyer B (1998): Voltammetry of solid microparticles immobilized on electrode surfaces. In: Bard AJ, Rubinstein I (eds) Electroanalytical chemistry, a series of advances 20: 1; (b) Grygar T, Marken F, Schröder U, Scholz F (2002) Coll Czech Chem Commun 67: 163
23. Mackay RA, Texter J (1992) Electrochemistry in colloids and dispersions. VCH, Weinheim
24. (a) Bond AM, Pfund VB (1992) J Electroanal Chem 335: 281; (b) Tanaka K, Tamamushi R (1995) J Electroanal Chem 380: 279
25. Murray R (1992) Molecular design of electrode surfaces. John Wiley, New York
26. Girault HH (1993) Modern aspects of electrochemistry No. 25. Plenum, New York
27. Armstrong FA, Heering HA, Hirst J (1997) Chem Soc Rev 26: 169
28. (a) Kim BH, Kim HJ, Hyun MS, Park DH (1999) J Microbiol Biotechnol 9: 127; (b) Compton RG, Perkin SJ, Gamblin DP, Davis J, Marken F, Padden AN, John P (2000) New J Chem 24: 179 and references cited therein
29. Gale RJ (1988) Spectroelectrochemistry: theory and practice. Plenum Press, New York
30. Gtierrez C, Melendres C (1990) Spectroscopic and diffraction techniques in interfacial electrochemistry. Nato ASI Series C: Mathematical and Physical Sciences, vol 320. Kluwer Academic, Drodrecht
31. Ward MD (1995) In: Rubinstein I (ed) Physical electrochemistry: principles, methods, and applications. Marcel Dekker, New York, Chap. 7
32. Abruna HD (1991) Electrochemical interfaces. VCH, Weinheim
33. Kissinger PT, Heinemann WR (1984) Laboratory techniques in electroanalytical chemistry. Marcel Dekker, New York, Chap. 6
34. (a) Janz GJ, Ives DJG (1961) Reference electrodes. Academic Press, New York; (b) Southampton electrochemistry group (1985) Instrumental methods in electrochemistry. Ellis Horwood, Chichester, Chap. 11
35. (a) Bond AM, Oldham KB, Snook GA (2000) Anal Chem 72: 3492; (b) Pavlishchuk VV, Addison AW (2000) Inorg Chim Acta 298: 97
36. Kissinger PT, Heinemann WR (1984) Laboratory techniques in electroanalytical chemistry. Marcel Dekker, New York, Chap. 15
37. Kiesele H (1981) Anal Chem 53: 1952
38. Yamamoto A (1986) Organotransition metal chemistry. John Wiley, New York, p 155
39. Reichardt C (1988) Solvents and solvent effects in organic chemistry 2nd edn. VCH, Weinheim, p 414
40. Way DM, Bond AM, Wedd AG (1997) Inorg Chem 36: 2826
41. Schmickler W (1996) Interfacial electrochemistry. Oxford University Press, Oxford
42. (a) Safford LK, Weaver MJ (1991) J Electroanal Chem 312: 69; (b) Nirmaier HP, Henze G (1997) Electroanalysis 9: 619
43. Norton JD, White HS, Feldberg SW (1990) J Phys Chem 94: 6772
44. Atkins PW (1998) Physical chemistry, 6th edn. Oxford University Press, Oxford, p 753
45. (a) Amatore C (1995) In: Rubinstein I (ed) Physical electrochemistry – principles, methods, and applications. Marcel Dekker, New York, p 191; (b) Amatore C, Maisonhaute E, Simonneau G (2000) Electrochem Commun 2: 81
46. Nicholson RS (1966) Anal Chem 38: 1406
47. Fisher AC (1996) Electrode dynamics. Oxford University Press, Oxford
48. Oldham KB, Myland JC (1994) Fundamentals of electrochemical science. Academic Press, London, p 237
49. Bard AJ, Faulkner RF (1980) Electrochemical methods. John Wiley, New York, p 238

50. (a) Randles JEB (1948) Trans Faraday Soc 44: 327; (b) Ševčik A (1948) Collect Czech Chem Commun 13: 349
51. Wilke CR, Chang P (1955) AIChE J 264
52. (a) Bard AJ, Faulkner RF (1980) Electrochemical methods. John Wiley, New York, p 523 and references cited therein; (b) Gileadi E (1993) Electrode kinetics for chemists, chemical engineers, and material scientists. VCH, Weinheim, p 420
53. (a) Bard AJ, Faulkner RF (1980) Electrochemical methods. John Wiley, New York, p 409; (b) Hubbard AT, Anson FC (1970) Electroanal Chem 4: 129
54. Brown AP, Anson FC (1977) Anal Chem 49: 1589
55. Ball JC, Marken F, Qiu FL, Wadhawan JD, Blythe AN, Schröder U, Compton RG, Bull SD, Davies SG (2000) Electro Analysis 12: 1017
56. Hubbard AT (1969) J Electroanal Chem 22: 165
57. Bockris JOM, Khan SUM (1993) Surface electrochemistry: a molecular level approach. Plenum, New York, p 223
58. Oldham KB, Myland JC (1994) Fundamentals of electrochemical science. Academic Press, London, p 373
59. Macdonald JR (1987) Impedance spectroscopy. John Wiley, New York
60. Coles BA, Compton RG, Larsen JP, Spackman RA (1996) Electro Analysis 8: 913
61. Tschuncky P, Heinze J (1995) Anal Chem 67: 4020
62. Bond AM, Feldberg SW, Greenhill HB, Mahon PJ, Colton R, Whyte T (1992) Anal Chem 64: 1014
63. Matsuda H, Ayabe Y (1955) Z Elektrochemie 59: 494
64. Aoki K (1993) Electroanalysis 5: 627
65. Neudeck A, Dittrich J (1991) J Electroanal Chem 313: 41
66. (a) Bond AM (1994) Analyst 119: R1; (b) Heinze J (1993) Angew Chem Int Ed Engl 32: 1268
67. Amatore C (1995) In: Rubinstein I (ed) Physical electrochemistry – principles, methods, and applications. Marcel Dekker, New York, p 163
68. (a) Paeschke M, Wollenberger U, Kohler C, Lisec T, Schnakenberg U, Hintsche R (1995) Anal Chim Acta 305: 126; (b) Aoki K (1990) J Electroanal Chem 284: 35
69. Fletcher S, Horne MD (1999) Electrochem Commun 1: 502
70. Scholz F, Hermes M (1999) Electrochem Commun 1: 345
71. (a) Keyes TE, Forster RJ, Jayaweera PM, Coates CG, McGarvey JJ, Vos JG (1998) Inorg Chem 37: 5925; (b) Bond AM, Marken F, Williams CT, Beattie DA, Keyes TE, Forster RJ, Vos JG (2000) J Phys Chem B 104: 1977
72. (a) Kounaves SP, O'Dea JJ, Chandresekhar P, Osteryoung J (1987) Anal Chem 59: 386; (b) Aoki K, Tokuda K, Matsuda H (1986) J Electroanal Chem 207: 25; (c) Aoki K, Maeda K, Osteryoung J (1989) J Electroanal Chem 272: 17
73. (a) Bard AJ, Faulkner RF (1980) Electrochemical methods. John Wiley, New York, p 92; (b) Vetter KJ (1967) Electrochemical kinetics. Academic Press, New York; (c) Fisher AC (1996) Electrode dynamics. Oxford University Press, Oxford
74. Oldham KB, Myland JC (1994) Fundamentals of electrochemical science. Academic Press, London, p 167
75. Cannon RD (1981) Electron transfer reactions. Butterworths, London.
76. Savéant JM (1993) Acc Chem Res 26: 455
77. Eberson L (1987) Electron transfer reactions in organic chemistry. Springer, Berlin Heidelberg New York
78. Oldham KB, Myland JC (1994) Fundamentals of electrochemical science. Academic Press, London, p 421
79. (a) Delahay P (1965) Double layer and electrode kinetics. John Wiley, New York, p 197; (b) Frumkin ANZ (1933) Physik Chem 164A: 121; (c) Parsons R (1961). The structure of the electric double layer and its influence on the rates of electrode reactions. In: Delahay P (ed) Advances in electrochemistry and electrochemical engineering, vol 1. Interscience, New York, p 1

80. (a) Schmickler W (1996) Interfacial electrochemistry. Oxford University Press, Oxford, p 68; (b) Chidsey CED (1991) Science 251: 919; (c) Savéant JM, Tessier D (1975) J Electroanal Chem 65: 57
81. Neudeck A, Marken F, Compton RG (1999) Electroanalysis 11: 1149
82. (a) Kaifer AE, Gomez-Kaifer M (1999) Supramolecular electrochemistry. Wiley-VCH, New York; (b) Aoki K (1996) J Electroanal Chem 419: 33
83. Geiger WE (1986) In: Zuckermann JJ (ed) Inorganic reactions and methods, vol 15. VCH, Weinheim, p 124
84. (a) Ammar F, Savéant JM (1973) J Electroanal Chem 47: 215; (b) Flanagan JB, Margel S, Bard AJ, Anson FC (1978) J Am Chem Soc 100: 4248
85. (a) Lowering DG (1974) J Electroanal Chem 50: 91; (b) Verplaetse H, Kiekens P, Temmerman E, Verbeck F (1980) J Electroanal Chem 115: 235
86. Smith WH, Bard AJ (1977) J Electroanal Chem 76: 19
87. (a) Evans DH, Lehmann MW (1999) Acta Chem Scand 53: 765; (b) Evans DH (1998) Acta Chem Scand 52: 194
88. Bard AJ, Faulkner RF (1980) Electrochemical methods. John Wiley, New York, p 371
89. Andrieux CP (1994) Pure Appl Chem 66: 2445
90. (a) Bard AJ, Faulkner RF (1980) Electrochemical methods. John Wiley, New York, p 35; (b) Rieger PH (1994) Electrochemistry, 2nd edn. Chapman and Hall, New York, p 270
91. Brown ER, Large RF (1971) In:Weissenberger A, Rossiter BW (eds) Techniques of chemistry – physical methods of chemistry. John Wiley, New York, p 476
92. (a) Nicholson RS, Shain I (1964) Anal Chem 36: 706; (b) Ahlberg E, Parker VD (1981) J Electroanal Chem 121: 73; (c) Neudeck A, Dittrich J (1989) J Electroanal Chem 264: 91
93. Andrieux CP, Savéant JM, Tallec A, Tardivel R, Tardy C (1997) J Am Chem Soc 119: 2420
94. McQuarrie DA, Simon JD (1997) Physical chemistry: a molecular approach. University Science Books, Sausalito, p 1095
95. Geiger WE (1985) Prog Inorg Chem 33: 2275
96. (a) Hubler P, Heinze J (1998) Ber Bunsenges Phys Chem Chem Phys 102: 1506; (b) Tschuncky P, Heinze J, Smie A, Engelmann G, Kossmehl G (1997) J Electroanal Chem 433: 223
97. (a) Volke J, Liska F (1994) Electrochemistry in organic synthesis. Springer, Berlin Heidelberg New York, p 118; (b) Simonet J (1991) In: Lund H, Baizer MM (eds) Organic electrochemistry. Marcel Dekker, New York, p 1217
98. (a) Andrieux CP, Hapiot P, Savéant JM (1990) Chem Rev 90: 723; (b) Savéant JM (1993) Acc Chem Res 26: 455
99. (a) Savéant JM, Vianello E (1963) Electrochim Acta 8: 905; (b) Savéant JM, Vianello E (1967) Electrochim Acta 12: 629
100. Marken F, Leslie WM, Compton RG, Moloney MG, Sanders E, Davies SG, Bull SD (1997) J Electroanal Chem 424: 25
101. (a) Orsini J, Geiger WE (1999) Organometallics 18: 1854; (b) Rebouillat S, Lyons MEG, Bannon T (1999) J Solid State Electrochem 3: 215; (c) Treimer SE, Evans DH (1998) J Electroanal Chem 449: 39
102. Albery WJ, Bell RP (1963) Proc Chem Soc 169
103. (a) Hershberger JW, Klingler RJ, Kochi JK (1983) J Am Chem Soc 105: 61; (b) Neto CC, Kim S, Meng Q, Sweigart DA, Chung YK (1993) J Am Chem Soc 115: 2077; (c) Rossenaar BD, Hartl F, Stufkens DJ, Amatore C, Maisonhaute E, Verpeaux JN (1997) Organometallics 16: 4675
104. Karp S (1968) J Phys Chem 72: 1082
105. (a) Gonzalez FJ (1998) Electro analysis 10: 638; (b) Neudeck A, Dittrich J (1989) J Electroanal Chem 264: 91
106. (a) Feldberg SW (1971) J Phys Chem 75: 2377; (b) Moulton RD, Chandler DJ, Arif AM, Jones RA, Bard AJ (1988) J Am Chem Soc 110: 5714; (c) Baeza A, Ortiz JL, Gonzalez J (1997) J Electroanal Chem 429: 121

107. Feldberg SW, Jeftic L (1972) J Phys Chem 76: 2439
108. (a) Feldberg SW (1969) Electroanal Chem 3: 199; (b) Astruc D (1995) Electron transfer and radical processes in transition metal chemistry. VCH, Weinheim, p 421 and references cited therein
109. Stulmann F, Domschke G, Bartl A, Neudeck A, Petr A, Omelka L (1997) Magn Reson Chem 35: 124
110. (a) Wooster TT, Geiger WE, Ernst RD (1995) Organometallics 14: 3455; (b) Bond AM, Colton R, Feldberg SW, Mahon PJ, Whyte T (1991) Organometallics 10: 3320; (c) Bond AM, Oldham KB (1983) J Phys Chem 87: 2492
111. Kano SK, Glass RS, Wilson GS (1993) J Am Chem Soc 115: 592
112. (a) Laviron E, Meunierprest R, Lacasse R (1994) J Electroanal Chem 375: 263; (b) Laviron E (1980) J Electroanal Chem 109: 57
113. Peirce DT, Geiger WE (1992) J Am Chem Soc 114: 6063, 7636
114. Tschuncky P, Heinze J (1993) Synth Metals 55: 1603
115. Scholz F, Schröder U, Gulaboski R (2005) Electrochemistry of immobilized particles and droplets. Springer, Heidelberg
116. Majda M (1992) Dynamics of electron transport in polymer assemblies of redox centers. In: Murray R (ed) Molecular design of electrode surfaces. John Wiley, New York, p 159
117. Shi CN, Anson FC (1999) J Phys Chem B 103: 6283
118. (a) Scholz F, Komorsky-Lovrić S, Lovrić M (2000) Electrochem Commun 2: 112; (b) Marken F, Webster RD, Bull SD, Davies SG (1997) J Electroanal Chem 437: 209
119. (a) Banks CE, Davies TJ, Evans RG, Hignett G, Wain AJ, Lawrence NS, Wadhawan JD, Marken F, Compton RG (2003) PCCP 5: 4053; (b) Scholz F, Gulaboski R (2005) Chem Phys Chem 6: 16
120. Katif N, Harries RA, Kelly AM, Fossey JS, James TD, Marken F (2009) J Solid State Electrochem 13: 1475
121. (a) Garcia E, Texter J (1994) J Coll Interface Sci 162: 262; (b) Dayalan E, Qutubuddin S, Texter J (1991) J Coll Interface Sci 143: 423
122. (a) Rusling JF, Zhou DL (1997) J Electroanal Chem 439: 89; (b) Campbell CJ, Haddleton DM, Rusling JF (1999) Electrochem Commun 1: 618
123. Bond AM, Colton R, Daniels F, Fernando DR, Marken F, Nagaosa Y, VanSteveninck RFM, Walter JN (1993) J Am Chem Soc 115: 9556

Chapter II.2
Pulse Voltammetry

Zbigniew Stojek

II.2.1 Introduction

The idea of imposing potential pulses and measuring the currents at the end of each pulse was proposed by Barker in a little-known journal as early as in 1958 [1]. However, the first reliable trouble-free and affordable polarographs offering voltammetric pulse techniques appeared on the market only in the 1970s. This delay was due to some limitations on the electronic side. In the 1990s, again substantial progress in electrochemical pulse instrumentation took place. This was related to the introduction of microprocessors, computers, and advanced software.

Pulse voltammetry was originally introduced for the dropping mercury electrode (DME) and therefore was initially called pulse polarography. At that time, it seemed reasonable, and very advantageous, to apply one pulse per one drop of the DME. Now, at the beginning of the twenty-first century, researchers rarely use the dropping mercury electrode. True, such an electrode offers a continually renewed mercury surface (since the drops are formed and grow one after another); however, the dropping mercury electrode complicates the measurements by generating large charging currents and a difficult-to-control convection at the electrode surface. This convection often leads to significantly enhanced currents. The dropping mercury electrode was quickly replaced by the static mercury drop electrode (SMDE), the hanging mercury drop electrode (HMDE), and the mercury film electrode (MFE). The dispensing of a new drop in an SMDE is controlled electronically and is done fast. Figure II.2.1 compares how the electrode area changes in time for a Hg drop in the dropping and the static mercury electrode. The application of specifically arranged potential pulses to the SMDE and electrodes made of metal, graphite, and other solid conducting materials results in pulse voltammetry.

The imposition of a potential pulse to the electrode leads in most experimental situations to a considerable improvement (increase) in the ratio of the charging and faradaic currents compared to that for linear scan voltammetry. This is because the

Z. Stojek (✉)
Department of Chemistry, University of Warsaw, Pasteura 1, 02-093 Warsaw, Poland
e-mail: stojek@chem.uw.edu.pl

F. Scholz (ed.), *Electroanalytical Methods*, 2nd ed.,
DOI 10.1007/978-3-642-02915-8_5,

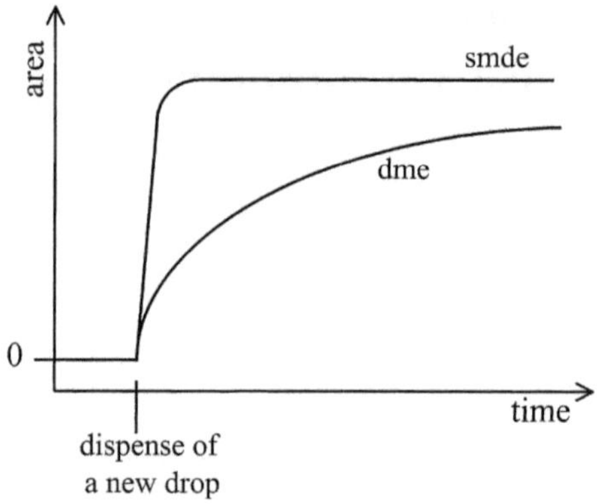

Fig. II.2.1 Mercury drop area plotted vs. time: *dme* dropping mercury electrode; *smde* static mercury electrode

faradaic current usually decreases with $1/t^{1/2}$, while the charging current decreases much faster (exponentially). In consequence, decreased lower limits of detection are obtained.

The following components can be distinguished in the total charging current:

(1) The current generated by the potential scan rate, v; this can be represented by Eq. (II.2.1):

$$I_{\text{cap,v}} = AC_{\text{d}}v \tag{II.2.1}$$

where A is the electrode area and C_{d} is the differential capacitance of the electrical double layer. This component is equal to zero as long as the pulse lasts, since $v = 0$.

(2) The current associated with the imposition of the potential pulse and the appropriate charging of the electrical double layer. This component always exists and can be described as

$$I_{\text{cap,p}} = k\,\mathrm{e}^{-t/R_{\text{s}}C_{\text{d}}} \tag{II.2.2}$$

where t is time measured from the moment of imposition of the pulse, k is a constant determined by the solution resistance and the imposed potential ($k = R_{\text{s}}/\Delta E$), and R_{s} is the solution resistance. This component of the overall charging current usually decreases very fast. Problems with discrimination of this current appear when $R_{\text{s}}C_{\text{d}}$ is large. In such a situation, the instrumental ohmic-potential-drop compensation circuit should be activated.

(3) In polarography, the continuous growth of the mercury drop brings about a corresponding charging current, $I_{\text{cap,gr}}$. This is described as follows:

$$I_{\text{cap,gr}} = 0.0056\,C_{\text{i}}(E_{\sigma=0} - E)\,m^{2/3}t^{-1/3} \tag{II.2.3}$$

where C_{i} is the integral capacity of the electrical double layer, m is the rate of mercury flow from the capillary, and t is the drop time.

(4) There are other components of the charging current which cannot be rigorously quantified. These are related to the changes in the electrical double-layer

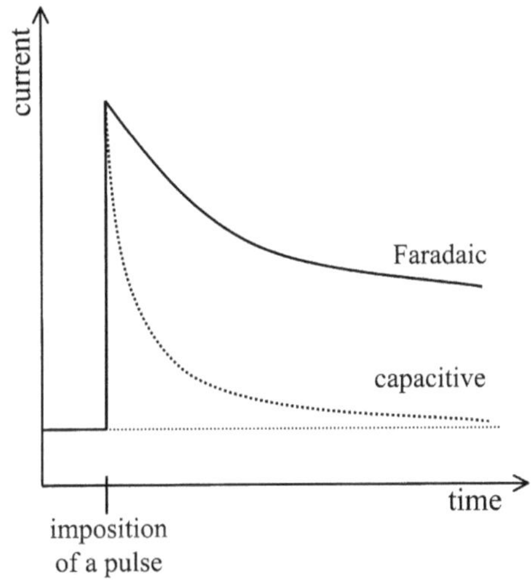

Fig. II.2.2 Faradaic and nonfaradaic currents flowing after the application of a potential pulse plotted vs. time

structure that include adsorption and desorption, reorientation, and the changes in the electrode area caused by the electrode reaction products.

Faradaic and nonfaradaic currents flowing after the application of a potential pulse are qualitatively illustrated in Fig. II.2.2. If the aim is to obtain a current possibly free of the charging component, the current sampling should be set for a time when the charging current is negligible. On the other hand, if one wants to expose the charging current in the absence of a faradaic reaction, the sampling should be done as soon as possible after the pulse imposition.

It is worth noting that, in practice, the current sampling is carried out over a period of time (sampling period) $t_2 - t_1$, where t_1 and t_2 are measured from the moment of imposition of the pulse. During that period the current is integrated and the mean value is sent to the instrument output. Ideally, $t_2 - t_1$ should be as short as possible. In some potentiostats this period is not longer than several microseconds, so the sampling can be treated as the point sampling. However, it is sometimes advantageous, especially in analytical chemistry, to prolong the sampling period to milliseconds, because, in such a way, the electrical noise can be substantially decreased. This is why in some polarographs/voltammographs this period is rather wide. In such a situation, to be able to calculate anything, the time of the sampling, t_s (or simply the equivalent pulse time, t_p), should be estimated. If each pulse is applied to an individual Hg drop or the pulses imposed to a solid electrode are separated by a sufficiently long rest time, the time of the sampling is considered to be the pulse time and can be calculated using the following formula:

$$\sqrt{\frac{1}{t_p}} = \frac{2}{\sqrt{t_1} + \sqrt{t_2}} \qquad \text{(II.2.4)}$$

The way the potential pulses are imposed on the electrode defines the voltammetric technique. There are many waveforms and procedures for current measuring published in the literature. Very often they differ in details. On the other hand, the same technique may be implemented in a slightly different way in various polarographs/voltammographs. The reader may also be confused by the manuals of various instrumentation makers, since the terminology is not ideally fixed. In this chapter the most popular and significant pulse techniques are discussed. Square-wave voltammetry, due to its specific character, is discussed in a separate chapter.

II.2.2 Staircase Voltammetry

This is the simplest pulse voltammetric technique; however, it is probably also the one most often used for a dynamic electrochemical examination of various compounds. The sequence of pulses in staircase voltammetry (SV) forms a potential staircase. An appropriate potential waveform is illustrated in Fig. II.2.3.

Staircase voltammetry is in fact a modified, discrete linear scan (or cyclic) voltammetry. The potential scan can be reversed in SV, similarly as it is done in cyclic voltammetry, and then a cyclic staircase voltammogram can be obtained. Staircase voltammograms are peaked-shaped the same as linear scan voltammograms. There are some differences between these voltammetries anyway. A linear scan (or cyclic) voltammogram forms a continuous current vs. potential curve, while each staircase voltammogram consists of a number of $i - E$ points. Also, the peak heights obtained under conditions of identical scan rates in linear scan and staircase voltammetries ($v = \Delta E/\Delta t$) may differ considerably.

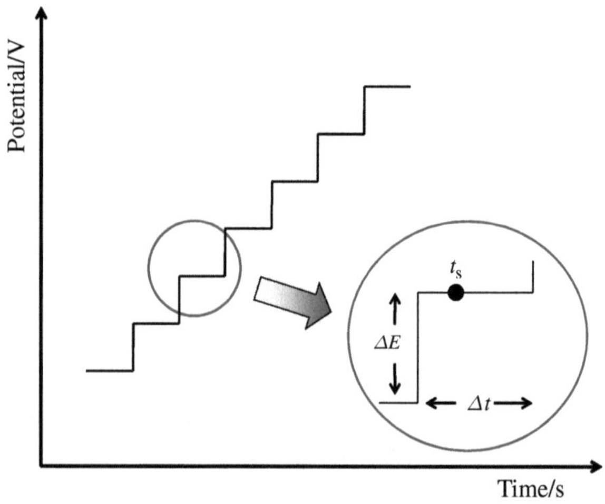

Fig. II.2.3 Potential waveform for *staircase voltammetry*. t_s time of sampling; Δt pulse (step) width; ΔE pulse (step) height

II.2.2.1 Equivalence of Staircase and Linear Scan Voltammetries

The theory for staircase voltammetry is more complex compared to the linear scan (or cyclic) theory. This is because more parameters must be taken into account in the derivation of the theoretical equations. One would be satisfied during the work with staircase voltammetry if the peak height obtained for $v = \Delta E/\Delta t$ is of identical height to LSV peak and correspondingly the simpler LSV theory could be used to interpret the staircase voltammetric data. There are two solutions to this problem. The first one is that the appropriate sampling time should be applied in staircase voltammetry. It has been theoretically estimated that this sampling time, expressed as a fraction of step width, Δt, should equal either 0.25 or 0.5 depending on the type of electrode process [2]. Another, more rigorous theoretical elaboration has led to the conclusion that, for reversible electrode processes, the current should be sampled at the fraction 0.36 of the step width [3]. The second way to arrive at the equivalence of staircase and linear scan (cyclic) voltammograms is to use a very small step height. Step heights of 1–2 mV are sufficiently small to yield identical peaks for identical scan rates ($v = \Delta E/\Delta t$). The problem of equivalence of staircase and linear scan voltammetries is particularly important for the work with wide potential ranges and potential step hights of several mV. A good idea is to realize that in several voltammographs based on microprocessors or working on-line with PCs, linear scan, and cyclic voltammetries are offered, and in fact they are staircase voltammetries.

II.2.3 Normal Pulse Voltammetry

A waveform for normal pulse voltammetry (NPV) [4, 5] is presented in Fig. II.2.4. The electrode is subjected to a train of increasing potential pulses. During the pulses, if the electrode potential is sufficiently close to or more negative than the formal potential of the system under investigation ($E_c^{\ominus'}$), product R is generated (reaction Ox + ne $\rightarrow$ Red is assumed here). Before the application of each consecutive pulse, the electrode is kept at the initial potential, E_i, for the waiting time t_w. At this potential, for a reversible electrode reaction, the product generated during the pulse is transformed to the substrate again, and the initial distribution of the substrate is reconstituted or renewed. For sufficiently long waiting times and irreversible electrode reactions, for which the product cannot be transformed back to the substrate, the initial distribution of the concentrations is also renewed, since the product can diffuse away from the solid electrode to the solution bulk. When the pulse potential cannot be sufficiently short (several ms) or the waiting time acceptably long (several s), a short period of stirring after the pulse application is helpful. The ability to renew the initial concentrations at the electrode surface after each pulse is important, since it simplifies substantially the theoretical formulas for the normal pulse waves. When a dropping mercury electrode (normal pulse polarography) or a static mercury drop electrode is used, this problem is simply solved by dislodging the old

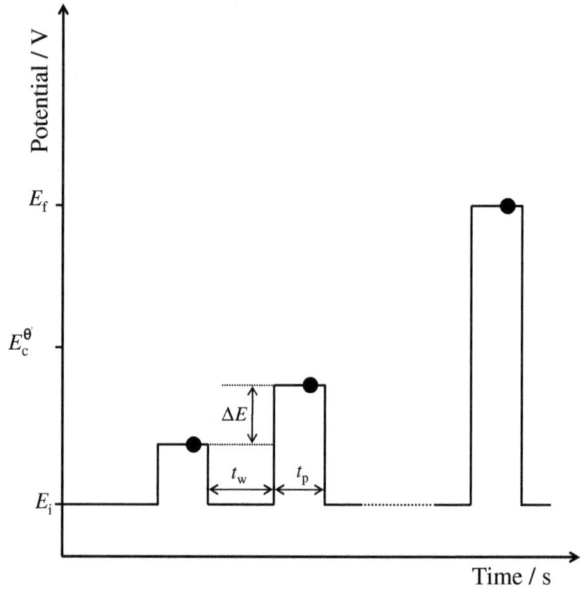

Fig. II.2.4 Normal pulse voltammetric waveform

drop after each pulse. At microelectrodes (of size in the μm range) the situation is also simple due to much faster transport of Ox and Red in the spherical diffusional field formed at the electrode surface [6].

The current in NPV is sampled most often near the end of the pulse; in Fig. II.2.4 the current sampling is represented by filled circles.

At successive potential pulses the current increases accordingly until the concentration of the substrate at the electrode surface approaches zero. Then the current reaches a plateau. A typical experimental normal pulse voltammogram is presented in Fig. II.2.5. [Tris-(9,10-dioxo-1-anthryl)]trisaminoacetyl amine (TDATAA) has been taken as an illustration. A corresponding cyclic voltammogram is added to the figure to serve as a reference. Apparently, TDATAA is reduced in two one-electron steps. It is easier to draw this conclusion in the case of NPV, since both waves are easily found to be of very similar height.

If the initial conditions are renewed after each potential pulse, then the wave height in NPV, $i_{\text{lim,NP}}$, will be described by the following formula:

$$i_{\text{lim,NP}} = \frac{nFAD^{1/2}C^*}{\pi^{1/2}t_{\text{p}}^{1/2}} \quad \text{(II.2.5)}$$

Formula (II.2.5) states that the NPV current is proportional to the bulk substrate concentration, C^*, the number of electrons transferred, n, the square root of the substrate diffusion coefficient, D, the electrode area, A, and is inversely proportional to the pulse time, t_{p}. In fact, instead of the pulse time, the formula should contain

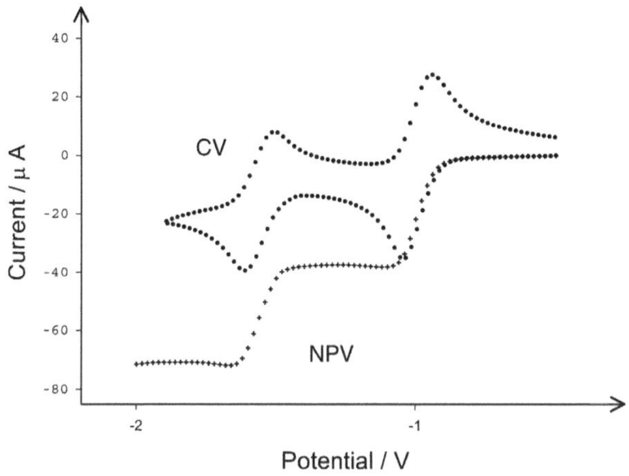

Fig. II.2.5 NPV (t_p = 20 ms) and CV (v = 100 mV/s) of 1 mM [tris-(9,10-dioxo-1-anthryl)]trisaminoacetyl amine at a 3-mm platinum electrode. Solution: 0.1 M tetrabutylammonium perchlorate in DMF. Both: Red and Ox are soluble. E is given vs. Ag/AgCl (0.1 M TBACl in methanol), $T = 25°C$

the sampling time, t_s; especially when the sampling is not performed exactly at the end of the pulse. Formula (II.2.5) is valid for electrodes of regular size (radius in the range of a few mm) and for processes where the transport of the substrate to the electrode surface is done only by diffusion. Chemical reactions that involve the substrate and either precede or follow the electron transfer will also lead to different currents. The height of NPV waves does not depend on the electron transfer rate, so this technique is considered as a very reliable one for the determination of diffusion coefficients of the examined compounds.

The potential at half height of the wave, $E_{1/2}$, is characteristic for the substrate examined. This potential does not vary much from the formal potential of the examined Ox/Red system. The relationship between the formal potential $E_c^{\ominus'}$ and the half-wave potential $E_{1/2}$, for reversible processes, is given in Eq. (II.2.6).

$$E_{1/2} = E_c^{\ominus'} + \frac{RT}{nF} \ln \left(\frac{D_{Ox}}{D_{Red}} \right)^{1/2} \tag{II.2.6}$$

To check whether the examined electrode reaction is reversible, a semi-logarithmic plot of E vs. $\ln\left[(i_{lim} - i)/i\right]$ can be drawn. For a reversible reaction, in the potential range around $E_{1/2}$, this plot should have a value of $59.1/n$ mV at 25°C. The corresponding formula giving the slope and the shape of a NP voltammogram is as follows:

$$E = E_{1/2} + \frac{RT}{nF} \ln \left(\frac{i_{lim} - i}{i} \right) \tag{II.2.7}$$

The equations for NPV of quasi-reversible and irreversible electrode processes are much more complex. The details can be found elsewhere [7].

A considerable advantage of NPV over other electrochemical methods is related to its potential waveform. The researcher can keep the electrode, for most of the experimental time, at a potential convenient for the experiment. The potentials at which the electrode process takes place are reached only during the several millisecond pulses. This may allow the researcher to limit the unwanted side processes, e.g., the precipitation and the filming of the electrode.

II.2.3.1 Influence of Adsorption

Normal pulse voltammetry is very sensitive to adsorption of both the substrate and the product at the electrode surface. If the substrate is adsorbed, a peak-shaped voltammogram is obtained. The stronger the adsorption or the lower the concentration of the substrate or the shorter the pulse time, the better defined is the peak. This is illustrated in Fig. II.2.6 by the normal pulse voltammograms of imidazoacridinone. This compound, at –0.4 V, is strongly adsorbed at the mercury surface, therefore the two consecutive reduction steps result in the formation of two peaks instead of two waves.

The adsorption of the product leads in NPV to the formation of a pre-wave. A good example of such an electrode process is the electro oxidation of a mercury

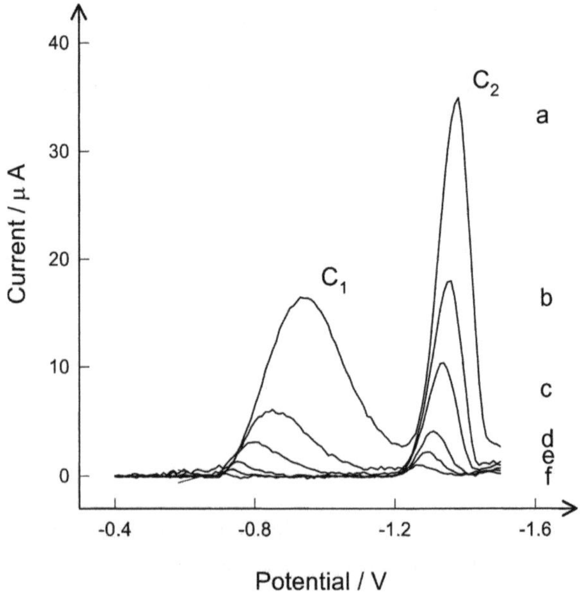

Fig. II.2.6 Normal pulse voltammograms at a hanging mercury drop electrode in 5×10^{-5} M imidazoacridinone (substrate adsorbs at electrode surface). Pulse time: *a* 2, *b* 5, *c* 10, *d* 25, *e* 50, and *f* 100 ms. Instrumental drop time: 1 s. (Adapted from [8] with permission)

electrode in the presence of EDTA, a strong ligand [9]. The product, HgEDTA, strongly adsorbs on mercury. The lower the concentration of EDTA or the shorter the pulse time, the higher (relative to the main wave) is the pre-wave.

II.2.4 Reverse Pulse Voltammetry

NPV does not allow direct examination of the products of the electrode reaction. To do this, a technique complementary to NPV may be employed. Reverse pulse voltammetry (RPV) is such a technique [5, 10, 11]. It is based on the waveform shown in Fig. II.2.7.

The main difference between NPP and RPV is that before the application of a pulse the electrode is kept at a potential corresponding to the NPV wave plateau. At this potential the product is generated from the substrate for a specified time t_g. During the consecutive pulses the product is transformed back into the primary substrate. In RPV the renewal of the initial conditions is straightforward only when a DME or an SMDE is employed. To obtain a renewal of the concentration profiles at solid electrodes either an extra waiting potential (and the corresponding waiting time, t_w) [12] or the disconnection of the working electrode followed by a stirring period [13] should be added to the waveform. In Fig. II.2.7, the addition of an extra waiting potential to the RPV waveform is shown. This potential is equal to the final

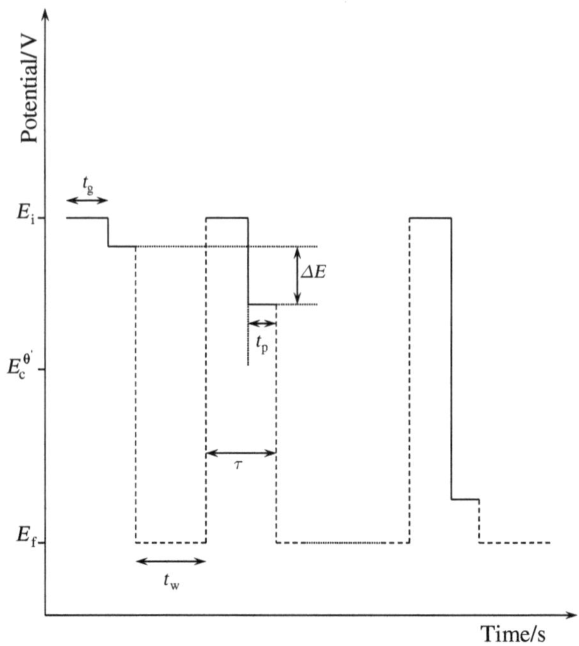

Fig. II.2.7 Potential waveform for RPV

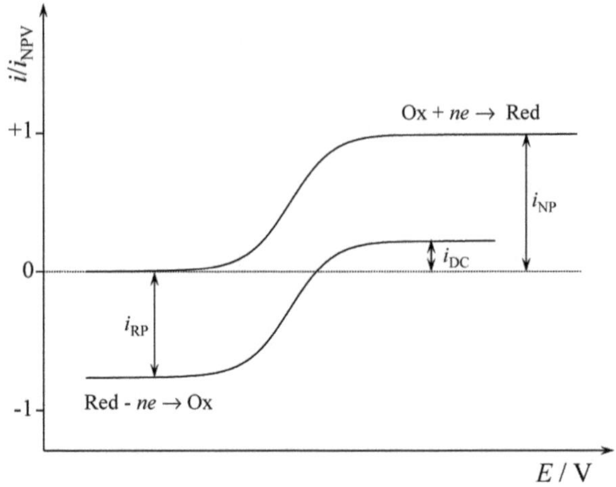

Fig. II.2.8 Typical RP voltammogram drawn against the corresponding NP voltammogram

pulse potential, where the product of the electrode reaction is transformed to the substrate.

A typical shape of a reverse pulse voltammogram is presented in Fig. II.2.8. The corresponding normal pulse voltammogram is added to the figure as a reference. The DC part of the RP voltammogram, i_{DC}, is positive as is the normal pulse current, i_{NP} (Ox + ne^{-} → Red). The reverse pulse current, i_{RP}, is of opposite sign, since the product of the electrode reaction is oxidized now.

If the working electrode is sufficiently large and the diffusion to its surface is planar, and the initial concentrations at the electrode surface are renewed after each pulse, the limiting DC current of the reverse pulse wave will be given by

$$i_{\mathrm{lim,DC}} = \frac{nFAD_{\mathrm{Ox}}^{1/2} C_{\mathrm{Ox}}^{*}}{\pi^{1/2} \tau_{\mathrm{p}}^{1/2}} \qquad \text{(II.2.8)}$$

and the limiting RP current is as follows:

$$-i_{\mathrm{lim,RP}} = \frac{nFAD_{\mathrm{Ox}}^{1/2} C_{\mathrm{Ox}}^{*}}{\pi^{1/2}} \left(\frac{1}{t_{\mathrm{p}}^{1/2}} + \frac{1}{\tau^{1/2}} \right) \qquad \text{(II.2.9)}$$

For a reaction where Red is the primary substrate and Ox is reduced in the reverse going pulses to Red, D_{Ox} and C_{Ox} should be replaced in Eqs. (II.2.8) and (II.2.9) by D_{Red} and C_{Red}.

A comparison of Eqs. (II.2.5), (II.2.8), and (II.2.9) allows us to write a very useful equation:

$$i_{\mathrm{lim,NP}} = i_{\mathrm{lim,DC}} - i_{\mathrm{lim,RP}} \qquad \text{(II.2.10)}$$

Another useful equation is that describing the ratio $i_{\lim\,\mathrm{mDC}}/i_{\lim,\mathrm{RP}}$:

$$-i_{\lim,\mathrm{RP}}/i_{\lim,\mathrm{DC}} = \left(n_{\mathrm{Red}}/n_{\mathrm{Ox}}\right)\left(\tau/t_{\mathrm{p}}\right)^{1/2} \qquad \text{(II.2.11)}$$

Equation (II.2.11) is a good approximation for $t_{\mathrm{p}} < 0.001\ \tau$.

If the experimental data do not fit Eqs. (II.2.10) and (II.2.11), this means that the electrode reaction in the first-generation step is not of stoichiometry 1:1, or simply the product of the first step is transferred in a chemical process to an electroinactive compound. The deviations may also be caused by the adsorption of either Ox or Red. Similar to NPV, the adsorption of Red at the electrode surface will make the RP voltammogram peak-shaped, and the adsorption of Ox will produce a pre-wave.

The entire RP voltammogram, i_{RP} vs. E, for an uncomplicated reversible process can be constructed using the formula:

$$-i_{\mathrm{RP}} = \frac{nFAD_{\mathrm{Ox}}^{1/2}C_{\mathrm{Ox}}^{*}}{\pi^{1/2}}\left[\left(1-\frac{1}{1+\xi\theta}\right)\frac{1}{t_{\mathrm{p}}^{1/2}}-\frac{1}{\tau^{1/2}}\right] \qquad \text{(II.2.12)}$$

where $\theta = \exp\left[nF\left(E-E^{\ominus}\right)/RT\right]$, and $\xi = \left(D_{\mathrm{Ox}}/D_{\mathrm{Red}}\right)^{1/2}$.

The shape of RP voltammograms is given by

$$E = E_{1/2} + \frac{RT}{nF}\ln\left(\frac{i_{\mathrm{lim,DC}} - i_{\mathrm{RP}}}{i_{\mathrm{RP}} - i_{\mathrm{lim,RP}}}\right) \qquad \text{(II.2.13)}$$

According to Eq. (II.2.13), a plot of E vs. $\ln\left(\frac{i_{\mathrm{lim,DC}}-i_{\mathrm{RP}}}{i_{\mathrm{RP}}-i_{\mathrm{lim,RP}}}\right)$ should give, for a reversible system, a slope equal to 2.303 *RT*/*nF* (59.1 mV at 25°C). Equation (II.2.13) is considered to be the best criterion for the reversibility of an electrode reaction.

More details on RPV can be found in the literature [7, 14].

II.2.5 Differential Pulse Voltammetry

A potential waveform for differential pulse voltammetry (DPV) is shown in Fig. II.2.9. In differential pulse polarography, a dropping mercury electrode is used and the old drop is dislodged after each pulse. Similar to NPV and RPV, simple theoretical formulas can only be obtained if the initial boundary conditions are renewed after each pulse. However, since the DPV technique is mainly used in electroanalysis, just one drop or a solid electrode is used to obtain a voltammogram. The importance of DPV in chemical analysis is based on its superior elimination of the capacitive/background current. This is achieved by sampling the current twice: once before pulse application and then at the end of the pulse. The current sampling is indicated by filled circles in Fig. II.2.9. The output from the potentiostat/voltammograph is equal to the difference in the two current values. The double

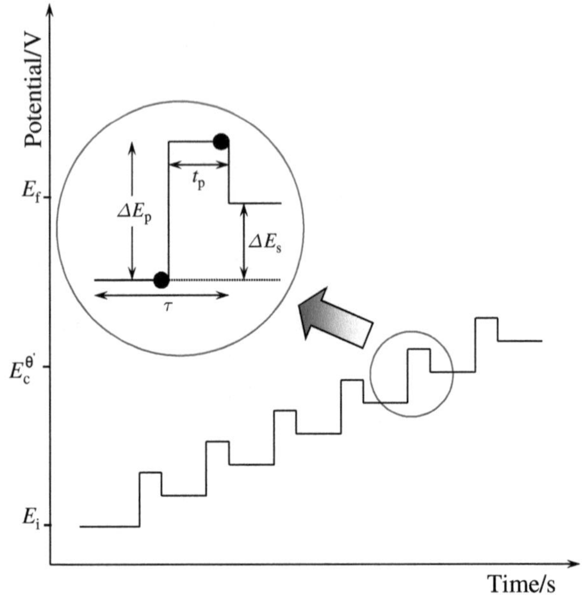

Fig. II.2.9 Potential program for differential pulse voltammetry

current sampling allows the analyst to detect the analytes present in the solution at a concentration as low as 0.05 μM. Another consequence of double sampling is that that the differential pulse voltammograms are peak-shaped.

For a reversible system, the peak height, Δi_p, of a differential pulse voltammogram is

$$\Delta i_p = \frac{nFAD_{Ox}^{1/2} C_{Ox}^*}{\pi^{1/2} t_p^{1/2}} \left(\frac{1-\sigma}{1+\sigma} \right) \quad \text{(II.2.14)}$$

where $\sigma = \exp\left(nF\Delta E_p/2RT\right)$. As ΔE_p decreases, the quotient $(1-\sigma)/(1+\sigma)$ in Eq. (II.2.14) diminishes and finally reaches zero. Due to double sampling of the current, the peak potential, E_p, precedes the formal potential, which can be written as

$$E_p = E_c^{\ominus'} + \frac{RT}{nF} \ln \left(\frac{D_{Red}}{D_{Ox}} \right)^{1/2} - \frac{\Delta E_p}{2} \quad \text{(II.2.15)}$$

The peak width at half height, $w_{1/2}$, for small values of ΔE_p turns out to be

$$w_{1/2} = 3.52\,RT/nF$$

which gives, for 25°C and n = 1, 2, and 3, the values of 90.4, 45.2, and 30.1 mV, respectively. A wider discussion of differential pulse voltammetry can be found in the literature [7, 14].

References

1. Barker GC, Gardner AW (1958) AEREC/R 2297, HMSO, London; Osteryoung RA, Osteryoung J (1981) Phil Trans R Lond A 302: 315
2. Bilewicz R, Wikiel K, Osteryoung J (1989) Anal Chem 1: 965
3. Penczek M, Stojek Z. Buffle J. (1989) J Electroanal Chem 270: 1
4. Osteryoung J (1983) J Chem Ed 60: 296
5. Osteryoung J, Schreiner M (1988) CRC Crit Rev Anal Chem 19: S1
6. Sinru L, Osteryoung J, O'Dea J, Osteryoung R (1988) Anal Chem 60: 1135
7. Galus Z (1994) Fundamentals of electrochemical analysis. Ellis Horwood and Polish Scientific Publishers PWN, New York
8. Cakala M, Mazerska Z, Donten M, Stojek Z (1999) Anal Chim Acta 379: 209
9. Stojek Z, Osteryoung J (1981) J Electroanal Chem 127: 57
10. Osteryoung J, Kirova-Eisner E (1980) Anal Chem 52: 62
11. Brumleve TR, Osteryoung J (1982) J Phys Chem 86: 1794
12. Stojek Z, Jaworski A (1992) Electroanalysis 4: 317
13. Karpinski Z (1988) Anal Chem 58: 2099
14. Bard AJ, Faulkner RF (2000) Electrochemical methods, 2nd edn. Wiley, New York

Chapter II.3
Square-Wave Voltammetry

Milivoj Lovrić

II.3.1 Introduction

Square-wave voltammetry (SWV) is one of the four major voltammetric techniques provided by modern computer-controlled electroanalytical instruments, such as Autolab and μAutolab (both EcoChemie, Utrecht), BAS 100 A (Bioanalytical Systems), and PAR Model 384 B (Princeton Applied Research) [1]. The other three important techniques are single scan and cyclic staircase, pulse, and differential pulse voltammetry (see Chap. II.2). All four are either directly applied or after a preconcentration to record the stripping process. The application of SWV boomed in the last decade, first because of the widespread use of the instruments mentioned above, second because of a well-developed theory, and finally, and most importantly, because of its high sensitivity to surface-confined electrode reactions. Adsorptive stripping SWV is the best electroanalytical method for the determination of electroactive organic molecules that are adsorbed on the electrode surface [2].

The theory and application of SWV are described in several reviews [2–7], and here only a brief account of the recent developments will be given. Contemporary SWV originates from the Kalousek commutator [8] and Barker's square-wave polarography [9, 10]. The Kalousek commutator switched the potential between a slowly varying ramp and a certain constant value to study the reversibility of electrode reactions [11–13]. Barker employed a low-amplitude symmetrical square wave superimposed on a ramp and recorded the difference in currents measured at the ends of two successive half-cycles, with the objective to discriminate the capacitive current [14–16]. SWV was developed by combining the high-amplitude, high-frequency square wave with the fast staircase waveform and by using computer-controlled instruments instead of analog hardware [17–26]. Figure II.3.1 shows the potential–time waveform of modern SWV. Each square-wave period occurs during one staircase period τ. Hence, the frequency of excitation signal is $f = \tau^{-1}$, and the pulse duration is $t_p = \tau/2$. The square-wave amplitude,

M. Lovrić (✉)
Rudjer Boskovic Institute, 10001 Zagreb, Croatia
e-mail: mlovric@irb.hr

F. Scholz (ed.), *Electroanalytical Methods*, 2nd ed.,
DOI 10.1007/978-3-642-02915-8_6, © Springer-Verlag Berlin Heidelberg 2010

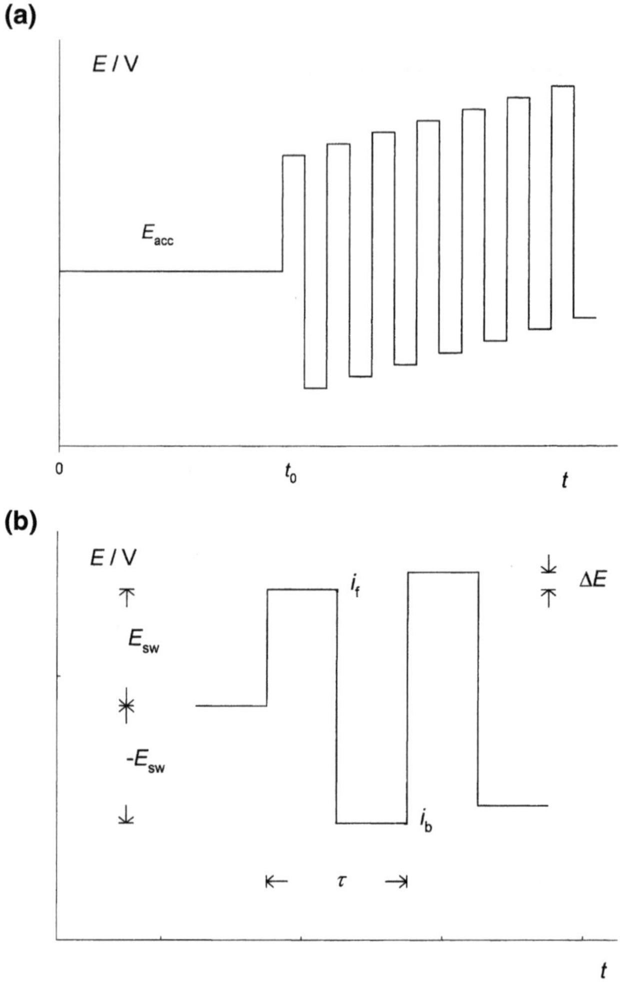

Fig. II.3.1 Scheme of the square-wave excitation signal. E_{acc} starting potential; t_0 delay time; E_{sw} SW amplitude; ΔE scan increment; τ SW period; i_f forward current; i_b backward current, and (•) points where they were sampled

E_{sw}, is one-half of the peak-to-peak amplitude, and the potential increment ΔE is the step height of the staircase waveform. The scan rate is defined as $\Delta E/\tau$. Relative to the scan direction, ΔE, forward and backward pulses can be distinguished. The currents are measured at the end of each pulse and the difference between the currents measured on two successive pulses is recorded as a net response. Additionally, the two components of the net response, i.e., the currents of the forward and backward series of pulses, respectively, can be displayed as well [6, 27–30]. The currents are plotted as a function of the corresponding potential of the staircase waveform.

II.3.2 Simple Reactions on Stationary Planar Electrodes

Figure II.3.2 shows the dimensionless square-wave net response $\Delta\Phi = \Delta i[nFSc^*(D_r f)^{1/2}]^{-1}$ of a simple, fast, and reversible electrode reaction

$$\text{Red} \rightleftarrows \text{Ox} + ne^- \quad \text{(II.3.1)}$$

where n is the number of electrons, F is the Faraday constant, S is the surface area of the electrode, c^* is the bulk concentration of the species Red, D_r is the diffusion coefficient of the species Red, f is the square-wave frequency and $E_{1/2}$ is the half-wave potential of the reaction (Eq. II.3.1). The response was calculated by using the planar diffusion model (see the Appendix, Eqs. II.3.24 and II.3.30). The dimensionless forward Φ_f and backward Φ_b components of the response are also shown in Fig. II.3.2. The net response ($\Delta\Phi = \Phi_f - \Phi_b$) and its components Φ_f and Φ_b consist of discrete current-staircase potential points separated by the potential increment ΔE [6, 17–19]. For a better graphical presentation, the points can be interconnected, as in Fig. II.3.2, but the line between two points has no physical significance. Hence, ΔE determines the density of information in the SWV response. However, the response depends on the product of the number of electrons and the potential increment (for the meaning of dimensionless potential φ_m^*, see Eqs. II.3.26 and II.3.30): $\varphi_m^* = F(nE_m - nE_{1/2})/RT$, and $nE_m = nE_{st} + nE_{sw}(1 \leq m \leq 25), nE_m =$

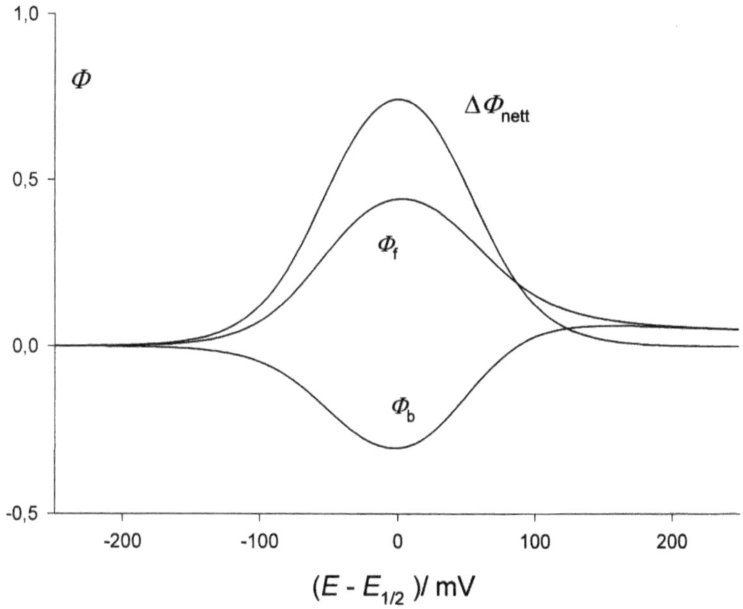

Fig. II.3.2 Square-wave voltammogram of fast and reversible redox reaction (II.3.1). $nE_{sw} = 50$ mV and $n\Delta E = 2$ mV. The net response ($\Delta\Phi_{nett}$) and its forward (Φ_f) and backward (Φ_b) components

$nE_{st} - nE_{sw}(26 \leq m \leq 50), nE_m = nE_{st} + nE_{sw} + n\Delta E(51 \leq m \leq 75), nE_m = nE_{st} - nE_{sw} + n\Delta E(76 \leq m \leq 100)$, etc. Besides, $E_{stair} = E_{st}(1 \leq m \leq 50)$ and $E_{stair} = E_{st} + \Delta E(51 \leq m \leq 100)$, etc. The larger the product $n\Delta E$, the larger will be the net response $\Delta\Phi$. This is because the experiment is performed on a solid electrode, or a single mercury drop, and the apparent scan rate is linearly proportional to the potential increment: $v = f\Delta E$. So, the response increases if the density of information is diminished (for the same frequency). Regardless of ΔE, there is always a particular value of $\Delta\Phi$ that is the highest of all. This is the peak current $\Delta\Phi_p$, and the corresponding staircase potential is the peak potential E_p. The latter is measured with the precision ΔE. In essence, there is no theoretical reason to interpolate any mathematical function between the two experimentally determined current–potential points. If ΔE is smaller, all $\Delta\Phi$ points, including $\Delta\Phi_p$, are smaller too. Frequently, the response is distorted by electronic noise and a smoothing procedure is necessary for its correct interpretation. In this case it is better if ΔE is as small as possible. By smoothing, the set of discrete points is transformed into a continuous current–potential curve, and the peak current and peak potential values can be affected slightly.

The dimensionless net peak current $\Delta\Phi_p$ primarily depends on the product nE_{sw} [31]. This is shown in Table II.3.1. With increasing nE_{sw} the slope $\partial\Delta\Phi_p/\partial nE_{sw}$ continuously decreases, while the half-peak width increases. The maximum ratio between $\Delta\Phi_p$ and the half-peak width appears for $nE_w = 50\,\text{mV}$ [6]. This is the optimum amplitude for analytical measurements. If $E_{sw} = 0$, the square-wave signal turns into the signal of differential staircase voltammetry, and $\Delta\Phi_p$ does not vanish [6, 32, 33].

The peak currents and potentials of the forward and backward components are listed in Table II.3.2. If the square-wave amplitude is not too small ($nE_{sw} > 10\,\text{mV}$), the backward component indicates the reversibility of the electrode reaction. In the

Table II.3.1 Dimensionless net peak currents of square-wave voltammograms of fast and reversible redox reaction (II.3.1)

$\Delta\Phi_p$			
nE_{sw}	$n\Delta E$/mV		
mV	2	4	6
0	0.0084	0.0160	0.0230
5	0.1043	0.1116	0.1183
10	0.1984	0.2053	0.2118
20	0.3744	0.3805	0.3863
30	0.5265	0.5316	0.5367
40	0.6505	0.6546	0.6588
50	0.7467	0.7499	0.7533
60	0.8186	0.8210	0.8237
70	0.8707	0.8726	0.8747
80	0.9077	0.9092	0.9108
90	0.9337	0.9348	0.9361
100	0.9517	0.9525	0.9535

case of reaction (II.3.1), this means that the reduction of the product Ox occurs. If $nE_{sw} < 10\,\text{mV}$, there is no minimum of the backward component. For the optimum amplitude, the separation of peak potentials of two components is 4 mV. The separation vanishes if $nE_{sw} = 70\,\text{mV}$, but increases to 18 mV if $nE_{sw} = 20\,\text{mV}$. At these amplitudes the peak potential of the anodic component is higher than the potential of the cathodic component, but, if $nE_{sw} \geq 80\,\text{mV}$, these potentials are inverted, as can be seen in Table II.3.2. The peak potentials of both components as well as of the net response are independent of the square-wave frequency, and this is the best criterion of the reversibility of reaction (II.3.1) [34, 35].

The net peak current depends linearly on the square root of the frequency:

$$\Delta i_p = nFSD_r^{1/2}\Delta\Phi_p f^{1/2}c^* \tag{II.3.2}$$

(where $\Delta\Phi_p$ depends on nE_{sw} and $n\Delta E$). The condition is that the instantaneous current is sampled only once at the end of each pulse, but then the response may appear noisy [36]. This procedure was assumed in the theoretical calculations presented in Fig. II.3.2 and Tables II.3.1 and II.3.2. In general, several instantaneous currents can be sampled at certain intervals during the last third, or some other portion of the pulse, and then averaged [1]. The average response corresponds qualitatively to an instantaneous current sampled in the middle of the sampling window [17, 19, 31]. The dimensionless peak current depends on the sampling procedure. The relationship between Δi_p and the square root of the frequency depends on the fraction of the pulse at which the current is sampled. This relationship is linear if the relative size of the sampling window is constant. If the absolute size of the sampling window is constant, its relative size increases and the pulse fraction decreases as the frequency is increased. So, the ratio $\Delta i_p/f^{1/2}$ increases with increasing frequency. If the relative size of the sampling window increases from 1% to 7% of the pulse duration, the dimensionless net peak current $\Delta\Phi_p$ increases by 10% [37].

Table II.3.2 Dimensionless peak currents and peak potentials of the forward and the backward components of square-wave voltammograms of fast and reversible redox reaction (II.3.1) ($n\Delta E = 2\,\text{mV}$)

nE_{sw}/mV	$\Phi_{p,f}$	$(E_{p,f} - E_{1/2})$/V	$\Phi_{p,b}$	$(E_{p,b} - E_{1/2})$/V
0	0.1201	0.030	–	–
5	0.1590	0.018	–	–
10	0.2011	0.012	–0.0201	–0.026
20	0.2812	0.008	–0.1023	–0.010
30	0.3494	0.006	–0.1815	–0.006
40	0.4035	0.004	–0.2491	–0.004
50	0.4440	0.002	–0.3035	–0.002
60	0.4731	0.002	–0.3456	–0.002
70	0.4933	0.000	–0.3774	0.000
80	0.5069	–0.002	–0.4009	0.002
90	0.5158	–0.004	–0.4182	0.002
100	0.5215	–0.006	–0.4308	0.004

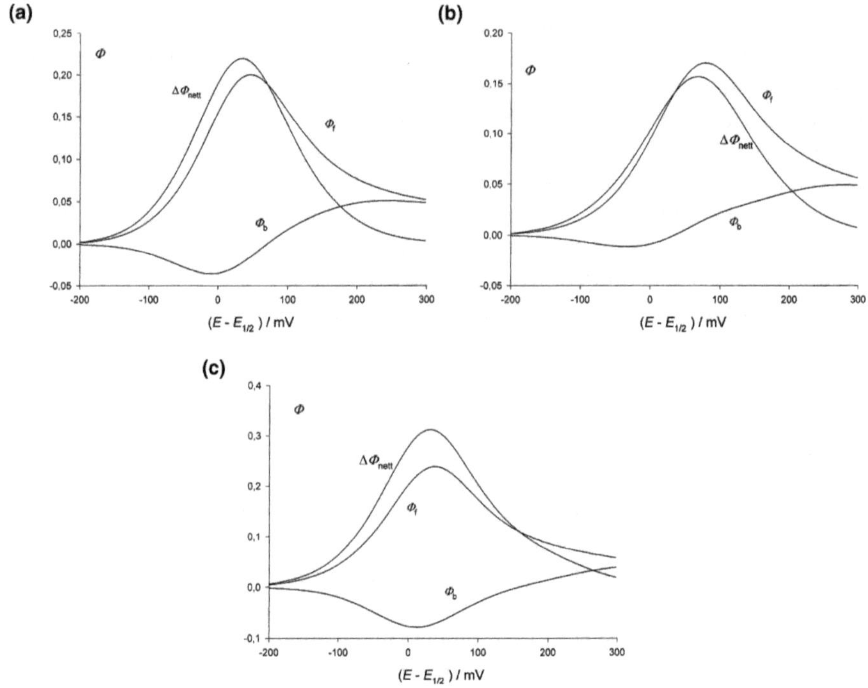

Fig. II.3.3 Square-wave voltammograms of reaction (II.3.1) controlled by electrode kinetics. $n\Delta E = 2$ mV, $nE_{\text{sw}} = 50$ mV (**a** and **b**) and 100 mV (**c**), $\alpha = 0.5$, $\lambda = 0.1$ (**a**) and 0.05 (**b** and **c**)

If reaction (II.3.1) is controlled by the electrode kinetics (see Chap. I.3), the square-wave response depends on the dimensionless kinetic parameter $\lambda = k_s(D_o f)^{-1/2}(D_o/D_r)^{\alpha/2}$ and the transfer coefficient α [38]. This relationship is shown in the Appendix (see Eqs. II.3.28, II.3.31, II.3.32, and II.3.33). In the quasi-reversible range ($-1.5 < \log\lambda < 0.5$), the dimensionless net peak current $\Delta\Phi_p$ decreases, while both the half-peak width and the net peak potential increase with diminishing λ. These dependencies are not linear, and vary with the transfer coefficient. The real net peak current Δi_p is not a linear function of the square root of frequency [39]. The change in the net response is caused by the transformation of the backward component under the influence of increased frequency. As can be seen in Fig. II.3.3, the minimum gradually disappears as if the amplitude is lessened (compare Figs. II.3.3a, b). The response is very sensitive to a change in the signal parameters, as can be seen by comparing Figs. II.3.3b, c. A complex simulation method based on Eqs. (II.3.31, II.3.32 and II.3.33) was developed and used for the estimation of λ and α parameters of the Zn^{2+}/Zn(Hg) electrode reaction [40–42].

The net current of a totally irreversible electrode reaction (Eq. II.3.1) is smaller than its forward component because the backward component is positive for all potentials (see Fig. II.3.4), regardless of the amplitude [43–45]. The ratio $\Delta i_p/f^{1/2}$ and the half-peak width are both independent of the frequency, but the net

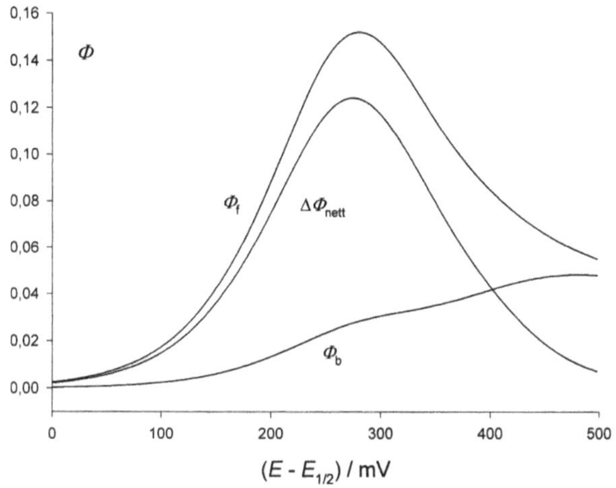

Fig. II.3.4 Square-wave voltammogram of irreversible reaction (II.3.1). $nE_{sw} = 50\,\text{mV}$, $n\Delta E = 2\,\text{mV}$, $\alpha = 0.5$ and $\lambda = 0.001$

peak potential is a linear function of the logarithm of frequency, with the slope $\partial E_p/\partial \log f = 2.3RT/2\alpha nF$ [6, 38].

II.3.3 Simple Reactions on Stationary Spherical Electrodes and Microelectrodes

The basic theory of SWV at a stationary spherical electrode is explained in the Appendix (see Eqs. II.3.34, II.3.35, II.3.36, II.3.37, II.3.38, II.3.39 and II.3.40). If the electrode reaction (Eq. II.3.1) is fast and reversible, the shape of the SWV response and its peak potential are independent of electrode geometry and size [46–48]. At a spherical electrode the dimensionless net peak current is linearly proportional to the inverse value of the dimensionless electrode radius $y = r^{-1}(D/f)^{1/2}$ [49, 50]. If $n\Delta E = 5\,\text{mV}$ and $nE_{sw} = 25\,\text{mV}$, the relationship is $\Delta\Phi_p = 0.46 + 0.45y$ [51]. Considering that the surface area of a hemispherical electrode is $S = 2\pi r^2$, the net peak current can be expressed as

$$\Delta i_p = 2.83\,nFc^* D^{1/2} r[1.03rf^{1/2} + D^{1/2}] \qquad \text{(II.3.3)}$$

Theoretically, if an extremely small electrode is used and low frequency is applied, so that $rf^{1/2} \ll D^{1/2}$, a steady-state, frequency-independent peak current should appear: $(\Delta i_p)_{ss} = 2.83\,nFc^* Dr$ [51]. However, even if the electrode radius is as small as 10^{-6} m, the frequency is only 10 Hz, and $D^{1/2}$ is as large as 5×10^{-5} m s$^{-1/2}$, this condition is not fully satisfied. Consequently, SWV measurements at hemispherical microelectrodes are unlikely to be performed under rigorously established steady-state conditions. Moreover, if the frequency is high and a hanging mercury drop electrode is used, the spherical effect is usually negligible ($y < 10^{-3}$). The relationship between $\Delta\Phi_p$ and $n\Delta E$ and nE_{sw} does not depend on electrode size [48]. The dimensionless net peak current of a reversible reaction (Eq. II.3.1) at an inlaid microdisk electrode is: $\Delta\Phi_p = 0.365 + 0.086\exp(-1.6y) + 0.58y$ (after recalculation to $n\Delta E$ and nE_{sw} as above) [47]. The theoretical steady-state, frequency-independent peak current at this electrode is: $(\Delta i_p)_{ss} = 1.83\,nFDc^*r$, where r is the radius of the disk. At a moderately small inlaid disk electrode, the average dimensionless peak current is: $\Delta\Phi_p = 0.452 + 0.47y$ [47]. Some electroanalytical applications of cylindrical [52, 53] and ring microelectrodes [54] have been described.

The dimensionless peak current of a totally irreversible electrode reaction is a function of the variable y; however, the relationship is not strictly linear. If $\alpha = 0.5$, it can be described with two asymptotes: $\Delta\Phi_p = 0.11 + 0.32y$ (for $y < 0.5$) and $\Delta\Phi_p = 0.15 + 0.24y$ (for $0.5 < y < 10$). The slopes and intercepts of these straight lines are linear functions of the transfer coefficient α. The responses of quasi-reversible electrode reactions are complex functions of both the electrode radius and the kinetics parameter $\kappa = k_s(Df)^{-1/2}$. Hence, no linear relationship between $\Delta\Phi_p$ and y was found [51].

II.3.4 Reactions of Amalgam-Forming Metals on Thin Mercury Film Electrodes

At a thin mercury film electrode, the dimensionless net peak current $\Delta\Phi_p$ of a reversible redox reaction $M^{n+} + ne^- \rightleftarrows M(Hg)$ depends on the dimensionless film thickness $\Lambda = l(f/D_r)^{1/2}$, where l is the real film thickness [55]. A classical voltammetric experiment was simulated by assuming that no metal atoms are initially present in the film. If $n\Delta E = 10\,\text{mV}$ and $nE_{sw} = 50\,\text{mV}$, $\Delta\Phi_p$ increases from 0.85 (for $\Lambda < 0.1$) to 1.12 (for $\Lambda = 1$) and decreases back to 0.74 for $\Lambda > 5$. The relationship between the real peak current Δi_p and the square root of the frequency is linear only if the parameter Λ remains either smaller than 0.1, or larger than 5 at all frequencies. At moderately thick films these conditions are usually not satisfied [56–58].

At a very thin film ($\Lambda < 0.1$), the real peak current of the reversible reaction (II.3.1) is linearly proportional to the frequency because $\Delta\Phi_p$ linearly depends on the parameter Λ [59, 60]. In reaction (II.3.1), it is assumed that only the species Red is initially present in the solution. This is the condition usually encountered in anodic stripping square-wave voltammetry [Red ≡ M(Hg)]. In the range $0.1 < \Lambda < 5$, $\Delta\Phi_p$ monotonously increases with Λ from 0.03 to 0.74, without a maximum for $\Lambda = 1$. The peak width changes from $99/n$ mV (for $\Lambda < 0.3$) to $124/n$ mV, for $\Lambda > 3$ [60, 61]. Simulations of SWV in the restricted diffusion space were extended to a thin layer cell [62]. The influence of electrode kinetics on direct and anodic stripping SWV on thin mercury film electrodes was analyzed recently [63–65].

SWV can be applied to systems complicated by preceding, subsequent, or catalytic homogeneous chemical reactions [38, 66–69]. Theoretical relationships between the measurable parameters, such as peak shifts, heights and widths, and the appropriate rate constants, were calculated and used for the extraction of kinetic information from the experimental data [66, 67].

II.3.5 Electrode Reactions Complicated by Adsorption of the Reactant and Product

SWV is a very sensitive technique partly because of its ability to discriminate against charging current [70–74]. However, a specific adsorption of reactant may significantly enhance SWV peak currents [75, 76]. Unlike alternating current voltammetry, SWV effectively separates a capacitive current from a so-called pseudocapacitance [77]. This is the basis for an electroanalytical application of SWV in combination with an adsorptive accumulation of analytes [78–81].

A redox reaction of surface-active reactants can be divided in two groups:

$$\mathrm{Ox} \rightleftarrows (\mathrm{Ox})_{\mathrm{ads}} + n\mathrm{e}^- \rightleftarrows (\mathrm{Red})_{\mathrm{ads}} \rightleftarrows \mathrm{Red} \tag{II.3.4}$$

$$\mathrm{Ox} \rightleftarrows (\mathrm{Ox})_{\mathrm{ads}} + n\mathrm{e}^- \rightleftarrows \mathrm{Red} \tag{II.3.5}$$

In the first group the product remains adsorbed on the electrode surface, while in the second group the product is not adsorbed [82]. The reactions of the majority of organic electroactive substances belong to the first group [83]. These are the so-called surface redox reactions. Examples of the second group of reactions are anion-induced adsorption and reduction of amalgam-forming metal ions on mercury electrodes [84–87]. These are mixed redox reactions.

The difference in responses of surface and mixed reactions is most pronounced if the reactions are fast and reversible. The square-wave stripping scan is preceded by a certain accumulation period during which the electrode is charged to the initial potential and the reactant is adsorbed on the electrode surface. The initial potential is rather high, so that only a minute amount of the reactant is reduced. During the stripping process, the equilibrium at the electrode surface is rhythmically disrupted. After each square-wave pulse the redox system tends to reestablish the Nernst equilibrium between Γ_{ox} and Γ_{red} (for the surface reaction), or Γ_{ox} and $c_{\mathrm{red}(x=0)}$ (for the mixed reaction). The current is a measure of the rate of this process (see the Appendix, Eqs. II.3.46 and II.3.47). Besides, the current is caused by the fluxes of dissolved Ox and Red species. If the equilibrium between Γ_{ox} and Γ_{red} is established at the beginning of the pulse, the current sampled at the end of the pulse does not originate from the reduction of initially accumulated reactant, but from the fluxes in the solution. This current is very small because, during the period of adsorptive accumulation, a thick diffusion layer of the reactant has developed and its flux is diminished. Hence, the response is smaller than in direct SWV, without adsorption [88]. In the case of a mixed reaction, the equilibrium between Γ_{ox} and $c_{\mathrm{red}\,(x=0)}$ is continuously disturbed by the diffusion of Red species from the electrode surface into the solution. For this reason the current sampled at the end of the pulse is proportional to the surface concentration of initially adsorbed reactant [89]. This initial surface concentration is proportional to the bulk concentration of the reactant and the square root of the duration of the accumulation in a solution that is not stirred [78]. The accumulation occurs during the delay time preceding the stripping scan and during the first period of the scan, before the reduction of the reactant:

$$t_{acc} = t_{delay} + (E_p - E_{acc})/f\Delta E \qquad \text{(II.3.6)}$$

where E_p and E_{acc} are the stripping peak potential and the accumulation potential, respectively.

The relationship between the stripping peak current of a fast and reversible mixed reaction and the square-wave frequency is a curve defined by $\Delta i_p = 0$, for $f = 0$, and an asymptote $\Delta i_p = kf + z$ [90]. The intercept z depends on the delay time and apparently vanishes when $t_{delay} > 30$ s. Consequently, the ratio $\Delta i_p / f$ may not be constant for all frequencies. This effect is caused by the additional adsorption during the first period of the stripping scan. The stripping peak potential of a reversible mixed reaction depends linearly on the logarithm of frequency [89]:

$$\partial E_p / \partial \log f = -2.3RT/2nF \qquad \text{(II.3.7)}$$

The peak current depends on the square-wave amplitude E_{sw} and the potential increment ΔE in the same way as in the case of the simple reaction (Eq. II.3.1) (see Table II.3.1). The half-peak width also depends on the amplitude and has no diagnostic value. However, the response of the reversible reaction (II.3.5) is narrower than the response of the reversible reaction (Eq. II.3.1). If $nE_{sw} = 50$ mV and $n\Delta E = 10$ mV, the half-peak widths are 100 mV and 125 mV, respectively [88].

Here it should be mentioned that for the reaction

$$\mathrm{Ox} + ne^- \rightleftarrows (\mathrm{Red})_{ads} \rightleftarrows \mathrm{Red} \qquad \text{(II.3.8)}$$

in which only the reactant Ox is initially present in the solution, but only the product Red is adsorbed, the square-wave peak current depends linearly on the square root of the frequency. The peak potential is a linear function of the logarithm of frequency, with the slope 2.3 $RT/2nF$ [90].

Under the influence of electrode kinetics, the surface reaction (Eq. II.3.4) depends on the dimensionless kinetic parameter $\kappa = k_s/f$ (Eq. II.3.57) and the dimensionless adsorption parameters $a_{ox} = K_{ox} f^{1/2} D_o^{-1/2}$ and $a_{red} = K_{red} f^{1/2} D_r^{-1/2}$ (Eqs. II.3.58 and II.3.59) [89, 91–93]. Equations (II.3.50) and (II.3.53) are complicated by the diffusion of the redox species Ox and Red and their adsorption equilibria. The kinetic effects can be investigated separately by analyzing a simplified surface reaction (Eq. II.3.64) that is a model of strong and totally irreversible

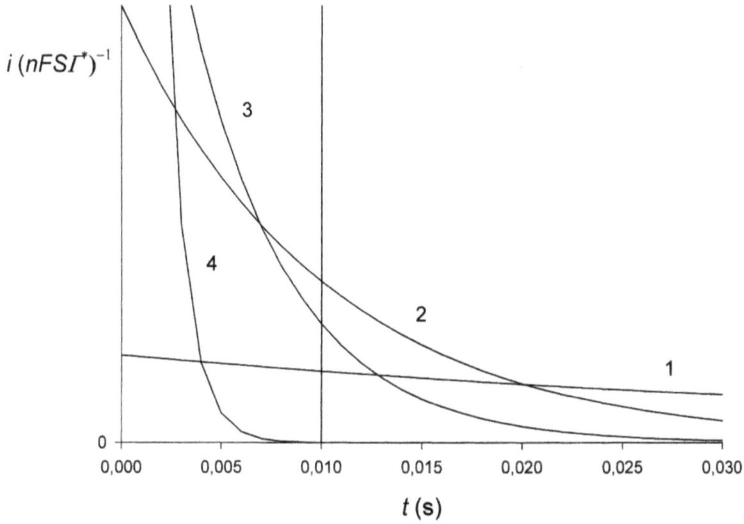

Fig. II.3.5 Chronoamperogram of reaction (II.3.64) (Eq. II.3.74) $E = E^{\ominus}$, $k_s\,/\,s^{-1} = 10$ (1), 50 (2), 100 (3) and 500 (4)

adsorption of an electroactive reactant [93, 94]. Under chronoamperometric conditions (for $E = E^{\ominus}$), the current depends exponentially on the product $k_s t$, where t is the time after the application of the potential pulse (see Eq. II.3.74). Figure II.3.5 shows the relationship between the current and the time for different values of the reaction rate constant. The vertical bar denotes the pulse duration $t_p = 10$ ms. If the k_s value is very large (curve 4, $k_s = 500\ s^{-1}$), the current quickly decreases and virtually vanishes before the end of the pulse. This is the response of a fast and reversible surface reaction as discussed above. The current caused by a slower reaction declines less rapidly. After 10 ms, the highest current corresponds to $k_s = 50\ s^{-1}$, while both faster ($k_s = 100\ s^{-1}$) and slower ($k_s = 10\ s^{-1}$) charge transfers cause lower currents. The first derivative of Eq. (II.3.74) shows that $k_{s,max} = (2t_p)^{-1}$. This is the rate constant of a kinetically controlled electrode reaction that gives a maximum response at the end of the pulse. Equations (II.3.75), (II.3.76), (II.3.77), (II.3.78), (II.3.79) and (II.3.80) show the general relationship between the dimensionless chronoamperometric response (Eq. II.3.71) and the dimensionless kinetic parameter $\lambda = k_s t$. At any electrode potential there is a certain λ_{max} giving the maximum response. The numerical analysis (Eq. II.3.69) of a SWV of a surface reaction (Eq. II.3.64) is shown in Fig. II.3.6. The dimensionless peak current $\Delta\Phi_p = \Delta i_p (nFS\Gamma^*_{ox} f)^{-1}$ is plotted as a function of the logarithm of the dimensionless kinetic parameter $\kappa = k_s / f$. The peak currents of quasi-reversible surface redox reactions are much higher than the peak currents of both reversible and totally irreversible surface reactions. The same relationship was found in SWV of the surface reaction (Eq. II.3.4) if $0.1 < a_{ox}/a_{red} \leq 10$ (see Eqs. II.3.50 and II.3.53). This is

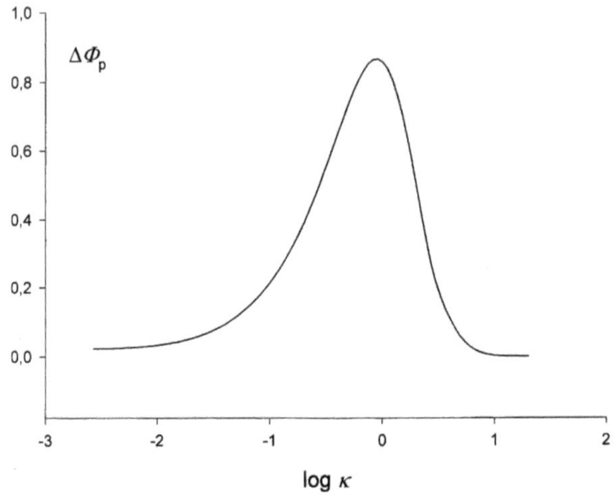

Fig. II.3.6 Theoretical quasi-reversible maximum in SWV. A dependence of the dimensionless net peak current $\Delta\Phi_p = \Delta i_p(nFS\Gamma^*_{ox}f)^{-1}$ on the logarithm of dimensionless kinetic parameter $\kappa = k_s/f$ (*Eq.* II.3.69). $n\Delta E = 2$ mV, $nE_{sw} = 50$ mV, $\alpha = 0.5$, and M = 50

called the quasi-reversible maximum in SWV [93, 95]. The critical dimensionless rate constant κ_{max} depends on the transfer coefficient α and the product nE_{sw}, but does not depend on the surface concentration of the adsorbed reactant if there are no interactions between the molecules of the deposit. Values of κ_{max} are listed in Table II.3.3 [96].

A variation in frequency changes the apparent reversibility of the surface reaction (Eq. II.3.4) [91, 92]. The reaction appears reversible if $k_s/f > 5$, and totally irreversible if $k_s/f < 10^{-2}$. The ratio of the real peak current and the corresponding frequency ($\Delta i_p/f$) increases with increasing frequency if $1 < \kappa < 5$, but it decreases if $10^{-2} < \kappa < 1$. Hence, the ratio $\Delta i_p/f$ may depend parabolically on the logarithm

Table II.3.3 Dependence of the critical kinetic parameter κ_{max} on the normalized square-wave amplitude nE_{sw} and the transfer coefficient α

κ_{max}					
α	nE_{sw}/mV				
	15	25	30	40	50
0.1	1.43	1.35	1.38	1.33	1.26
0.2	1.32	1.30	1.25	1.17	1.08
0.3	1.31	1.26	1.20	1.10	0.97
0.4	1.29	1.20	1.16	1.04	0.90
0.5	1.28	1.19	1.13	1.01	0.88
0.6	1.27	1.18	1.13	1.02	0.89
0.7	1.26	1.22	1.17	1.04	0.94
0.8	1.25	1.24	1.19	1.12	1.04
0.9	1.25	1.27	1.30	1.26	1.19

of frequency. The characteristic frequency f_{max} of the maximum ratio $(\Delta i_p/f)_{max}$ is related to the standard reaction rate constant by the equation: $k_s = \kappa_{max} f_{max}$, where κ_{max} depends on the experimental conditions (see Table II.3.3). If the transfer coefficient α is not known, an average value of κ_{max} can be used [96]. The kinetic parameters of electrode reactions of adsorbed alizarin red S [97], europium(III)–salicylate complex [98], probucole [99], azobenzene [96], 6-propyl-2-thiouracil [100], and indigo [101] were determined by this method.

The response of the fast and reversible surface reaction (Eq. II.3.4) splits into two peaks if $nE_{sw} > 40$ mV [89, 91, 92, 102]. The splitting is shown in Fig. II.3.7 a for the reaction: $(\mathrm{Red})_{ads} \rightleftarrows (\mathrm{Ox})_{ads} + ne^-$. It was observed in SWV of adsorbed methylene blue [89], cytochrome *c* [92], azobenzene [91], and alizarin red S [102]. The separation between the anodic and cathodic components of the response is a consequence of the current sampling procedure. The ratio between the cathodic and anodic peak currents of the net response depends on the transfer coefficient [102]:

$$\Delta\Phi_{p,c}/\Delta\Phi_{p,a} = 5.64\exp(-3.46\alpha) \tag{II.3.9}$$

If $\alpha = 0.5$ these two peaks are equal. The peak potentials of the net response are independent of the frequency, regardless of the amplitude, if $\kappa > 5$.

Figure II.3.7b–d shows the changes in the cathodic and anodic components of the net response in the range of the quasi-reversible maximum. These currents are very sensitive to a change in frequency and allowed the evaluation of kinetic parameters of azobenzene [91], cytochrome *c* [92], and Cu(II)-oxine complexes [103] by fitting experimental data with theoretical curves. In addition, a theory of a two-step surface reaction was developed and applied to the kinetics of the reduction of adsorbed 4-(dimethylamino)azobenzene [104].

The SWV response of the mixed reaction (Eq. II.3.5) influenced by the electrode kinetics depends on the complex dimensionless parameter $K_{kin} = k_s K_{ox} r_s^{-1/2} D^{-1/4} f^{-3/4}$, where k_s and r_s are defined by the equation:

$$i/nFS = k_s\exp(-\alpha\varphi)[\Gamma_{ox} - r_s(c_{red})_{x=0}\exp\varphi] \tag{II.3.10}$$

and K_{ox} is the adsorption constant (Eq. II.3.43). The dimensionless peak current $\Delta\Phi_p = \Delta i_p(nFSK_{ox}c^*_{ox}f)^{-1}$ is independent of the kinetic parameter if either log $K_{kin} > 1.5$ (reversible reactions) or log $K_{kin} < -1.5$ (totally irreversible reactions). In the range $-1.5 < \log K_{kin} < 1.5$, the quasi-reversible maximum appears [105–107]. The critical kinetic parameter, $K_{kin,max}$, depends on the product nE_{sw} and the transfer coefficient α [107]. It can be used for an estimation of rate constants of mixed reactions [98].

A special case of mixed reactions is an accumulation of insoluble mercuric and mercurous salts on the mercury electrode surface and their stripping off by SWV:

$$\mathrm{HgL(s)} \rightleftarrows \mathrm{Hg(l)} + \mathrm{L^{2-}(aq)} + 2e^- \tag{II.3.11}$$

$$\mathrm{Hg_2X_2(s)} \rightleftarrows 2\mathrm{Hg(l)} + 2\mathrm{X^-} + 2e^- \tag{II.3.12}$$

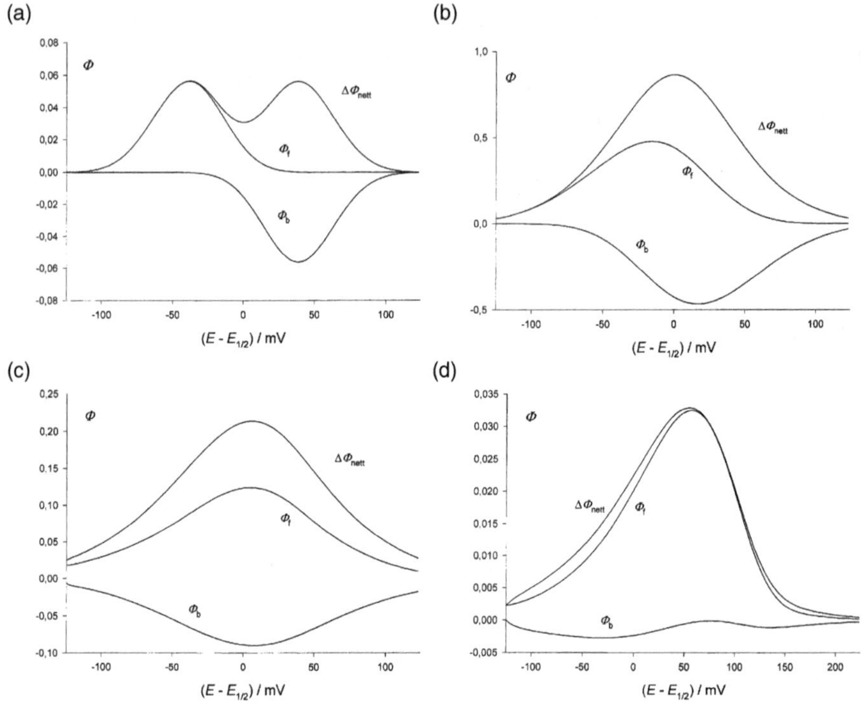

Fig. II.3.7 Theoretical square-wave voltammograms of a kinetically controlled reaction (Eq. II.3.64). $n\Delta E = 2$ mV, $nE_{\mathrm{sw}} = 50$ mV, $\alpha = 0.5$ and $\kappa = 5$ (**a**), 0.9 (**b**), 0.1 (**c**) and 0.01 (**d**)

The quasi-reversible maxima in SWV of first-order [106, 107] and second-order [108] mixed reactions were analyzed. If the ligand L^{2-} in Eq. (II.3.11) remains adsorbed on the electrode surface, the mixed reaction turns into a surface reaction and the response changes accordingly [100].

SWV responses of totally irreversible surface and mixed reactions are identical [78, 89, 109, 110]. The real peak current is linearly proportional to the frequency, while the peak potential depends linearly on the logarithm of frequency, with the slope:

$$\partial E_{\mathrm{p}}/\partial \log f = -2.3RT/\alpha nF \tag{II.3.13}$$

The half-peak width is independent of the amplitude if $nE_{\mathrm{sw}} > 20$ mV [109]:

$$\Delta E_{\mathrm{p}/2} = (63.5 \pm 0.5)/\alpha n \,\mathrm{mV} \tag{II.3.14}$$

II.3.6 Applications of Square-Wave Voltammetry

SWV has been applied in numerous electrochemical and electroanalytical measurements [2, 4, 6, 7]. Apart from the investigation of charge transfer kinetics of dissolved zinc ions [40, 42] and adsorbed organic species mentioned above [91, 92, 96–101, 103, 110], the mechanisms of redox reactions of titanium(III), iron(II) [66] and adsorbed metal complexes [77, 84–87] were analyzed. Electroanalytical application of SWV can be divided into direct and stripping measurements. Analytes measured directly, without accumulation, were Bi(III), Cu(II), Pb(II), Tl(I), In(III), Cd(II), Zn(II), Fe(III/II)oxalate [18], Ni(II) [36], tert-butyl hydroperoxide and *N*-acetylpenicillamine thionitrite [44]. The stripping measurements were based either on the accumulation of amalgams [18, 111, 112] and metal deposits on solid electrodes [52, 113], or on the adsorptive accumulation of organic substances [114] and metal complexes [115]. Anodic stripping SWV was applied to thin mercury film covered macro electrodes [116] and micro electrodes [52, 117]. Some examples of stripping SWV with adsorptive accumulation include the analyses of cimetidine [78], nicotinamide adenine dinucleotide [79, 80], berberine [81, 109], azobenzene [83], sulfide [106], and cysteine [108]. Other examples can be found in Table II.3.4.

Finally, SWV was used for the detection of heavy metals in thin-layer chromatography [118] and various organic substances in high-performance liquid chromatography [119].

Table II.3.4 Some elements and compounds measured by square-wave voltammetry

Reactant	Footnote	Reactant	Footnote
Bismuth	F1	Captopril	F28
Cadmium	F2	Cocaine	F29
Chromium	F3	Cytochrome	F30
Copper	F4	Daminozide	F31
Europium	F5	Dimethoate	F32
Germanium	F6	Dopamine	F33
Gold	F7	Famotidine	F34
Indium	F8	Fluorouracil	F35
Iodine	F9	Folic acid	F36
Lead	F10	Hemoglobin	F37
Manganese	F11	Hydroxypyrene	F38
Mercury	F12	Melatonin	F39
Molybdenum	F13	Mifepristone	F40
Nickel	F14	Molinate	F41
Ruthenium	F15	Nitrophenol	F42
Selenium	F16	Nitroprusside	F43
Thallium	F17	Orthochlorophenol	F44
Uranium	F18	Quinoxaline	F45
Zinc	F19	Serotonin	F46
Adenosine	F20	Sulfadimetoxol	F47
Ametryne	F21	Sulfaquinoxaline	F48

Table II.3.4 (continued)

Reactant	Footnote	Reactant	Footnote
Amisulpride	F22	Sunset yellow	F49
Azidothymidine	F23	Tetramethrin	F50
Azithromycin	F24	Toxogonin	F51
Benzaldehyde	F25	Uric acid	F52
Benzoylecgonine	F26	Venlafaxine	F53
Caffeine	F27	Warfarin	F54

(F1) Komorsky-Lovrić Š (1988) Anal Chim Acta 204: 161 (F2) Yantasee W, Lin Y, Fryxell GE, Busche BJ (2004) Anal Chim Acta 502: 207 (F3) Boussemart M, van den Berg CMG, Ghaddaf M (1992) Anal Chim Acta 262: 103 (F4) Hardcastle JL, Compton RG (2002) Electroanalysis 14: 753 (F5) Zelić M (2003) Croat Chem Acta 76: 241 (F6) Sohn SC, Park YJ, Joe KS (1997) J Korean Chem Soc 41: 590 (F7) Riley JP, Wallace GG (1991) Electroanalysis 3: 191 (F8) Zelić M, Mlakar M, Branica M (1994) Anal Chim Acta 289: 299 (F9) Wong GTF, Zhang LS (1992) Talanta 39: 355 (F10) Richter EM, Pedrotti JJ, Angnes L (2003) Electroanalysis 15: 1871 (F11) Komorsky-Lovrić Š (1998) Croat Chem Acta 71: 263 (F12) Walcarius A, Devoy J, Bessiere J (2000) J Solid State Electrochem 4: 330 (F13) Quentel F, Mirčeski V (2004) Electroanalysis 16: 1690 (F14) Morfobos M, Economou A, Voulgaropoulos A (2004) Anal Chim Acta 519: 57 (F15) Prakash R, Tyagi B, Ramachandraiah G (1997) Indian J Chem A 36: 201 (F16) Tan SH, Kounaves SP (1998) Electroanalysis 10: 364 (F17) Zen J-M, Wu J-W (1997) Electroanalysis 9: 302 (F18) Mlakar M (1993) Anal Chim Acta 276: 367 (F19) Demetriades D, Economou A, Voulgaropoulos A (2004) Anal Chim Acta 519: 167 (F20) Flores JR, Alvarez JMF (1992) Electroanalysis 4: 347 (F21) Cabral MF, de Souza D, Alves CR, Machado SAS (2003) Eclet Quim 28: 41 (F22) Özkan SA, Uslu B, Sentürk Z (2004) Electroanalysis 16: 231 (F23) Vacek J, Andrysik Z, Trnkova L, Kizek R (2004) Electroanalysis 16: 224 (F24) Farghaly OAE-M, Mohamed NAL (2004) Talanta 62: 531 (F25) Saska M, Sturrock PE (1983) Anal Chim Acta 155: 243 (F26) Komorsky-Lovrić Š, Gagić S, Penovski R (1999) Anal Chim Acta 389: 219 (F27) Zen J-M, Ting Y-S (1998) Analyst 123: 1145 (F28) Parham H, Zargar B (2005) Talanta 65: 776 (F29) Pavlova V, Mirčeski V, Komorsky-Lovrić Š, Petrovska-Jovanović S, Mitrevski B (2004) Anal Chim Acta 512: 49 (F30) Bianco P, Lattuca C (1997) Anal Chim Acta 353: 53 (F31) Ianniello RM (1987) Anal Chim Acta 193: 81 (F32) Hernandez P, Ballesteros Y, Galan F, Hernandez L (1994) Electroanalysis 6: 51 (F33) Zen J-M, Chen I-L (1997) Electroanalysis 9: 537 (F34) Mirčeski V, Jordanoski B, Komorsky-Lovrić Š (1998) Portugaliae Electrochim Acta 16: 43 (F35) Mirčeski V, Gulaboski R, Jordanoski B, Komorsky-Lovrić Š (2000) J Electroanal Chem 490: 37 (F36) Cakir S, Atayman I, Cakir O (1997) Mikrochim Acta 126: 237 (F37) Liu H, Hu N (2003) Anal Chim Acta 481: 91 (F38) Castro AA, Wagener ALR, Farias PAM, Bastos MB (2004) Anal Chim Acta 521: 201 (F39) Beltagi AM, Khashaba PY, Ghoneim MM (2003) Electroanalysis 15: 1121 (F40) Rodriguez J, Berzas JJ, Castaneda G, Rodriguez N (2004) Electroanalysis 16: 661 (F41) Oliveira Brett AM, Garrido EM, Lima JLFC, Delerue-Matos C (1997) Portugaliae Electrochim Acta 15: 315 (F42) Pedrosa VA, Codognoto L, Avaca LA (2003) J Braz Chem Soc 14: 530 (F43) Carapuca HM, Simao JEJ, Fogg AG (1998) J Electroanal Chem 455: 93 (F44) Manisankar P, Prabu HG (1995) Electroanalysis 7: 594 (F45) Barros AA, Rodrigues JA, Almeida PJ, Rodrigues PG, Fogg AG (1999) Anal Chim Acta 385: 315 (F46) Zen J-M, Chen I-L, Shih Y (1998) Anal Chim Acta 369: 103 (F47) Berzas JJ, Rodrigues J, Lemus JM, Castaneda G (1997) Electroanalysis 9: 474 (F48) Berzas JJ, Rodriguez J, Lemus JM, Castaneda G (1993) Anal Chim Acta 273: 369 (F49) Berzas Nevado JJ, Rodrigues Flores J, Villasenor Llerena MJ (1997) Talanta 44: 467 (F50) Hernandez P, Galan-Estella F, Hernandez L (1992) Electroanalysis 4: 45 (F51) Komorsky-Lovrić Š (1990) J Electroanal Chem 289: 161 (F52) Zen J-M, Chen P-J (1997) Anal Chem 69: 5087 (F53) Lima JLFC, Loo DV, Delerue-Matos C, Roque Da Silva AS (1999) Farmaco 54: 145 (F54) Ghoneim MM, Tawfik A (2004) Anal Chim Acta 511: 63

Appendix

(A) For a stationary planar diffusion model of a simple redox reaction (Eq. II.3.1) the following differential equations and boundary conditions can be formulated:

$$\frac{\partial c_{\mathrm{red}}}{\partial t} = D_{\mathrm{r}} \frac{\partial^2 c_{\mathrm{red}}}{\partial x^2} \tag{II.3.15}$$

$$\frac{\partial c_{\mathrm{ox}}}{\partial t} = D_{\mathrm{o}} \frac{\partial^2 c_{\mathrm{ox}}}{\partial x^2} \tag{II.3.16}$$

$$t = 0, x \geq 0\text{: } c_{\mathrm{red}} = c^*, c_{\mathrm{ox}} = 0 \tag{II.3.17}$$

$$t > 0, x \to \infty,\ c_{\mathrm{red}} \to c^*, c_{\mathrm{ox}} \to 0 \tag{II.3.18}$$

$$x = 0\text{: } D_{\mathrm{r}} \left(\frac{\partial c_{\mathrm{red}}}{\partial x} \right)_{x=0} = \frac{i}{nFS} \tag{II.3.19}$$

$$D_{\mathrm{o}} \left(\frac{\partial c_{\mathrm{ox}}}{\partial x} \right)_{x=0} = -\frac{i}{nFS} \tag{II.3.20}$$

If reaction (II.3.1) is fast and reversible, the Nernst equation has to be satisfied:

$$(c_{\mathrm{ox}})_{x=0} = (c_{\mathrm{red}})_{x=0} \exp(\varphi) \tag{II.3.21}$$

$$\varphi = \frac{nF}{RT}(E - E^{\ominus}) \tag{II.3.22}$$

If reaction (II.3.1) is kinetically controlled, the Butler-Volmer equation applies:

$$\frac{i}{nFS} = -k_{\mathrm{s}} \exp(-\alpha\varphi)[(c_{\mathrm{ox}})_{x=0} - (c_{\mathrm{red}})_{x=0} \exp(\varphi)] \tag{II.3.23}$$

where c_{red} and c_{ox} are the concentrations of the reduced and oxidized species, respectively. D_{r} and D_{o} are the corresponding diffusion coefficients, k_{s} is the standard rate constant, α is the transfer coefficient, $E^{\ominus}$ is the standard potential, x is the distance from the electrode surface, t is the time variable, and the other symbols are explained below Eq. (II.3.1) above.

The solution of Eqs. (II.3.15), (II.3.16), (II.3.17), (II.3.18), (II.3.19), (II.3.20), (II.3.21) and (II.3.22) is an integral equation [120]:

$$\int_0^t \Phi^*[\pi(t - \tau)]^{-1/2} \mathrm{d}\tau = \exp(\varphi^*)[1 + \exp(\varphi^*)]^{-1} \tag{II.3.24}$$

where

$$\Phi^* = i[nFSc^* D_r^{1/2}]^{-1} \tag{II.3.25}$$

$$\varphi^* = nF(E - E_{1/2})/RT \tag{II.3.26}$$

$$E_{1/2} = E^{\ominus} + RT[\ln(D_r/D_o)]/2nF \tag{II.3.27}$$

The solution for the kinetically controlled reaction is [120]:

$$\Phi^* = -\lambda^* \exp(-\alpha\varphi^*)\left[1 + \exp(\varphi^*)\right] \int_0^t \Phi^* [\pi(t-\tau)]^{-1/2} \mathrm{d}\tau + \lambda^* \exp\left[(1-\alpha)\varphi^*\right] \tag{II.3.28}$$

The convolution integrals in Eqs. (II.3.24) and (II.3.28) can be solved by the method of numerical integration proposed by Nicholson and Olmstead [38, 121]:

$$\int_0^t \Phi^* [\pi(t-\tau)]^{-1/2} \mathrm{d}\tau = 2(d/\pi)^{1/2} \sum_{j=1}^{m} \Phi_j^* S_{m-j+1} \tag{II.3.29}$$

where d is the time increment, $t = md$, Φ_j^* is the average value of the function Φ^* within the jth time increment, $S_k = k^{1/2} - (k-1)^{1/2}$ and $S_1 = 1$. Each square-wave half-period is divided into 25 time increments: $d = (50f)^{-1}$. By this method, Eq. (II.3.24) is transformed into the system of recursive formulae:

$$\Phi_m = 5(\pi/2)^{1/2} \exp(\varphi_m^*)\left[1 + \exp(\varphi_m^*)\right]^{-1} - \sum_{j=1}^{m-1} \Phi_j S_{m-j+1} \tag{II.3.30}$$

where $\Phi = i[nFSc^*(D_r f)^{1/2}]^{-1}, \varphi_m^* = nF(E_m - E_{1/2})/RT, m = 1,2,3,\ldots \mathrm{M}$ and $\mathrm{M} = 50(E_{\text{fin}} - E_{\text{st}})/\Delta E$. The potential E_m changes from $E_{\text{stair}} = E_{\text{st}}$ to $E_{\text{stair}} = E_{\text{fin}}$ according to Fig. II.3.1. The recursive formulae for the kinetically controlled reaction are [38, 40, 41, 44]:

$$\Phi_m = Z_1 - Z_2 \sum_{j=1}^{m-1} \Phi_j S_{m-j+1} \tag{II.3.31}$$

$$Z_1 = \frac{\lambda \exp[(1-\alpha)\varphi_m^*]}{1 + \frac{\lambda\sqrt{2}}{5\sqrt{\pi}}[\exp(-\alpha\varphi_m^*) + \exp((1-\alpha)\varphi_m^*)]} \tag{II.3.32}$$

$$Z_2 = \frac{\frac{\lambda\sqrt{2}}{5\sqrt{\pi}}[\exp(-\alpha\varphi_m^*) + \exp((1-\alpha)\varphi_m^*)]}{1 + \frac{\lambda\sqrt{2}}{5\sqrt{\pi}}[\exp(-\alpha\varphi_m^*) + \exp((1-\alpha)\varphi_m^*)]} \tag{II.3.33}$$

where

$$\lambda = \frac{k_s}{\sqrt{D_o f}}\left(\frac{D_o}{D_r}\right)^{\frac{\alpha}{2}}$$

is a dimensionless kinetic parameter.

(B) On a stationary spherical electrode, a simple redox reaction

$$\mathrm{Ox} + ne^- \leftrightarrows \mathrm{Red} \tag{II.3.34}$$

can be mathematically represented by the well-known integral equation [120]:

$$\Phi = \frac{k_s}{(Df)^{1/2}} \exp(-\alpha\varphi)[1 - f^{1/2}(1 + \exp(\varphi))I^o] \tag{II.3.35}$$

$$\Phi = i(nFSc_{ox}^*)^{-1}(Df)^{-1/2} \tag{II.3.36}$$

$$I^o = \int_0^t \Phi[\pi(t-u)]^{-1/2}\mathrm{d}u - \frac{D^{1/2}}{r}\int_0^t \Phi \exp[D(t-u)r^{-2}]\mathrm{erfc}[D^{1/2}r^{-1}(t-u)^{1/2}]\mathrm{d}u \tag{II.3.37}$$

where r is the radius of the spherical electrode and c_{ox}^* is the bulk concentration of the oxidized species. The meanings of all other symbols are as above. It is assumed that both the reactant and product are soluble, that only the oxidized species is initially present in the solution, and that the diffusion coefficients of the reactant and product are equal. For numerical integration, Eq. (II.3.35) can be transformed into a system of recursive formulae [51]:

$$\Phi_m = \frac{-\frac{D^{1/2}}{rf^{1/2}} - (1 + \exp(\varphi_m))\sum_{i=1}^{m-1}\Phi_i S_{m-i+1}}{\frac{D}{k_s r}\exp(\alpha\varphi_m) + S_1(1 + \exp(\varphi_m))} \tag{II.3.38}$$

$$S_1 = 1 - \exp(Df^{-1}r^{-2}N^{-1})\mathrm{erfc}(D^{1/2}f^{-1/2}r^{-1}N^{-1/2}) \tag{II.3.39}$$

$$\begin{aligned} S_k = {} & \exp\left[Df^{-1}r^{-2}N^{-1}(k-1)\right]\mathrm{erfc}\left[D^{1/2}f^{-1/2}r^{-1}N^{-1/2}(k-1)^{1/2}\right] \\ & - \exp\left(Df^{-1}r^{-2}N^{-1}k\right)\mathrm{erfc}\left(D^{1/2}f^{-1/2}r^{-1}N^{-1/2}k^{1/2}\right) \end{aligned} \tag{II.3.40}$$

where N is the number of time increments in each square-wave period. The ratio $k_s\, r/D$ is the dimensionless standard charge transfer rate constant of reaction (II.3.34) and the ratio $rf^{1/2}/D^{1/2}$ is the dimensionless electrode radius.

(C) A surface redox reaction (II.3.4) on a stationary planar electrode is represented by the system of differential equations (II.3.15) and (II.3.16), with the following initial and boundary conditions [89]:

$$t = 0, x \geq 0: c_{ox} = c^*_{ox}, c_{red} = 0,\ \Gamma_{ox} = \Gamma_{red} = 0 \tag{II.3.41}$$

$$t > 0: x \to \infty:\ c_{ox} \to c^*_{ox},\ c_{red} \to 0 \tag{II.3.42}$$

$$x = 0: K_{ox}(c_{ox})_{x=0} = \Gamma_{ox} \tag{II.3.43}$$

$$K_{red}(c_{red})_{x=0} = \Gamma_{red} \tag{II.3.44}$$

$$i/nFS = k_s \exp(-\alpha\varphi)[\Gamma_{ox} - \exp(\varphi)\Gamma_{red}] \tag{II.3.45}$$

$$D_o(\partial c_{ox}/\partial x)_{x=0} = d\Gamma_{ox}/dt + i/nFS \tag{II.3.46}$$

$$D_r(\partial c_{red}/\partial x)_{x=0} = d\Gamma_{red}/dt - i/nFS \tag{II.3.47}$$

$$\varphi = nF(E - E^{\ominus}_{\Gamma_{ox}/\Gamma_{red}})/RT \tag{II.3.48}$$

$$E^{\ominus}_{\Gamma_{ox}/\Gamma_{red}} = E^{\ominus} + (RT/nF)\ln(K_{red}/K_{ox}) \tag{II.3.49}$$

where Γ_{ox} and Γ_{red} are the surface concentrations of the oxidized and reduced species, respectively, and K_{ox} and K_{red} are the constants of linear adsorption isotherms. The solution of Eqs. (II.3.15) and (II.3.16) is an integral equation:

$$i/nFS = k_s \exp(-\alpha\varphi) \left\{K_{ox}c^*_{ox}\left[1 - \exp\left(D_o t K_{ox}^{-2}\right)\operatorname{erfc}\left(D_o^{1/2} t^{1/2} K_{ox}^{-1}\right)\right] - I_{ox} - I_{red}\exp(\varphi)\right\} \tag{II.3.50}$$

$$I_{ox} = \int_0^t (i/nFS)\exp\left[D_o(t-\tau)K_{ox}^{-2}\right]\operatorname{erfc}\left[D_o^{1/2}(t-\tau)^{1/2}K_{ox}^{-1}\right]\mathrm{d}\tau \tag{II.3.51}$$

$$I_{red} = \int_0^t (i/nFS)\exp\left[D_r(t-\tau)K_{red}^{-2}\right]\operatorname{erfc}\left[D_r^{1/2}(t-\tau)^{1/2}K_{red}^{-1}\right]\mathrm{d}\tau \tag{II.3.52}$$

For numerical integration, Eqs. (II.3.50), (II.3.51) and (II.3.52) are transformed into a system of recursive formulae [93]:

$$\Phi_m = \frac{\kappa\exp(-\alpha\varphi_m)\left[1 - \exp\left(a_{ox}^{-2}mN^{-1}\right)\operatorname{erfc}\left(a_{ox}^{-1}m^{1/2}N^{-1/2}\right) - SS_1 + SS_2\right]}{1 + \kappa\exp(-\alpha\varphi_m)\left[2(N\pi)^{-1/2}(a_{ox} + a_{red}\exp(\varphi_m)) - a_{ox}^2 M_1 - a_{red}^2\exp(\varphi_m)P_1\right]} \tag{II.3.53}$$

$$SS_1 = 2(N\pi)^{-1/2}\left[a_{ox} + a_{red}\exp(\varphi_m)\right]\sum_{j=1}^{m-1}\Phi_j S_{m-j+1} \tag{II.3.54}$$

$$SS_2 = \sum_{j=1}^{m-1} \Phi_j \left[a_{\text{ox}}^2 M_{m-j+1} + a_{\text{red}}^2 \exp\left(\varphi_m\right) P_{m-j+1} \right] \tag{II.3.55}$$

$$\Phi = i \left(nFSK_{\text{ox}} c_{\text{ox}}^* f \right)^{-1} \tag{II.3.56}$$

$$\kappa = k_{\text{s}}/f \tag{II.3.57}$$

$$a_{\text{ox}} = K_{\text{ox}} f^{1/2} D_{\text{o}}^{-1/2} \tag{II.3.58}$$

$$a_{\text{red}} = K_{\text{red}} f^{1/2} D_{\text{r}}^{-1/2} \tag{II.3.59}$$

$$d = N^{-1} f^{-1} \tag{II.3.60}$$

$$S_k = k^{1/2} - (k-1)^{1/2} \tag{II.3.61}$$

$$M_1 = 1 - \exp\left(a_{\text{ox}}^{-2} N^{-1} \right) \text{erfc} \left(a_{\text{ox}}^{-1} N^{-1/2} \right)$$

$$\begin{aligned} M_k = {} & \exp\left[a_{\text{ox}}^{-2} (k-1) N^{-1} \right] \text{erfc}\left[a_{\text{ox}}^{-1} (k-1)^{1/2} N^{-1/2} \right] \\ & - \exp\left[a_{\text{ox}}^{-2} k N^{-1} \right] \text{erfc}\left[a_{\text{ox}}^{-1} k^{1/2} N^{-1/2} \right] \end{aligned} \tag{II.3.62}$$

$$P_1 = 1 - \exp\left(a_{\text{red}}^{-2} N^{-1} \right) \text{erfc} \left(a_{\text{red}}^{-1} N^{-1/2} \right)$$

$$\begin{aligned} P_k = {} & \exp\left[a_{\text{red}}^{-2} (k-1) N^{-1} \right] \text{erfc}\left[a_{\text{red}}^{-1} (k-1)^{1/2} N^{-1/2} \right] \\ & - \exp\left[a_{\text{red}}^{-2} k N^{-1} \right] \text{erfc}\left[a_{\text{red}}^{-1} k^{1/2} N^{-1/2} \right] \end{aligned} \tag{II.3.63}$$

(D) In a simplified approach to the surface redox reaction, the transport of Ox and Red in the solution is neglected. This assumption corresponds to a totally irreversible adsorption of both redox species [94]:

$$(\text{Ox})_{\text{ads}} + n\text{e}^- \rightleftarrows (\text{Red})_{\text{ads}} \tag{II.3.64}$$

The current is determined by Eq. (II.3.45), with the initial and boundary conditions:

$$t = 0, \Gamma_{\text{ox}} = \Gamma_{\text{ox}}^*, \Gamma_{\text{red}} = 0 \tag{II.3.65}$$

$$t > 0 : \Gamma_{\text{ox}} + \Gamma_{\text{red}} = \Gamma_{\text{ox}}^* \tag{II.3.66}$$

$$\text{d}\Gamma_{\text{ox}}/\text{d}t = -i/nFS \tag{II.3.67}$$

$$\text{d}\Gamma_{\text{red}}/\text{d}t = i/nFS \tag{II.3.68}$$

The solution of Eq. (II.3.45) is a system of recursive formulae:

$$\phi_{\mathrm{m}} = \frac{\kappa \exp(-\alpha\varphi_m)\left[1 - N^{-1}(1+\exp(\varphi_m))\sum_{j=1}^{m-1}\Phi_j\right]}{1+\kappa\exp(-\alpha\varphi_m)N^{-1}(1+\exp(\varphi_m))} \tag{II.3.69}$$

$$\Phi = \frac{i}{nFS\Gamma_{\mathrm{ox}}^{*}f} \tag{II.3.70}$$

The kinetic parameter κ is defined by Eq. (II.3.57).

Under chronoamperometric conditions (E = const.), the solution of Eq. (II.3.45) is

$$\Phi = \lambda \exp(-\alpha\varphi)\exp[-\lambda\exp(-\alpha\varphi)(1+\exp(\varphi))] \tag{II.3.71}$$

$$\Phi = it(nFS\Gamma_{\mathrm{ox}}^{*})^{-1} \tag{II.3.72}$$

$$\lambda = k_{\mathrm{s}}t \tag{II.3.73}$$

If $\varphi = 0$, Eq. (II.3.71) is reduced to

$$i/nFS\Gamma_{\mathrm{ox}}^{*} = k_{\mathrm{s}}\exp(-2k_{\mathrm{s}}t) \tag{II.3.74}$$

The maximum chronoamperometric response is defined by the first derivative of Eq. (II.3.71):

$$\partial\Phi/\partial\lambda = 0 \tag{II.3.75}$$

$$\lambda_{\max} = \exp(\alpha\varphi)[1+\exp(\varphi)]^{-1} \tag{II.3.76}$$

The second condition is:

$$\partial\lambda_{\max}/\partial\varphi = 0 \tag{II.3.77}$$

with the result:

$$\exp(\varphi_{\max}) = \frac{\alpha}{1-\alpha} \tag{II.3.78}$$

$$\lambda_{\max,\max} = \alpha^{\alpha}(1-\alpha)^{1-\alpha} \tag{II.3.79}$$

$$\Phi_{\max} = \frac{1-\alpha}{e} \tag{II.3.80}$$

This derivation shows that, for any electrode potential E, there is a certain dimensionless kinetic parameter $\lambda_{\max}$ which gives the highest response (Eq. II.3.76). The maximum of $\lambda_{\max}$ (Eq. II.3.79) is a parabolic function of the transfer coefficient: $0.5 \leq \lambda_{\max,\max} < 1$, for $0 < \alpha < 1$. If $\alpha = 0.5$, then $\lambda_{\max,\max} = 0.5$ and

$\Phi_{max} = (2e)^{-1}$. This is in the agreement with Eq. (II.3.74). From the condition $\partial i/\partial k_s = 0$, it follows that $k_{s,max} = (2t)^{-1}$ and $(i/nFS\Gamma^*_{ox})_{max} = (2et)^{-1}$.

References

1. Autolab, Installation guide, EcoChemie BV, Utrecht, 1993; BAS 100A, Operation guide, Bioanalytical Systems, West Lafayette, 1987; Model 384 B, Operating manual, EG & G Princeton Applied Research, Princeton, 1983
2. Mirčeski V, Komorsky-Lovrić Š, Lovrić M (2007) Square-Wave Voltammetry. Springer, Berlin Heidelberg New York
3. de Souza D, Machado SAS, Avaca LA (2003) Quim Nova 26: 81
4. de Souza D, Codognoto L, Malagutti AR, Toledo RA, Pedrosa VA, Oliveira RTS, Mazo LH, Avaca LA, Machado SAS (2004) Quim Nova 27: 790
5. Osteryoung JG, Osteryoung RA (1985) Anal Chem 57: 101A
6. Osteryoung J, O'Dea JJ (1986) Square-wave voltammetry. In: Bard AJ (ed) Electro analytical chemistry, vol 14. Marcel Dekker, New York, p 209
7. Eccles GN (1991) Crit Rev Anal Chem 22: 345
8. Kalousek M (1948) Collect Czech Chem Commun 13: 105
9. Barker GC, Jenkins IL (1952) Analyst 77: 685
10. Barker GC (1958) Anal Chim Acta 18: 118
11. Ishibashi M, Fujinaga T (1952) Bull Chem Soc Jpn 25: 68
12. Kinard WF, Philp RH, Propst RC (1967) Anal Chem 39: 1557
13. Radej J, Ružić I, Konrad D, Branica M (1973) J Electroanal Chem 46: 261
14. Barker GC, Gardner AW, Williams MJ (1973) J Electroanal Chem 42: App. 21
15. Kalvoda R, Holub I (1973) Chem Listy 67: 302
16. Igolinski VA, Kotova NA (1973) Elektrokhimiya 9: 1878
17. Ramaley L, Krause MS Jr (1969) Anal Chem 41: 1362
18. Krause MS Jr, Ramaley L (1969) Anal Chem 41: 1365
19. Christie JH, Turner JA, Osteryoung RA (1977) Anal Chem 49: 1899
20. Ramaley L, Surette DP (1977) Chem Instrum 8: 181
21. Buchanan EB Jr, Sheleski WJ (1980) Talanta 27: 955
22. Yarnitzky C, Osteryoung RA, Osteryoung J (1980) Anal Chem 52: 1174
23. Anderson JA, Bond AM (1983) Anal Chem 55: 1934
24. Lavy-Feder A, Yarnitzky C (1984) Anal Chem 56: 678
25. Jayaweera P, Ramaley L (1986) Anal Instrum 15: 259
26. Wong KH, Osteryoung RA (1987) Electrochim Acta 32: 629
27. Ramaley L, Tan WT (1981) Can J Chem 59: 3326
28. Fatouros N, Simonin JP, Chevalet J, Reeves RM (1986) J Electroanal Chem 213: 1
29. Chen X, Pu G (1987) Anal Lett 20: 1511
30. Krulic D, Fatouros N, Chevalet J (1990) J Electroanal Chem 287: 215
31. Aoki K, Maeda K, Osteryoung J (1989) J Electroanal Chem 272: 17
32. Lovrić M (1995) Croat Chem Acta 68: 335
33. Krulic D, Fatouros N, El Belamachi MM (1995) J Electroanal Chem 385: 33
34. Molina A, Serna C, Camacho L (1995) J Electroanal Chem 394: 1
35. Brookes BA, Ball JC, Compton RG (1999) J Phys Chem B 103: 5289
36. Zachowski EJ, Wojciechowski M, Osteryoung J (1986) Anal Chim Acta 183: 47
37. Lovrić M (1994) Annali Chim 84: 379
38. O'Dea JJ, Osteryoung J, Osteryoung RA (1981) Anal Chem 53: 695
39. Elsner CI, Rebollo NL, Dgli WA, Marchiano SL, Plastino A, Arvia AJ (1994) ACH-Models Chem 131: 121
40. O'Dea JJ, Osteryoung J, Osteryoung RA (1983) J Phys Chem 87: 3911

41. O'Dea JJ, Osteryoung J, Lane T (1986) J Phys Chem 90: 2761
42. Go WS, O'Dea JJ, Osteryoung J (1988) J Electroanal Chem 255: 21
43. Ivaska AV, Smith DE (1985) Anal Chem 47: 1910
44. Nuwer MJ, O'Dea JJ, Osteryoung J (1991) Anal Chim Acta 251: 13
45. Fatouros N, Krulic D (1998) J Electroanal Chem 443: 262
46. O'Dea JJ, Wojciechowski M, Osteryoung J, Aoki K (1985) Anal Chem 57: 954
47. Whelan DP, O'Dea JJ, Osteryoung J, Aoki K (1986) J Electroanal Chem 202: 23
48. Aoki K, Tokuda K, Matsuda H, Osteryoung J (1986) J Electroanal Chem 207: 25
49. Ramaley L, Tan WT (1987) Can J Chem 65: 1025
50. Fatouros N, Krulic D, Lopez-Tenes M, El Belamachi MM (1996) J Electroanal Chem 405: 197
51. Komorsky-Lovrić Š, Lovrić M, Bond AM (1993) Electroanalysis 5: 29
52. Singleton ST, O'Dea JJ, Osteryoung J (1989) Anal Chem 61: 1211
53. Murphy MM, O'Dea JJ, Osteryoung J (1991) Anal Chem 63: 2743
54. Tallman DE (1994) Anal Chem 66: 557
55. Kounaves SP, O'Dea JJ, Chandrasekhar P, Osteryoung J (1986) Anal Chem 58: 3199
56. Wikiel K, Osteryoung J (1989) Anal Chem 61: 2086
57. Kumar V, Heineman W (1987) Anal Chem 59: 842
58. Kounaves SP, Deng W (1991) J Electroanal Chem 306: 111
59. Penczek M, Stojek Z (1986) J Electroanal Chem 213: 177
60. Kounaves SP, O'Dea JJ, Chandrasekhar P, Osteryoung J (1987) Anal Chem 59: 386
61. Wechter C, Osteryoung J (1989) Anal Chem 61: 2092
62. Aoki K, Osteryoung J (1988) J Electroanal Chem 240: 45
63. Brookes BA, Compton RG (1999) J Phys Chem B 103: 9020
64. Ball JC, Compton RG (1998) J Phys Chem B 102: 3967
65. Agra-Gutierrez C, Ball JC, Compton RG (1998) J Phys Chem B 102: 7028
66. Zeng J, Osteryoung RA (1986) Anal Chem 58: 2766
67. O'Dea JJ, Wikiel K, Osteryoung J (1990) J Phys Chem 94: 3628
68. Molina A (1998) J Electroanal Chem 443: 163
69. Fatouros N, Krulic D (1998) J Electroanal Chem 456: 211
70. Turner JA, Christie JH, Vuković M, Osteryoung RA (1977) Anal Chem 49: 1904
71. Barker GC, Gardner AW (1979) J Electroanal Chem 100: 641
72. Stefani S, Seeber R (1983) Annali Chim 73: 611
73. Zhang J, Guo SX, Bond AM, Honeychurch MJ, Oldham KB (2005) J Phys Chem B 109: 8935
74. Jadreško D, Lovrić M (2008) Electrochim Acta 53: 8045
75. Barker GC, Bolzan JA (1966) Z Anal Chem 216: 215
76. Ramaley L, Dalziel JA, Tan WT (1981) Can J Chem 59: 3334
77. Komorsky-Lovrić Š, Lovrić M, Branica M (1988) J Electroanal Chem 241: 329
78. Webber A, Shah M, Osteryoung J (1983) Anal Chim Acta 154: 105
79. Webber A, Shah M, Osteryoung J (1984) Anal Chim Acta 157: 1
80. Webber A, Osteryoung J (1984) Anal Chim Acta 157: 17
81. Komorsky-Lovrić Š (1987) J Electroanal Chem 219: 281
82. Komorsky-Lovrić Š, Lovrić M (1989) Fresenius Z Anal Chem 335: 289
83. Xu G, O'Dea JJ, Mahoney LA, Osteryoung JG (1994) Anal Chem 66: 808
84. Komorsky-Lovrić Š, Lovrić M, Branica M (1989) J Electroanal Chem 266: 185
85. Zelić M, Branica M (1991) J Electroanal Chem 309: 227
86. Zelić M, Branica M (1992) Electroanalysis 4: 623
87. Zelić M, Branica M (1992) Anal Chim Acta 262: 129
88. Lovrić M, Branica M (1987) J Electroanal Chem 226: 239
89. Lovrić M, Komorsky-Lovrić Š (1988) J Electroanal Chem 248: 239
90. Komorsky-Lovrić Š, Lovrić M, Branica M (1992) J Electroanal Chem 335: 297
91. O'Dea JJ, Osteryoung JG (1993) Anal Chem 65: 3090

92. Reeves JH, Song S, Bowden EF (1993) Anal Chem 65: 683
93. Komorsky-Lovrić Š, Lovrić M (1995) J Electroanal Chem 384: 115
94. Lovrić M (1991) Elektrokhimiya 27: 186
95. Komorsky-Lovrić Š, Lovrić M (1995) Anal Chim Acta 305: 248
96. Komorsky- Lovrić Š, Lovrić M (1995) Electrochim Acta 40: 1781
97. Komorsky-Lovrić Š (1996) Fresenius J Anal Chem 356: 306
98. Lovrić M, Mlakar M (1995) Electroanalysis 7: 1121
99. Mirčeski V, Lovrić M, Jordanoski B (1999) Electroanalysis 11: 660
100. Mirčeski V, Lovrić M (1999) Anal Chim Acta 386: 47
101. Komorsky-Lovrić Š (2000) J Electroanal Chem 482: 222
102. Mirčeski V, Lovrić M (1997) Electroanalysis 9: 1283
103. Garay F, Solis V, Lovrić M (1999) J Electroanal Chem 478: 17
104. O'Dea JJ, Osteryoung JG (1997) Anal Chem 69: 650
105. Lovrić M, Komorsky-Lovrić Š, Bond AM (1991) J Electroanal Chem 319: 1
106. Lovrić M, Pižeta I, Komorsky-Lovrić Š (1992) Electroanalysis 4: 327
107. Mirčeski V, Lovrić M (1999) Electroanalysis 11: 984
108. Mirčeski V, Lovrić M (1998) Electroanalysis 10: 976
109. Lovrić M, Komorsky-Lovrić Š, Murray RW (1988) Electrochim Acta 33: 739
110. O'Dea JJ, Ribes A, Osteryoung JG (1993) J Electroanal Chem 345: 287
111. Ostapczuk P, Valenta P, Nürnberg HW (1986) J Electroanal Chem 214: 51
112. Tercier M-L, Buffle J, Graziottin F (1998) Electroanalysis 10: 355
113. Zen J-M, Ting YS (1996) Anal Chim Acta 332: 59
114. Yarnitzky C, Smyth WF (1991) Int J Pharm 75: 161
115. Bobrowski A, Zarebski J (2000) Electroanalysis 12: 1177
116. Wang J, Tian B (1992) Anal Chem 64: 1706
117. Emons H, Baade A, Schoning MJ (2000) Electroanalysis 12: 1171
118. Petrovic SC, Dewald HD (1996) J Planar Chromatogr 9: 269
119. Hoekstra JC, Johnson DC (1999) Anal Chim Acta 390: 45
120. Galus Z (1994) Fundamentals of electrochemical analysis. Ellis Horwood, New York, Polish Scientific Publishers PWN, Warsaw
121. Nicholson RS, Olmstead ML (1972) Numerical solutions of integral equations. In: Matson JS, Mark HB, MacDonald HC (eds) Electrochemistry: calculations, simulations and instrumentation, vol 2. Marcel Dekker, New York, p 119

Chapter II.4
Chronocoulometry

György Inzelt

II.4.1 Introduction

In 1834 Faraday suggested two fundamental laws of electrolysis. According to Faraday, the amount of material deposited or evolved (m) during electrolysis is directly proportional to the current (I) and the time (t), i.e., on the quantity of electricity (Q) that passes through the solution (first law). The amount of the product depends on the equivalent mass of the substance electrolyzed (second law). (In fact, Faraday's laws are based on two fundamental laws, i.e., on the conservation of matter and the conservation of charge.) Accordingly,

$$m = \frac{M}{nF}Q = \frac{MIt}{nF} \tag{II.4.1}$$

where Q in the amount of charge consumed during the electrochemical transformation, n is the charge number of the electrochemical cell reaction, I is the current, and t is the duration of electrolysis.

If the current efficiency is 100%, i.e., the total charge is consumed only by a well-defined electrode reaction, the measurement of charge provides an excellent tool for both qualitative and quantitative analyses. For instance, knowing m and Q, M/n can be obtained, which is characteristic to a given substance and its electrode reaction.

By knowing M and n the amount of the substance in the solution can be determined. This method is known as *coulometry*. It is also possible to generate a reactant by electrolysis in a well-defined amount and then it will enter a reaction with a component of the solution. It is used in *coulometric titration* where the end-point is detected in a usual way, e.g., by using an indicator.

Chronocoulometry belongs to the family of step techniques. The essential features of chronoamperometry have already been discussed in Sect. I.3.2. Instead of following the variation of current with time after application of a potential step perturbation, it is possible to detect the amount of the charge passed as a function of

G. Inzelt (✉)
Department of Physical Chemistry, Eötvös Loránd University, 1518 Budapest 112, Hungary
e-mail: inzeltgy@chem.elte.hu

F. Scholz (ed.), *Electroanalytical Methods*, 2nd ed.,
DOI 10.1007/978-3-642-02915-8_7,

time, i.e., to carry out chronocoulometric measurements. Chronocoulometry gives practically the same information that is provided by chronoamperometry, since it is just an integrated form of the current–time response; however, chronocoulometry offers important experimental advantages.

First, unlike the current response that quickly decreases the measured signal usually increases with time, and hence the later parts of the transient can be detected more accurately. Second, a better signal-to-noise ratio can be achieved. Third, contributions of charging/discharging of the electrochemical double layer and any pseudocapacitance on the surface (charge consumed by the electrode reaction of adsorbed species) to the overall charge passed as a function of time can be distinguished from those due to the diffusing electroreactants.

II.4.2 Fundamental Theoretical Considerations

In the case of the electrochemical oxidation of a species R

$$R \to O^+ + e^- \tag{II.4.2}$$

under steady-state conditions, the current–potential curve shown in Fig. II.4.1 is obtained (see Sect. I.3.1 for details).

At potential E_1 practically no current flows, while at E_2 the current is limited by diffusion (more precisely by the rate at which the reactant is supplied to the electrode surface). If the experimental conditions are arranged so that species R are transported by linear diffusion (flat electrode, unstirred solution), the current that flows at any time after application of the potential step from E_1 to E_2 will obey the Cottrell equation [cf. Chap. I.3; Eq. (I.3.64)].

The time integral of the Cottrell equation – since $Q = \int_0^t I \mathrm{d}t$ – gives the cumulative charge passed in the course of oxidation R:

$$Q_{\mathrm{diff}}(t) = \int_0^t nFAD_{\mathrm{R}}^{1/2} c_{\mathrm{R}}^* \pi^{1/2} \mathrm{d}t = 2nFAD_{\mathrm{R}}^{1/2} c_{\mathrm{R}}^* \pi^{-1/2} t^{1/2} \tag{II.4.3}$$

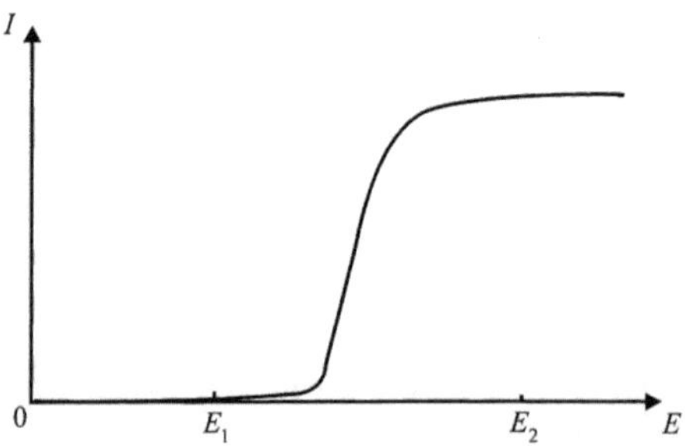

Fig. II.4.1 Steady-state current–potential curve for oxidation of a reactant (R) at an electrode. At E_1 practically no current flows, at E_2 the current is limited by mass transport

However, at least one additional current component has to be taken into account, because of the charging of the double layer while stepping the potential from E_1 to E_2. The equation for the time dependence of the capacitive current has been given in Sect. I.3.2.1 [Eq. (I.3.61)].

According to Sect. I.3.2.1, Eq. (I.3.61), after the application of a potential step of magnitude $E = E_2 - E_1$, the exponential decay of the current with time depends on the double-layer capacitance and the solution resistance, i.e., on the time constant $\tau = R_{\mathrm{s}}C_{\mathrm{d}}$. Consequently, if we assume that C_{d} is constant and the capacitor is initially uncharged ($Q = 0$ at $t = 0$), for the capacitive charge (Q_{c}) we obtain

$$Q_{\mathrm{c}} = EC_{\mathrm{d}}\left(1 - \mathrm{e}^{-t/R_{\mathrm{s}}C_{\mathrm{d}}}\right) \tag{II.4.4}$$

The typical waveform of a step experiment, as well as the respective chronoamperometric and chronocoulometric curves when species R is electrochemically inactive at E_1 but is oxidized at a diffusion-limited rate at E_2, are shown schematically in Fig. II.4.2. The charging current and Q_{c} vs. time curves are also displayed. The latter curves can be determined by repeating the experiment in the absence of R, i.e., in pure supporting electrolyte.

If R is adsorbed at the electrode surface at E_1, the adsorbed amount of R will also be oxidized when the potential is stepped to E_2.

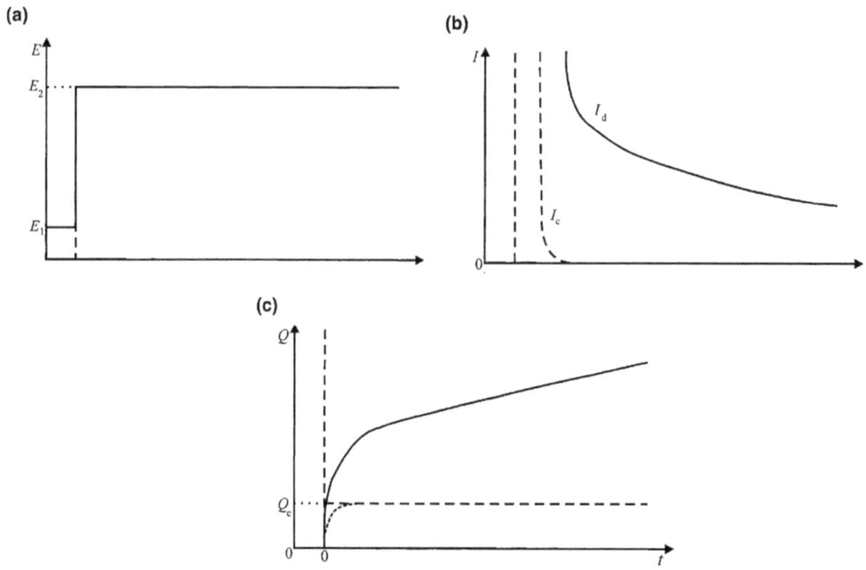

Fig. II.4.2 Typical waveform of (**a**) a potential step experiment and the respective (**b**) chronoamperometric and (**c**) chronocoulometric responses. E_1 and E_2 are as given in Fig. II.4.1. *Dashed curves* for the capacitive current (**b**) and capacitive charge (**c**) obtained when the experimental was repeated in pure supporting electrolyte

This process is usually very quick compared to the slow accumulation of R by diffusion. Thus, the total charge in the presence of adsorbed reactant can be given as follows:

$$Q(t) = Q_{\text{diff}}(t) + Q_{\text{c}}(t) + Q_{\text{ads}}(t) \quad \text{(II.4.5)}$$

From Q_{ads} the adsorbed amount of $R(\Gamma_{\text{R}})$ can be estimated, since

$$Q_{\text{ads}} = nF\Gamma_{\text{R}} \quad \text{(II.4.6)}$$

According to Eq. (II.4.3), the plot of Q_{diff} vs. $t^{1/2}$ should be linear and the slope proportional to the concentration of the reactant, as well as to n, A and $D^{1/2}$. This behavior is shown in Fig. II.4.3 and has been observed experimentally in a large number of cases. The intercept of Q vs. $t^{1/2}$ plot gives Q_{c} or $Q_{\text{c}} + Q_{\text{ads}}$ (in the case of adsorption of R). In the latter case, Q_{c} can be estimated from the charge passed in the same experiment performed with supporting electrolyte only. It is evident that at short times Q vs. $t^{1/2}$ curves are not linear, since neither the charging process nor the oxidation of the adsorbed amount of the reactant is instantaneous, albeit under well-designed experimental conditions (see next section) this period (dotted line in Fig. II.4.3) is less than 1 ms.

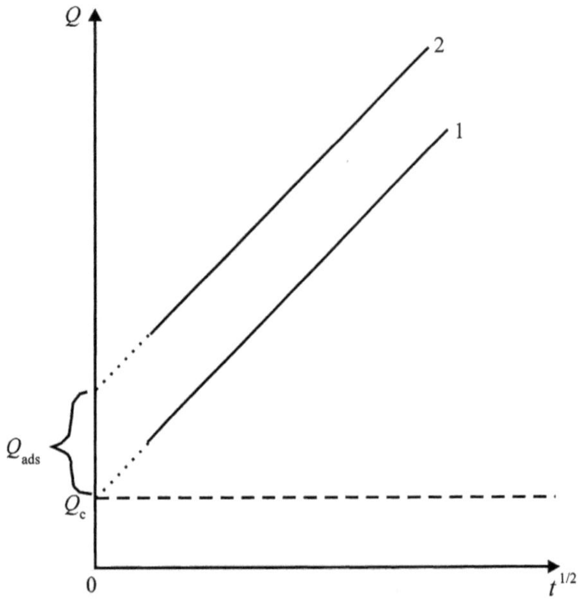

Fig. II.4.3 Chronocoulometric charge vs. (time)$^{1/2}$ plots. Line 1 results if the reactant is not adsorbed; line 2 results for adsorbed reactant with the same parameters (c^*_{R}, D_{R}). The *dotted extensions* of lines 1 and 2 indicate that the intercepts of chronocoulometric plots are usually obtained by linear extrapolation from the shortest times at which reliable data are available. The *dashed, horizontal line* represents the charge response in the absence of reactant

II.4.3 Practical Problems

In order to obtain a straight line in the Q vs. $t^{1/2}$ plot for a relatively long period of time – which is the precondition of straightforward data evaluation – the decay of the current belonging to the double-layer charging should be very fast. It follows from Eq. (II.4.4) that both the uncompensated solution resistance and the double-layer capacitance should be low. For instance, if $R_s = 1\,\Omega$ and $C_d = 20\,\mu\text{F}$, $\tau = 20\,\mu\text{s}$, i.e., the double-layer charging is 95% complete in 60 μs. However, in a highly resistive media, e.g., $R_s = 1000\,\Omega$ and large area electrode $C_d = 200\,\mu\text{F}$, $t = 200\,\text{ms}$, hence a charging current will flow for almost 1 s. It certainly will cause problems in the data evaluation, because of the distortion of the reliable part of the chronoamperometric response. At long times ($t > 20$ s) not only the relative change of I or Q with time will be small but, due to any spontaneous convection, the strict conditions of unstirred solution for the applicability of the Cottrell equation will be violated. Therefore, if possible, R_s should be decreased and a small area electrode (because C_d is proportional to A) should be used.

The application of a working electrode of small A and a supporting electrolyte in order to lower R_s is of importance also to maintain the potential control. It is especially crucial at the shortest times when high current flows in fast transient experiments. The potentiostat must have an adequate power reserve. It has to be able to supply necessary currents (even if they are demanded only momentarily), and it must be able to force those currents through the cell. At high currents, the output voltage of the potentiostat is mostly dropped across the solution resistance, and the voltage requirement can easily exceed 100 V. If the power of the potentiostat is not enough, it supplies a maximum constant current, i.e., it works as a galvanostat. In the case of high ohmic potential drop ($E_\Omega = IR_s$) and anodic current, the true potential of the working electrode will be less positive than the nominal value by that amount adjusted on the potentiostat. The opposite holds for a cathodic current. In some cases the resistance effect can be overcome by using larger potential step amplitudes, because, in this way, even if the effective potential was less than that was set on the potentiostat, we will still work in the region of diffusion-limited current. The optional setting of the value of E_2 is limited by the start of another electrode reaction, e.g., the decomposition of the solvent molecules.

The adsorbed reactant will be oxidized or reduced essentially instantaneously, because it is already present on the electrode surface. It is the situation when monolayer or submonolayer adsorption prevails ($\Gamma < 10^{-10}$ mol cm^{-2}). On the other hand, in the case of a large pseudocapacitance (e.g., any conductor covered with a molecular layer that contains more than one monolayer-equivalent of electroactive centers; typical examples are the polymer film electrodes), Q_c may be rather high and the current–time decay can reflect the diffusion rate of the charge carriers through the surface layer, thus at shorter times the decay of the current should conform to the Cottrell equation.

At long times, when $(Dt)^{1/2} \geq L = \Gamma/c^*$, where L is the film thickness, the concentration within the surface film impacts on the film-solution boundary, the chronoamperometric current will be less than that predicted by the Cottrell equation, and a finite diffusion relationship

$$I = \frac{nFAD^{1/2}c^*}{\pi^{1/2}t^{1/2}}\left[1 + 2\sum_{m=1}^{\infty}(-1)^m \exp\left(-\frac{m^2L^2}{Dt}\right)\right] \quad \text{(II.4.7)}$$

becomes appropriate.

Consequently, the chronocoulometric curve can be given as follows:

$$\frac{Q}{Q_T} = 1 - \frac{8}{\pi^2}\sum_{m=1}^{\infty}\left(\frac{1}{2m-1}\right)^2 \exp\left[-(2m-1)^2\pi^2\frac{Dt}{L^2}\right] \quad \text{(II.4.8)}$$

where Q_T is the total charge that can be consumed by the electroactive surface film ($Q_T = Q_{ads}$). For 2% accuracy, it is enough to consider the first member of the summation ($m = 1$), hence

$$\frac{Q}{Q_T} = 1 - \frac{8}{\pi^2}\exp\left(-\pi\frac{Dt}{L^2}\right) \quad \text{(II.4.9)}$$

It follows that, in the presence of a thick ($L > 100$ nm, $\Gamma > 10^{-8}$ mol cm^{-2}), electrochemically active surface layer, a superposition of $Q_{diff} + Q_{ads}$ can only be measured in the time window from milliseconds to some seconds. If the chronocoulometric response of the electroactive film is measured alone – in contact with inert supporting electrolyte – Cottrell-type response can be obtained usually for thick films only, because, at short times ($t < 0.1 - 1$ ms), the potential of the electrode is not established while, at longer times ($t >$ 1–10 ms), the finite diffusion conditions will prevail and I will exponentially decrease with time.

As mentioned, Q_c can be determined separately performing the same step experiment in pure supporting electrolyte. It is of importance to note that adsorption usually influences the interfacial capacitance; thus if R or O is adsorbed, the capacitive components determined in the absence or presence of the adsorbing species, respectively, will differ from each other. For films – especially in the case of porous conducting polymer – C_d evaluated for metal/electrolyte and metal/film/electrolyte systems, respectively, may differ by orders of magnitude.

II.4.4 Double-Step Chronocoulometry

The application of a double step, i.e., a reversal of the potential to its initial value E_1 from E_2, is a powerful tool in identifying adsorption phenomena, in obtaining information on the kinetics of coupled homogeneous reactions and for the determination of the capacitive contribution. Figure II.4.4 depicts the potential–, current–, and charge–time responses obtained in such a double potential-step experiment for the case when the product O^+ is reduced again during the second step from E_2 to E_1. The chronocoulometric curve obtained during the first potential step is, of course,

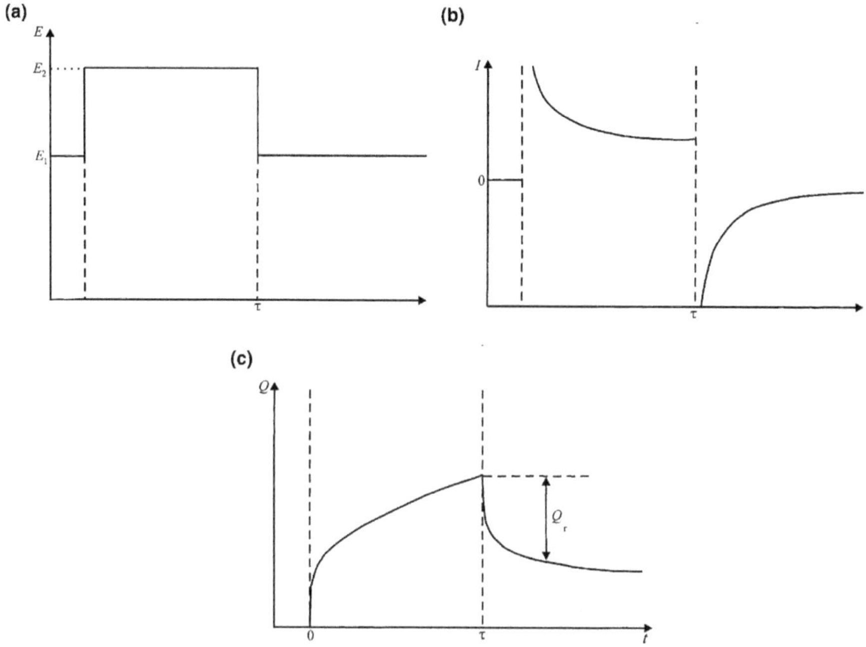

Fig. II.4.4 Temporal behavior of (**a**) potential, (**b**) current, and (**c**) charge in double-potential step coulometry. The meanings of τ and Q_r are given in the text

identical to that obtained in the single-step experiment discussed previously. If the magnitude of the reversal step is also large enough to ensure diffusion control, the chronoamperometric and chronocoulometric responses for $t > \tau$ – where t is the duration of the first step – are given by the following equations:

$$I_{\text{diff}}(t > \tau) = \frac{nFAD_R^{1/2}c_R^*}{\pi^{1/2}}\left[\frac{1}{(t-\tau)^{1/2}} - \frac{1}{t^{1/2}}\right] \quad \text{(II.4.10)}$$

$$Q_{\text{diff}}\,(t > \tau) = 2nFA\,D_R^{1/2}c_R^*\pi^{-1/2}\left(t^{1/2} - (t-\tau)^{1/2}\right) \quad \text{(II.4.11)}$$

It is important to note that there is no capacitive contribution because the net potential change is zero.

The quantity of charge consumed in the reversal step is the difference $Q\,(\tau) - Q\,(t > \tau)$ since the second step actually withdraws the charge injected during the first step:

$$Q_{\text{diff,r}}\,(t > \tau) = 2nFA\,D_R^{1/2}c_R^*\pi^{1/2}\left(t^{1/2} + (t-\tau)^{1/2} - t^{1/2}\right) \quad \text{(II.4.12)}$$

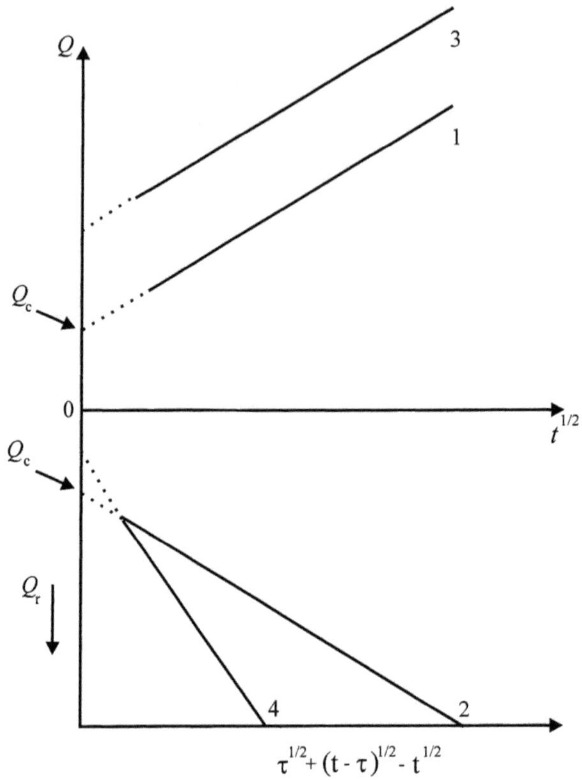

Fig. II.4.5 Chronocoulometric plots for double potential step experiments. Lines 1 and 2 correspond to the case when no adsorption of the reactant or product occurs. Lines 3 and 4 depict the linear responses when the reactant but not the product of the electrode reaction is adsorbed

By plotting $Q(t < \tau)$ vs. $t^{1/2}$ and $Q_r\,(t > \tau)$ vs. $\left[\tau^{1/2} - (t - \tau)^{1/2} - t^{1/2}\right]$, two straight lines should be obtained. In the absence of adsorption the intercept is Q_c and the two intercepts are equal, as seen in Fig. II.4.5 (lines 1 and 2).

In principle, the two linear plots also possess the same slope magnitude. It should be mentioned that the concentration profiles at the start of the second step may not correspond to the conditions of the ideal Cottrell response, especially in the case of adsorption.

In cases where the reactant but not the product of electrode reaction is adsorbed, the intercept of the chronocoulometric plot for the reverse step provides a direct measure of Q_c in the presence of adsorbed reactant, while the difference between the intercepts is $nFA\Gamma_R$, i.e., it gives the amount of the adsorbed reactants. This situation is also displayed in Fig. II.4.5 (lines 3 and 4). Under favorable conditions, intercepts of coulometric plots can be measured with a precision of ca. 0.5 μC cm^{-2}, which corresponds to an uncertainty in Γ of ca. 5×10^{-12} mol cm^{-2} for a one-electron process.

Double-step chronocoulometry is also extremely useful for characterizing coupled homogeneous reactions. Any deviation from the coulometric responses described by Eqs. (II.4.3) and (II.4.11) – providing that diffusion control prevails – implies a chemical complication. For example, O^+ rapidly reacts with a component of the solution, and this homogeneous chemical reaction results in the formation of an electrochemically inactive species. $Q_{\text{diff}}\,(t > \tau)$ falls less quickly than expected or, at complete conversion within the timescale of the experiment, no backward reaction is seen at all. A quick examination of this effect can be carried out by the evaluation of the ratio of $Q_{\text{diff}}\,(t = 2\tau)\,/Q\,(t = \tau)$. For stable systems this ratio is between 0.45 and 0.55.

Responses for various mechanisms have been discussed in the literature [1–5].

II.4.5 Effect of Heterogeneous Kinetics on Chronocoulometric Responses

Chronocoulometric responses may be governed wholly or partially by the charge transfer kinetics. It has been reported that, in some cases, the diffusion-limited situation cannot be reached, e.g., due to the insufficient power of the potentiostat and the inherent properties of the system, especially at the beginning of the potential step. On the other hand, in many cases, potential steps of smaller magnitudes than that required for diffusion control are applied in order to study the electrode kinetics.

If the potential amplitude of the step is less than $E_2 - E_1$, i.e., the step is made to any potential in the rising portion of the voltammogram, either charge transfer control or mixed kinetic-diffusional control prevails, i.e., the concentration of the reactant is not zero at the electrode surface. Still, $c_{\text{R}}\,(x = 0) < c_{\text{R}}^*$, so R does diffuse to the electrode surface where it is oxidized. Since the difference between the bulk and surface concentration is smaller than in the case of mass transport control, less material arrives at the surface per unit time, and the currents for corresponding times are smaller than in the diffusion-limited situation. Nevertheless, the concentration gradient for R still increases, which means that the current still decreases with time, albeit to a smaller extent. The current–time curves when the potential is stepped from $E_2 > E_{21} > E_{22}$ are shown in Fig. II.4.6.

If we sample the current at some fixed time (t_{s}), then, by plotting $I(t_{\text{s}})$ vs. E_1, E_{22}, E_{21}, E_2, etc., a current–potential curve is obtained that resembles the waveshape of a steady-state voltammogram. This kind of experiment is called *sampled-current voltammetry*. In fact, it serves as the basis for the theoretical description of dc polarography and several other pulse techniques. For the mathematical description, the flux equations [see Chap. I.3, Eq. (I.3.30)] should be combined with the respective charge transfer relationships [see Chap. I.3, Eqs. (I.3.5), (I.3.12), and (I.3.13)].

When initially only R is present, the following expression can be derived:

$$I(t) = nFAk_{\text{ox}}c_{\text{R}}^* \exp\left(k_{\text{ox}}^2/D_R t\right) \operatorname{erfc}\left(k_{\text{ox}}t^{1/2}/D_{\text{R}}^{1/2}\right) \qquad \text{(II.4.13)}$$

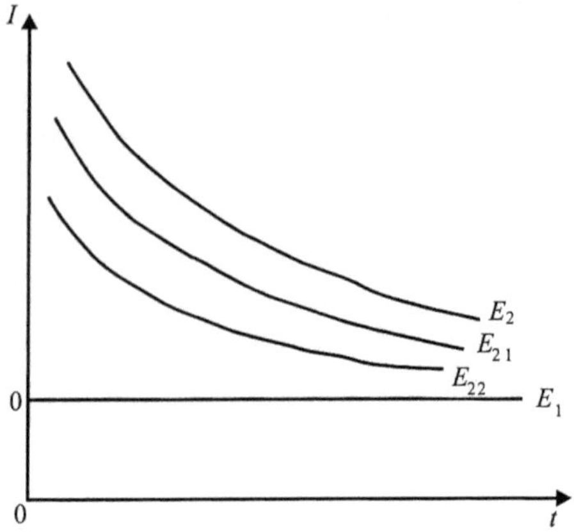

Fig. II.4.6 Current–time curves for different step magnitudes. E_1 is the initial potential where no current flows, while E_2 is the potential in the mass transport limited region. $E_2 > E_{21} > E_{22} > E_1$, and E_{21} and E_{22} correspond to potentials in the rising portion of the voltammogram where mixed charge transfer–mass transport control prevails

If both O and R are present, an even more complex equation is obtained:

$$I(t) = nFA(k_{\mathrm{ox}}c_{\mathrm{R}}^{*} - k_{\mathrm{red}}c_{\mathrm{R}}^{*})\exp(H^2t)\,\mathrm{erfc}\,(Ht^{1/2}) \tag{II.4.14}$$

where

$$H = \frac{k_{\mathrm{red}}}{D_{\mathrm{O}}^{1/2}} + \frac{k_{\mathrm{ox}}}{D_{\mathrm{R}}^{1/2}} \tag{II.4.15}$$

If the potential is stepped to a potential at the foot of the voltammetric wave, Eq. (II.4.13) or Eq. (II.4.15) can be linearized, since, in this case, k_{ox}, k_{red}, hence H, are small. Equation (II.4.15), for example, is simplified to

$$I(t) = nFAk_{\mathrm{ox}}c_{\mathrm{R}}^{*}\left(1 - \frac{2Ht^{1/2}}{\pi^{1/2}}\right) \tag{II.4.16}$$

Then plotting $I(t)$ vs. $t^{1/2}$ and extrapolating the linear plot to $t = 0$, k_{ox} can be obtained from the intercept.

The chronocoulometric responses can be derived from the current transients by integration from $t = 0$. In the case of Eq. (II.4.14), the integration gives the following expression:

$$Q\,(t) = \frac{nFAk_{\mathrm{ox}}c_{\mathrm{R}}^{*}}{H^2}\left[\exp(H^2t)\,\mathrm{erfc}\,(Ht^{1/2}) + \frac{2Ht^{1/2}}{\pi^{1/2}} - 1\right] \tag{II.4.17}$$

For $Ht^{1/2} > 5$, the first term in the brackets can be neglected, hence Eq. (II.4.17) takes the limiting form:

$$Q\,(t) = nFAk_{\mathrm{ox}}c_{\mathrm{R}}^{*}\left(\frac{2t^{1/2}}{H\pi^{1/2}} - \frac{1}{H^2}\right) \tag{II.4.18}$$

In this case, the plot of $Q(t)$ vs $t^{1/2}$ is linear and from the intercept on the $t^{1/2}$-axis ($t_i^{1/2}$) H can be estimated, since

$$H = 0.5\,\pi^{1/2} t_{\mathrm{i}}^{-1/2} \tag{II.4.19}$$

From the linear slope, $2nFAk_{\mathrm{ox}}c_{\mathrm{R}}^{*}RH^{-1}\pi^{1/2}$, k_{ox} can be calculated.

At high values of E, H approaches $k_{\mathrm{ox}}/D_{\mathrm{R}}^{1/2}$, and then the Cottrell relationship comes into play.

Equations (II.4.17) and (II.4.18) do not include contributions from double-layer charging or adsorbed species. It is advisable to design the experimental conditions in such a way that these effects are negligible compared to the diffusive component. It can be achieved, for instance, by using high concentration of the reactant.

The application of chronocoulometry is advantageous also for the evaluation of rate constants because the extrapolation from a later time domain is more accurate. The integration preserves the information of the kinetic effect that appears at the early period of the step experiment.

Several coulometric and pulse techniques are used in electroanalytical chemistry. Rather low detection limits can be achieved, and kinetic and transport parameters can be deduced with the help of these fast and reliable techniques. Since nowadays the pulse sequences are controlled and the data are collected and analyzed using computers, different pulse programs can easily be realized. Details of a wide variety of coulometric and pulse techniques, instrumentation and applications can be found in the following literature: controlled current coulometry [6], techniques, apparatus and analytical applications of controlled potential coulometry [7], coulostatic pulse techniques [8], normal pulse voltammetry [9], differential pulse voltammetry [9], and square-wave voltammetry [10].

The application of pulse techniques at microelectrodes has also been reviewed [11–13].

References

1. Bard AJ, Faulkner LR (2001) Electrochemical methods, fundamentals and applications, 2nd edn. John Wiley, New York
2. Rieger PH (1987) Electrochemistry. Prentice Hall, Oxford
3. Galus Z (1994) Fundamentals of electrochemical analysis, 2nd edn. Harwood, Chichester
4. Delahay P (1954) New instrumental methods in electrochemistry. Wiley, New York
5. Macdonald DD (1977) Transient techniques in electrochemistry. Plenum, New York
6. Janata J, Mark HB Jr (1969) Application of controlled-current coulometry to reaction kinetics. In: Bard AJ (ed) Electroanalytical chemistry, vol 3. Marcel Dekker, New York, pp 1–56
7. Harrar JE (1975) Techniques, apparatus, and analytical applications of controlled-potential coulometry. In: Bard AJ (ed) Electroanalytical chemistry, vol 8. Marcel Dekker, New York, pp 2–167
8. van Leeuwen HP (1982) Coulostatic pulse techniques. In: Bard AJ (ed) Electroanalytical chemistry, vol 12. Marcel Dekker, New York, pp 159–238
9. Bond AM (1980) Modern polarographic methods in analytical chemistry. Marcel Dekker, New York

10. Osteryoung J, O'Dea J (1986) Square-wave voltammetry. In: Bard AJ (ed) Electroanalytical chemistry, vol 14. Marcel Dekker, New York, pp 209–308
11. Montenegro MI, Querios MA, Daschbach JL (ed) (1991) Microelectrodes: theory and applications. Proc NATO ASI. Kluwer, Dordrecht
12. Amatore C (1995) Electrochemistry at ultramicroelectrodes. In: Rubinstein I (ed) Physical electrochemistry. Marcel Dekker, New York, pp 131–208
13. Heinze J (1993) Angew Chem Int Ed Engl 32: 1268

Chapter II.5
Electrochemical Impedance Spectroscopy

Utz Retter and Heinz Lohse

II.5.1 Introduction

Non-steady-state measuring techniques are known to be extremely suitable for the investigation of the electrode kinetics of more complex electrochemical systems. Perturbation of the electrochemical system leads to a shift of the steady state. The rate at which it proceeds to a new steady state depends on characteristic parameters (reaction rate constants, diffusion coefficients, charge transfer resistance, double-layer capacity). Due to non-linearities caused by the electron transfer, low-amplitude perturbation signals are necessary. The small perturbation of the electrode state has the advantage that the solutions of relevant mathematical equations used are transformed in limiting forms that are normally linear. Impedance spectroscopy represents a powerful method for investigation of electrical properties of materials and interfaces of conducting electrodes. Relevant fields of application are the kinetics of charges in bulk or interfacial regions, the charge transfer of ionic or mixed ionic–ionic conductors, semiconducting electrodes, the corrosion inhibition of electrode processes, investigation of coatings on metals, characterisation of materials and solid electrolyte as well as solid-state devices.

II.5.2 Definitions, Basic Relations, the Kramers–Kronig Transforms

If a monochromatic alternating voltage $U(t) = U_{\mathrm{m}} \sin(\omega t)$ is applied to an electrode then the resulting current is $I(t) = I_{\mathrm{m}} \sin(\omega t - \vartheta)$ where ϑ is the phase difference between the voltage and the current and U_{m} and I_{m} are the amplitudes of the sinusoidal voltage and current, respectively. Then the impedance is defined as

$$Z = U(t)/I(t) = |Z|\,\mathrm{e}^{\mathrm{j}\vartheta} = Z' + \mathrm{j}Z'' \qquad \text{(II.5.1)}$$

U. Retter (✉)
Balzerstr. 45, 12683 Berlin, Germany
e-mail: Utz.Retter@gmx.de

F. Scholz (ed.), *Electroanalytical Methods*, 2nd ed.,
DOI 10.1007/978-3-642-02915-8_8,

with

$$j = (-1)^{1/2} \quad \text{(II.5.2)}$$

where Z' and Z'' are the real and imaginary part of Z, respectively.

Impedance Z and admittance Y are related as follows:

$$Y = 1/Z = Y' + jY'' \quad \text{(II.5.3)}$$

Now let us sum up the following definitions: resistance: Z' (also R), reactance: Z'' (also X), magnitude of impedance: $|Z|$, conductance: Y' (also G), susceptance: Y'' (also B), quality factor: $Q = |Z''/Z'| = |Y''/Y'|$, dissipation factor: $D = 1/Q = |\tan\delta|$, loss angle: δ.

$$|Z|^2 = (Z')^2 + (Z'')^2 \quad \text{(II.5.4)}$$

$$\delta = \arctan(Z'/Z'') \quad \text{(II.5.5)}$$

Frequently used procedures in modelling include the conversion of a parallel circuit to a series one and the conversion of a series circuit to a parallel one.

The series circuit and the parallel circuit must be electrically equivalent. This means the dissipation factors $|\tan\delta|$ must be the same and the absolute values $|Z| = 1/|Y|$.

The following equations enable these procedures to be performed [1]:

$$\begin{aligned} &\tan\delta = \cot\vartheta = Z'/Z'' = -Y'/Y'' \text{ and} \\ &|Z|^2 = 1/|Y|^2 = (Z')^2 + (Z'')^2 = 1/\left((Y')^2 + (Y'')^2\right) \end{aligned} \quad \text{(II.5.6)}$$

For an RC circuit with components R_s, C_s (series) and R_{par}, C_{par} (parallel), the following relations are valid:

$$\begin{aligned} Z' &= R_s, \\ Z'' &= -1/(\omega C_s), \\ Y' &= 1/R_{par}, \\ Y'' &= \omega C_{par} \end{aligned} \quad \text{(II.5.7)}$$

$$R_{par} = R_s\left(1 + 1/\tan^2\delta\right) \quad \text{(II.5.8)}$$

$$C_{par} = C_s/\left(1 + \tan^2\delta\right) \quad \text{(II.5.9)}$$

The Kramers–Kronig frequency domain transformations enable the calculation of one component of the impedance from another or the determination of the phase angle from the magnitude of the impedance or the polarization resistance R_p from the imaginary part of the impedance. Furthermore, the Kramers–Kronig (KK) transforms allow the validity of an impedance data set to be checked. Precondition for the application of KK transforms is, however, that the impedance must be finite-valued for the limits $\omega \to 0$ and $\omega \to \infty$, and must be a continuous and finite-valued function at all intermediate values.

Let us now sum up the KK transforms:

$$Z'(\omega) - Z'(\infty) = (2/\pi) \int_0^\infty \left[\left(xZ''(x) - \omega Z''(\omega) \right) / \left(x^2 - \omega^2 \right) \right] \mathrm{d}x \tag{II.5.10}$$

$$Z'(\omega) - Z'(0) = (2\omega/\pi) \int_0^\infty \left[\left((\omega/x)\, Z''(x) - Z''(\omega) \right) / \left(x^2 - \omega^2 \right) \right] \mathrm{d}x \tag{II.5.11}$$

$$Z''(\omega) = -\,(2\omega/\pi) \int_0^\infty \left[\left(Z'(x) - Z'(\omega) \right) / \left(x^2 - \omega^2 \right) \right] \mathrm{d}x \tag{II.5.12}$$

$$\vartheta\,(\omega) = (2\omega/\pi) \int_0^\infty \left[(\ln |Z\,(x)|) / \left(x^2 - \omega^2 \right) \right] \mathrm{d}x \tag{II.5.13}$$

$$R_\mathrm{p} = (2/\pi) \int_0^\infty \left[\left(Z''(x) \right) / x \right] \mathrm{d}x \tag{II.5.14}$$

II.5.3 Measuring Techniques

The measuring principle is simple. On an electrochemical system in equilibrium a small signal acts (time-dependent potential or current). The response of the system is measured then. "Small signal" means the perturbation of the system is so low that the response is linear, i.e. harmonic generation and frequency mix products can be neglected. The signal can be a single sinus wave or consist of a sum of such waves with different amplitudes, frequencies and phases (e.g. single potential or current step, pulse-shaped signals, noise). In the majority of cases, electrochemical systems are linear at signal amplitudes of 10 mV or less. On application of a sum signal, the effective value of the signal must keep that condition. The overall equivalent circuit at high frequencies can be assumed as a series combination of the linear solution resistance and the predominantly capacitive interface. Then, only a part of the amplitude applied to the electrochemical cell is applied to the interface because most of the potential drop occurs at the solution resistance. Therefore, the signal amplitude can be chosen higher without violating the linearity conditions. One can check the validity of the linearity condition by checking the independence of the impedance on the test signal amplitude.

The input signal can be a single frequency, a discrete number of frequencies (e.g. computer generated) or a theoretical unlimited spectrum of frequencies (white noise). Primarily, it seems of advantage to apply a large number of frequencies and measure the response simultaneously. However, the electrochemical system generates noise and, because of the linearity condition, the signal amplitudes are very small. Therefore, a signal averaging must be used for the response signal. This is very time consuming, mainly for low frequencies and high impedances.

In time-domain measurements the test signal has a time-dependent shape (ramp function, triangle or square pulse) and the time dependence of the system response is measured. From the time dependence of the signal response, information can be obtained about the system parameters but their extraction is very complicated or impossible for non-trivial systems. Data transition from the time domain into the frequency domain and back can be made with transform methods. Commercial measuring systems use fast Fourier transformation (FFT). The use of FFT is recommended in the low frequency region (10^{-3} – 10^{2} Hz) because the cycle duration of the highest signal frequency is large in comparison to the conversion time of the precision analog-to-digital converters (ADC) and FFT speed. Digital signal filtering is here superior to analog filters. The measuring methods for system response are physically equivalent (same limitations in noise-bandwidth) but different methods can be suited more or less in dependence of the system under investigation and the hard- and software used in the measuring system.

Frequency analysis (measuring in frequency domain) can be used over a very large frequency range (10^{-3} – 10^{7} Hz). Normally, a single frequency signal is used and the amplitude and phase shift, or real and imaginary parts of response signal, is measured [2]. Figure II.5.1 shows a block diagram of a potentiostatic frequency response analyser.

In commercial impedance analysers, commonly the polarization and the test signals are generated separately. The polarization is a large-scale signal and the requirements on electrical circuits as amplifiers and digital-to-analog converters (DAC) are: high constant system parameters in time and temperature and precise amplitude and low noise characteristic over the whole potential region (e.g. ± 5 V in 1 mV steps). The system for generating the test signal (small signal) must be

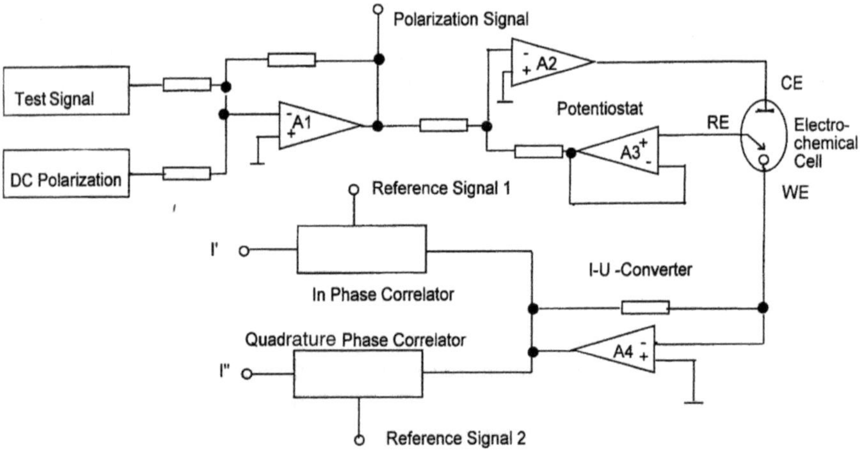

Fig. II.5.1 Block diagram of a potentiostatic frequency response analyser

linear over the entire frequency and amplitude scale and the noise amplitude must be low too. The sum of signals at the output of the operations amplifier A1 is the polarization signal of the electrochemical cell. The amplifiers A2 and A3 together with the cell compose a potentiostat. The potential of the reference electrode (RE) is compared with the polarization signal and the control loop with A2 changes the potential at the counter electrode (CE) until the potential at RE is equal the sum from the test signal and the DC polarization signal. An important problem for measuring the impedance behaviour at high frequencies is the influence of parasitic elements and the phase shift connected with the potentiostatic control of the system. In the worst case the system can oscillate. However, before oscillations occur, the phase shifts and variations of the amplitude of the response signals give incorrect impedance values. Then, a precise dummy cell with well-known parameters similar to the cell impedance should be measured to examine it. For cases without or very small direct current, a two-electrode configuration can be used to avoid problems with phase shifts. Then the RE is connected to the CE and the amplifier A3 can be replaced with a resistor.

The decoupling of ac and dc signals is another way to avoid phase shifts and amplitude deformations at high frequencies. The cut-off frequency of the potentiostat is then far lower than the cycle duration of the ac signal [3]. In Fig. II.5.1 the output potential of the amplifier A4 is proportional to the cell current (*I*-*U* converter). Here, the input potential of A4 is controlled to be zero (virtual earth) and there are no additional parasitic impedance elements in the measuring circuit. In the case of high cell currents and for very high frequencies, the *I*-*U* converter can show non-ideal behaviour. Then, the current can be measured by determining the potential drop on a small resistance between the working electrode and the earth. The output signal of A4 is applied to a phase sensitive detector. Here, it is compared with reference signals 1 and 2 in phase or 90° shifted to the test signal, respectively. As a result, the output signals I' and I'' are proportional to the in-phase and quadrature components of the cell current. Some analysers determine the amplitude and phase of the response signal. The results are equivalent and the inherent problems are the same. At high frequencies and/or if I' and I'' are very different (phase angle $\approx 0°$ or $\approx -90°$), the phase discriminators show a "cross talk" effect, e.g. the component with the high amplitude influences the other. Generally, the precision of the phase angle is lower at higher frequencies.

An impedance spectrum is the result of a sweep about the selected frequency band. Narrow band measuring (filters or lock-in amplifiers) is able to improve the accuracy and sensitivity but it is time consuming at low frequencies. For precise measurements (suppressing of transients), a signal averaging at least five periods is recommended.

Broadband perturbing signals used in connection with frequency transformation (mostly FFT algorithm) can allow a faster measuring in time-varying systems. However, an increase in speed decreases the accuracy and deteriorates the signal-to-noise ratio. To improve this ratio signal accumulation must be performed and the gain in measuring time is lost. A special method to measure the impedance or admittance of an electrochemical system is to compare its signal response with the response of a combination of calibrated impedance elements (use of audio- and radio-frequency bridges or substitution methods). Here, the accuracy of the method is only affected by the elements and the sensitivity and stability of the amplitude and phase measuring. The handicap here is that the method is very time consuming. Time domain techniques mostly use step and ramp signals.

For electrochemical impedance spectroscopy a number of excellent commercial measuring systems exist. In some cases it is favourable to use precision impedance analysers designed specially for accurate impedance measurements of electronic components and materials in a broad-frequency scale.

II.5.4 Representation of the Impedance Data

The impedance data will be represented in complex plane plots as $-Z''$ vs. Z' (normally called the Nyquist diagrams), Y'' vs. Y' diagrams and derived quantities as the modulus function $M = \mathrm{j}\omega C_{\mathrm{C}} Z = M' + \mathrm{j}M''$ (C_{C} is the capacity of the empty cell) and the complex dielectric constant $\varepsilon = 1/M = \varepsilon' - \mathrm{j}\varepsilon''$ depictions. Frequently used terms are as follows: the complex plane plot of the frequency normalized admittance components Y'/ω vs. Y''/ω specially for an investigation of non-faradaic processes (the frequency acts as parameter) and the Bode plot for the representation of the magnitude of impedance $|Z|$ and the phase angle ϑ vs. log ν, (ν being the measuring frequency). It should be noted that the representation form and any mathematical transformation can never improve the quality of the data fitting, i.e. the fitting should always start with the experimental data.

II.5.5 Equivalent Circuits

Any electrochemical cell can be represented in terms of an equivalent electrical circuit that comprises a combination of resistances, capacitances or inductances as well as mathematical components. At least the circuit should contain the double-layer capacity, the impedance of the faradaic or non-faradaic process and the high-frequency resistance. The equivalent circuit has the character of a model, which more or less precisely reflects the reality. The equivalent circuit should not involve too many elements because then the standard errors of the corresponding parameters become too large (see Sect. II.5.7), and the model considered has to be assessed as not determined, i.e. it is not valid.

II.5.6 The Constant Phase Element

For an ideally polarized electrode, the impedance consists of the double-layer capacity C_{d} and the solution resistance R_{Ω} in series. In the impedance plane plot, a straight vertical line results intersecting the Z'-axis at $Z' = R_{\Omega}$. At solid electrodes, especially due to contamination and roughness, a straight line can be observed intersecting the Z'-axis at R_Ω at an angle smaller than 90° to the real axis. The corresponding phase angle $\vartheta_C = -(1-\beta)\,\pi/2$ is assumed to be independent of frequency, i.e. a "constant phase angle" occurs [4, 5] and, consequently, the impedance follows as

$$Z = R_\Omega + Z_C = R_\Omega\left[1 + (j\omega\tau)^{-(1-\beta)}\right] \tag{II.5.15}$$

where $\tau = R_\Omega C_d$. The dimensionless parameter β ranges between 0 and 1. For $\beta = 0$, the well-known case of a series connection of R_Ω and C_d follows. If the electrode consists of R_Ω in series with an impedance Z_C with a constant-phase angle (the "constant-phase element", CPE), C_d will be replaced by the CPE. As a consequence, the corresponding admittance changes from a semicircular arc to a depressed semicircular arc.

$$Y = 1/Z = (1/R_\Omega)\left[1 - 1/\left(1 + (j\omega\tau)^{(1-\beta)}\right)\right] \tag{II.5.16}$$

When a charge transfer proceeds at the electrode, the equivalent circuit consists of C_d and the charge-transfer resistance R_{ct} in parallel. Therefore, the corresponding Nyquist impedance plot represents a semicircular arc. Analogous to the case just considered above, a replacement of C_d by the CPE leads to a change from a semicircular arc to a depressed semicircular arc.

A full discussion of the distribution of relaxation times as the origin of constant-phase elements is available in the literature [5].

In another report [4], an error in the interpretation of the CPE is pointed out. On the one hand, the double-layer capacity is replaced by the CPE, i.e. the CPE is a property of the double layer itself. On the other, the CPE is discussed as originating from surface inhomogeneities.

II.5.7 Complex Non-Linear Regression Least-Squares (CNRLS) for the Analysis of Impedance Data

Let us consider a set of data Z_i' and Z_i''. The measurements were performed at the angular frequencies $\omega_i\,(i = 1\ldots K)$. The theoretical values are denoted by $Z_{it}\,(\omega_i, P_1, P_2 \ldots P_m)$, where $P_1, P_2 \ldots P_m$ are parameters to be determined and m is the number of parameters. Such parameters can be rate constants, the charge transfer resistance, the double-layer capacity or the high-frequency resistance. The aim of the complex least-squares analysis consists of minimising the sum S:

$$S = \sum_{i=1}^{K}\left\{\left[Z_i' - Z_{it}'\,(\omega_i, P_1, P_2, \ldots P_m)\right]^2 + \left[Z_i'' - Z_{it}''\,(\omega_i, P_1, P_2, \ldots P_m)\right]^2\right\} \tag{II.5.17}$$

For the minimizing procedure one uses normally the Marquardt algorithm [6, 7].

Good starting values of the parameter play an important role here. Otherwise the least-squares sum may converge to local minima instead of the absolute minimum or the sum may even diverge. One should not blindly accept a model and tentatively assume that it is correct. There are the following criteria to assess the validity of a model, which must be fulfilled simultaneously:

(a) Small relative standard deviations of the parameters, smaller than 30% [5]. Otherwise the corresponding parameters have to be removed from the model. One should not believe "The more parameters the better." Often a simplification of the model leads to success, or other models should be tested.
(b) Small relative residuals for the data points are demanded [8]. The overall standard deviation of the fit, $(S/2K)^{1/2}$, divided by the mean value of the measuring values, should not exceed 10% for a model. In some cases, there are several models with comparable standard error of the fit and of the parameters. In this case, additional dependencies have to be investigated. For instance, if the potential dependence of a parameter alone does not allow a decision as to which model is valid, then the concentration dependence should also be investigated. To sum up this section: non-linear regression is a necessary mathematical quality control of models. Abuse of non-linear regression can only lead to a modern sort of (electro)alchemy, absurdly based on high-tech measurements.

II.5.8 Commercial Computer Programs for Modelling of Impedance Data

Several commercial computer programs for the modelling of impedance data currently exist. In the first operation, an adequate equivalent circuit can be created. The standard impedance elements used here are: resistance, capacity, inductance, constant-phase element, Warburg impedance (semi-infinite linear diffusion), Warburg impedance (semi-infinite hemispherical diffusion), finite-length diffusion for transmissive and reflective boundary conditions, impedance of porous electrodes, impedance for the case of rate control by a homogeneous chemical reaction. In the second operation, the validity of the impedance data is checked by the KK rule check (see Sect. II.5.2). As the third operation, the model parameters will be adapted to the measured data using the complex non-linear regression least-squares (CNRLS) fit. The last operation involves the representation of the experimental data and the optimized calculated data using plot diagrams. Here, most often the Bode plot and the Nyquist impedance plot are taken.

II.5.9 Charge Transfer at the Electrode – the Randles Model

A quasi-reversible charge transfer is considered with R_{ct} the charge-transfer resistance and j_0 the exchange current density:

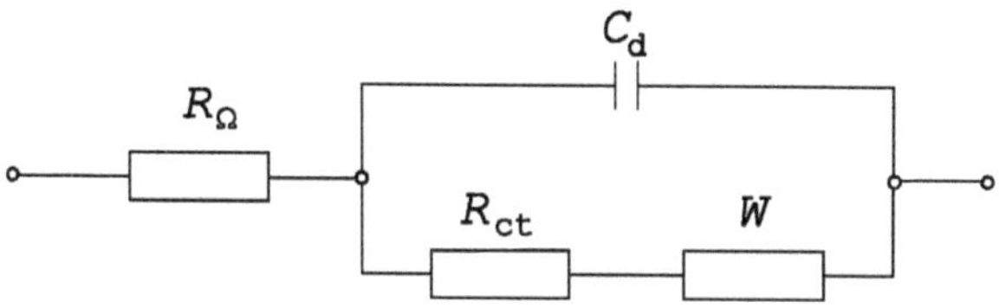

Fig. II.5.2 Randles equivalent circuit

$$\text{Ox} + ne \rightarrow \text{Red} \tag{II.5.18}$$

Let the species Ox and Red with concentrations c_O and c_R diffuse in the solution in the direction perpendicular to the electrode. D_O and D_R are the diffusion coefficients. The initial conditions demand that the solution is homogeneous and that the concentrations are equal to c_O^* and c_R^* for $t = 0$. Outside the Nernst layer the concentrations are equal to c, the concentrations in the bulk. At the electrode surface, the fluxes of Ox and Red are identical and equal to the normalized faradaic current I_F/nFA (A: electrode area). The charge-transfer resistance R_{ct} is defined as:

$$1/R_{ct} = (\delta I_F/\delta E)_c \tag{II.5.19}$$

$$k_f = k_f^0 \exp\left[-\alpha\,(nFE/RT)\right] \tag{II.5.20}$$

$$k_b = k_b^0 \exp\left[(1-\alpha)\,(nF/RT)\,E\right] \tag{II.5.21}$$

$$1/R_{ct} = \left(An^2F^2/RT\right)\left[ak_f c_O^* + (1-\alpha)\,k_b c_R^*\right] \tag{II.5.22}$$

$$R_{ct} = RT/\,(nFj_0) \tag{II.5.23}$$

where α is the apparent cathodic transfer coefficient and j_0 the exchange current density.

A quasi-reversible charge transfer is considered with the equivalent circuit shown in Fig. II.5.2.

Let us first consider the Randles model for higher frequencies. R_Ω is the high-frequency series resistance or electrolyte resistance and C_d the double-layer capacity.

$$\begin{aligned} Z' &= R_\Omega + R_{ct}/\left(1+\omega^2C_d^2R_{ct}^2\right); \\ Z'' &= -\omega C_d R_{ct}^2/\left(1+\omega^2C_d^2R_{ct}^2\right) \end{aligned} \tag{II.5.24}$$

Rearranging, the equation gives

$$\left(Z' - R_\Omega - R_{ct}/2\right)^2 + \left(Z''\right)^2 = R_{ct}^2/4 \tag{II.5.25}$$

$$Z'' = -\left(Z' - R_\Omega\right)(\omega R_{ct} C_d) \tag{II.5.26}$$

This is the equation of a circle with its centre on the Z'-axis at $Z' = R_\Omega + 0.5\,R_{ct}$ and radius 0.5 R_{ct}. From the intersection point of the semicircle with the R_s-axis, one obtains the value R_Ω + R_{ct} and the high-frequency ohmic component is equal to R_Ω. Let ω_{max} be the frequency at which -Z'' exhibits a maximum vs Z'. For the maximum of – Z'' it follows that:

$$\omega_{max} R_{ct} C_d = 1 \qquad \text{(II.5.27)}$$

From the frequency at which – Z'' exhibits a maximum, C_d can be estimated.

$$C_d = 1/\left(R_{ct}\omega_{max}\right) \qquad \text{(II.5.28)}$$

The following problem arises in the interpretation of such semicircles in the complex plane impedance plots: every parallel combination of a constant resistance and constant capacity leads to a semicircle in the Nyquist plot of the impedance. To verify a charge transfer, for instance, the potential dependence of the charge-transfer resistance should be investigated to demonstrate the Butler–Volmer potential dependence of the exchange current.

Let us now consider a semi-infinite linear diffusion of charged particles from and to the electrode. The Faraday impedance is defined as the sum of the charge-transfer resistance R_{ct} and the Warburg impedance W corresponding to the semi-infinite diffusion of the charged particles

$$Z_F = R_{ct} + W \qquad \text{(II.5.29)}$$

$$W = \left[R_{ct}K_W/\left(2\omega\right)^{1/2}\right](1 - \mathrm{j}) \qquad \text{(II.5.30)}$$

$$K_W = \left(k_{ox}/\left(D_O\right)^{1/2}\right) + \left(k_{red}/\left(D_R\right)^{1/2}\right) \qquad \text{(II.5.31)}$$

Here, k_{ox}, k_{red} are the rate constants for oxidation and reduction, respectively, and D_O and D_R are the diffusion coefficients of the oxidized and reduced reactants:

$$\sigma = K_W R_{ct}/\left(2\right)^{1/2} \qquad \text{(II.5.32)}$$

where σ is the Warburg coefficient. The Nyquist plot of Z_W is a straight line at an angle of 45° to the real axis.

The complete Randles model includes mixed control by diffusion and charge transfer control.

The corresponding equations for Z' and Z'' are [9, 10]

$$Z' = R_\Omega + \left(R_{ct} + \sigma\omega^{-1/2}\right)/Q_R \qquad \text{(II.5.33)}$$

$$Z'' = -\left(\omega C_d\left(R_{ct} + \sigma\omega^{-1/2}\right)^2 + \sigma^2 C_d + \sigma\omega^{-1/2}\right)/Q_R \qquad \text{(II.5.34)}$$

$$Q_R = \left(\sigma\omega^{1/2}C_d + 1\right)^2 + \omega^2 C_d^2\left(R_{ct} + \sigma\omega^{-1/2}\right)^2 \qquad \text{(II.5.35)}$$

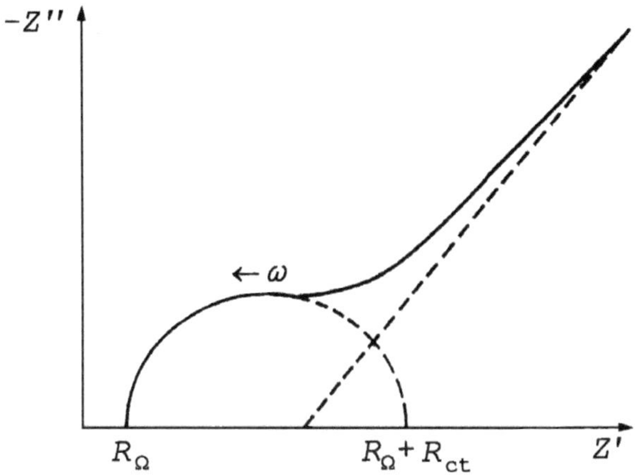

Fig. II.5.3 Scheme of the impedance of the Randles equivalent circuit in the complex impedance plane (Nyquist plot)

Figure II.5.3 represents the Nyquist plot of the Randles impedance with the semicircle at higher frequencies and the straight line at an angle of 45° to the real axis at lower frequencies.

In the case of complex expressions for the impedance for more complicated electrochemical reactions, the calculations of the real and imaginary component can be very complicated. Then, it is much easier to split the whole calculation into elementary steps. We denote this method as cumulative calculation of the cell impedance. As an example, let us take again the Randles model.

The Warburg impedance is (see Eqs. II.5.30, II.5.31, and II.5.33)

$$W' = \sigma\omega^{-1/2}; \quad W'' = -\sigma\omega^{-1/2} \tag{II.5.36}$$

In the first step, the charge-transfer resistance will be added

$$Z_1{}' = W' + R_{\mathrm{ct}}; \quad Z_1{}'' = W'' \tag{II.5.37}$$

The second step involves a conversion from a series to a parallel circuit:

$$\text{Series in parallel: } \left(Z_1{}', Z_1{}''\right) \rightarrow \left(1/R_{\mathrm{par2}}, C_{\mathrm{par2}}\right) \tag{II.5.38}$$

In the third step, the double-layer capacity C_{d} is added:

$$C_{\mathrm{par2}} + C_{\mathrm{d}} = C_{\mathrm{par3}}; \quad R_{\mathrm{par2}} = R_{\mathrm{par3}} \tag{II.5.39}$$

The fourth step includes a conversion from a parallel to a series circuit:

$$\text{Parallel in series: } \left(R_{\mathrm{par3}}, C_{\mathrm{par3}}\right) \rightarrow \left(R_{\mathrm{s4}}, 1/\omega C_{\mathrm{s4}}\right) \tag{II.5.40}$$

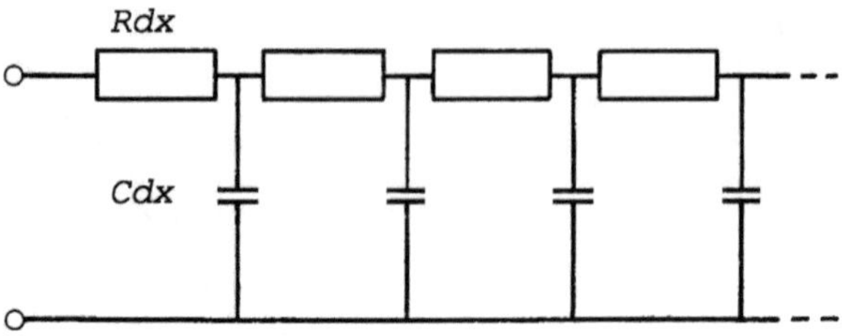

Fig. II.5.4 Resistive-capacitive semi-infinite transmission line – the equivalent circuit for semi-infinite diffusion. R and C are normalized to unit length

In the fifth step, the high-frequency series resistance will be added:

$$R_{s4} + R_{\Omega} = R_{s5}; \quad 1/\omega C_{s4} = 1/\omega C_{s5} \tag{II.5.41}$$

Now a direct modelling of the experimentally measured impedance data is possible due to comparison of R_{s5} with $R_{s,exp}$ and C_{s5} with $C_{s,exp}$.

The use of subroutines $(Z'; Z'') \rightarrow (Y'/\omega; Y''/\omega)$ and $(Y'/\omega; Y''/\omega) \rightarrow (Z'; Z'')$ makes the calculations much easier because now the calculations are reduced to repeating conversions of series components in parallel ones, and vice versa, and additions of the relevant parameters.

Equation (II.5.36) shows that the Warburg impedance cannot be represented as a series combination of frequency-independent elements in an equivalent circuit. This is possible, however, by a semi-infinite resistive-capacitive transmission line with a series resistance R per unit length and a shunt capacity C per unit length (Fig. II.5.4).

II.5.10 Semi-infinite Hemispherical Diffusion for Faradaic Processes

The Warburg impedance for hemispherical diffusion corresponds to that of planar diffusion; however, with a resistance R'_{par} in parallel [10]. $1/R'_{par}$ is inversely proportional to the mean size of the active centres. $1/R'_{par}$ can be determined by representing $1/R_{par}$ vs. $(\omega)^{1/2}$. This gives a straight line with the intersection point for $\omega \rightarrow 0$ equal to $1/R'_{par}$. According to Vetter [11], hemispherical diffusion occurs if the electrode surface is energetically inhomogeneous and the diffusion wavelength $l_d = (2D/\omega)^{1/2}$ is larger than r_a, the mean size of the active centres, but smaller than r_i, the mean size of the inactive centres. So an inhibiting film at the electrodes with proper sizes of pores leads to a hemispherical diffusion of the reacting ion. Indeed, such an effect was verified for the $Tl^{+}/Tl(Hg)$ electrode reaction in the presence of adsorbed tribenzylamine (TBA) condensed film [12]. The evaluation of the impedance data resulted in the sizes of active and inactive centres of the electrode, which are, for 0.5×10^{-4} M TBA, for instance, equal to 0.7 μm and 3.4 μm, respectively.

II.5.11 Diffusion of Particles in Finite-Length Regions – the Finite Warburg Impedance

The infinite-length Warburg impedance obtained from solution of Fick's second law for one-dimensional diffusion considers the diffusion of a particle in semi-infinite space. Diffusion in the finite-length region is much more real. The thickness of the electrode or of a diffusion layer plays an important role then. In the case of a supported electrolyte, the thickness of the Nernst diffusion layer determines the finite length. In the case of an unsupported electrolyte, the finite length of solid electrolytes is decisive. Let the diffusion length be much less than the region available for diffusion then the case of infinite-length Warburg impedance is realized. When the diffusion length approaches the thickness of the diffusion region for decreasing frequencies, the shape of the complex plane impedance changes from 45° straight-line behaviour. Unhindered disappearance of diffusion at the far end due to contact with a conductor leads to a parallel combination of capacity C_l and diffusion resistance R_d at low frequencies, i.e. to a semicircle in the complex plane impedance plot (Fig. II.5.5).

From C_l, R_d and the diffusion coefficient D, the thickness l_d of the diffusion region can be obtained according to

$$C_l R_d = l_d^2 / 3D \tag{II.5.42}$$

This is denoted as diffusion for the transmissive boundary condition. The corresponding complete impedance is [5]

$$Z_{Wtr} = R_d \left[\tanh\left((js)^{1/2} \right) / (js)^{1/2} \right] \tag{II.5.43}$$

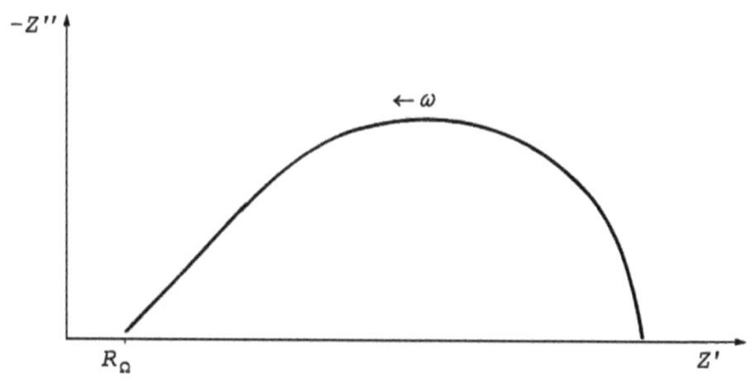

Fig. II.5.5 Nyquist impedance plot due to finite length diffusion with a transmissive boundary condition

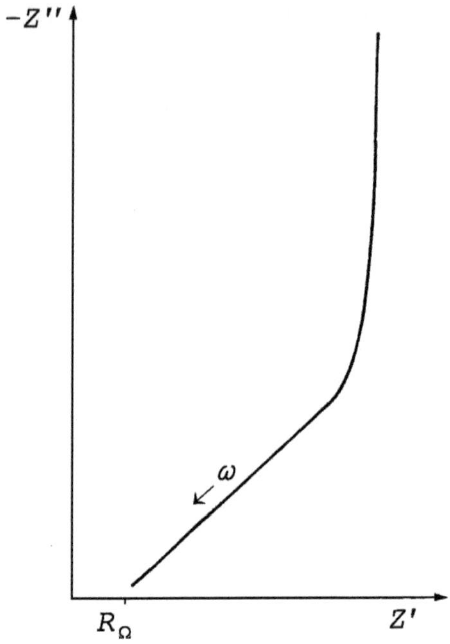

Fig. II.5.6 Nyquist impedance plot due to finite-length diffusion with a reflective boundary condition

with

$$s = l_{\mathrm{d}}^{2}\omega/D \qquad \text{(II.5.44)}$$

An experimental example can be found in [13]. The diffusion coefficients of electrons and potassium ions in copper(II) hexacyanoferrate(II) composite electrodes were determined using impedance spectroscopy. Composite electrodes are mixtures of graphite and copper hexacyanoferrate (Cu hcf) powder embedded in paraffin. The diameter of the Cu hcf particles amounted to about 30 μm. The diffusion region of electrons is limited by the size of these particles. For the first time, a diffusion of electrons in hexacyanoferrates could be detected. The diffusion coefficient obtained was $0.1\ \mathrm{cm^2\ s^{-1}}$.

Let us now assume an open-circuit condition at the far end of the transmission line, i.e. no direct current can flow in the actual system. This is defined as diffusion in the case of the reflective boundary condition. At the far end complete blocking of diffusion occurs. This results in a vertical line at low frequencies in the Nyquist plot corresponding to a capacity only (Fig. II.5.6). Here, at very low frequencies, resistance R_{d} and capacity C_{l} are in series.

The complete expression for the Warburg impedance corresponding to finite diffusion with reflective boundary condition is [5]

$$Z_{\mathrm{Wr}} = (s/(\omega C_{\mathrm{l}}))\left[\operatorname{ctnh}\left((\mathrm{j}s)^{1/2}\right)/(\mathrm{j}s)^{1/2}\right] \qquad \text{(II.5.45)}$$

II.5.12 Homogeneous or Heterogeneous Chemical Reaction as Rate-Determining Step

According to Vetter [11, § 72], an electrochemical process can also be controlled by a chemical reaction in the bulk (homogeneous reaction) or at the interface (heterogeneous reaction). The corresponding expressions for the real and imaginary components of the impedance are presented for both cases in the paper by Vetter [11].

II.5.13 Porous Electrodes

For porous electrodes, an additional frequency dispersion appears. First, it can be induced by a non-local effect when a dimension of a system (for example, pore length) is shorter than a characteristic length (for example, diffusion length), i.e. for diffusion in finite space. Second, the distribution characteristic may refer to various heterogeneities such as roughness, distribution of pores, surface disorder and anisotropic surface structures. De Levie used a transmission-line-equivalent circuit to simulate the frequency response in a pore where cylindrical pore shape, equal radius and length for all pores were assumed [14].

The impedance for pores is similar to that for diffusion in finite space [15]. Penetration into pores increases with decreasing frequency. The pore length determines the maximal possible penetration depth. The pore length plays the role of thickness in the case of finite space diffusion with reflective boundary condition. The diffusion is blocked at the end of the pore. At high frequencies, a straight line in the Nyquist plot follows with an angle of 45° to the real axis. For double-layer charging only, a vertical line in the Nyquist plot is predicted at low frequencies. For a charge transfer in the pores in addition to the double-layer charging, the low frequency part of the impedance corresponds to a semicircle.

Double-layer charging of the pores only (non-faradaic process) and inclusion of a pore-size distribution leads to complex plane impedance plots, as in Fig. II.5.7, i.e. at high frequencies, a straight line results in an angle of 45° to the real axis and, at lower frequencies, the slope suddenly increases but does not change to a vertical line [16].

II.5.14 Semiconductor Electrodes

The general scheme for a semiconductor electrode takes into account a two-step charge-transfer process [17]. One step corresponds to the transfer of electrons and ions through the Helmholtz layer. Let Z_H be the corresponding impedance that is in parallel to C_H, the capacity of the Helmholtz layer and let Z_1 be the impedance of this parallel combination. The other step exists due to the localization of charges in surface states or intermediates. Here the corresponding impedance, Z_{sc}, is in

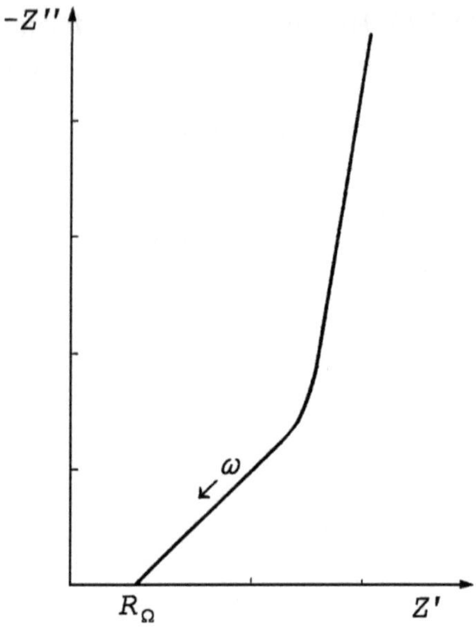

Fig. II.5.7 Nyquist impedance plot due to a porous electrode with log-normal distribution of the pore sizes and with double-layer charging only

parallel to C_{sc}, the capacity of the depletion layer. Let Z_2 be the impedance of this parallel combination. The total impedance of the semiconductor electrode, Z_{tot}, results as $Z_{tot} = Z_1 + Z_2$. The quantity $(C_{sc})^{-2}$ plotting vs. the potential V gives the Mott–Schottky plot. From this one can determine the flat-band potential of a semiconducting electrode [17].

II.5.15 Kinetics of Non-Faradaic Electrode Processes

Let us now consider non-faradaic processes, i.e. kinetics of adsorption/desorption at electrodes without a charge transfer [1]. Here, the charge density q_M is a function of both the electrode potential and the degree of coverage of the adsorbed substance. Sufficiently low frequencies result in the "thermodynamic" or "low-frequency capacity" C_{LF}, which is the sum of the "high-frequency capacity" $C_{HF} = (\mathrm{d}q_M/\mathrm{d}E)_{\Gamma,c}$ and a capacity, which is mainly determined by the potential dependence of the degree of coverage Θ. This capacity leads to the occurrence of maxima in the capacity-potential dependence, i.e. at the potential of the adsorption/desorption maxima is the potential dependence of the degree of coverage maximal. If the adsorption process is too slow to follow the potential changes (for higher frequencies), non-capacitive behaviour can be observed, i.e. an adsorption admittance occurs. There are three main mechanisms of adsorption kinetics: first a diffusion control, second a control by the adsorption exchange rate and third a mixed control by diffusion and adsorption exchange.

Mixed adsorption–diffusion control was considered by Lorenz and Möckel [18], and they derived the following equations for the frequency-normalized admittance:

$$1/\omega R_{\text{par}} = (C_{\text{LF}} - C_{\text{HF}}) \left((\omega\tau_{\text{D}}/2)^{1/2} + \omega\tau_{\text{A}} \right) \Big/ \left[\left((\omega\tau_{\text{D}}/2)^{1/2} + \omega\tau_{\text{A}} \right)^2 + \left((\omega\tau_{\text{D}}/2)^{1/2} + 1 \right)^2 \right] \quad \text{(II.5.46)}$$

$$C_{\text{par}} = C_{\text{HF}} + (C_{\text{LF}} - C_{\text{HF}}) \left((\omega\tau_{\text{D}}/2)^{1/2} + 1 \right) \Big/ \left[\left((\omega\tau_{\text{D}}/2)^{1/2} + \omega\tau_{\text{A}} \right)^2 + \left((\omega\tau_{\text{D}}/2)^{1/2} + 1 \right)^2 \right] \quad \text{(II.5.47)}$$

$$\tau_{\text{D}} = (\text{d}\Gamma/\text{d}c)_{\text{E}}^2 / D \quad \text{(II.5.48)}$$

$$\tau_{\text{A}} = (\text{d}\Gamma/\text{d}v)_{\text{E},c} \quad \text{(II.5.49)}$$

If the validity of the Frumkin adsorption isotherm is assumed, it follows that [19]

$$\tau_{\text{D}} = \Gamma_{\text{m}}^2 \Theta^2 (1 - \Theta)^2 / \left[c^2 D (1 - 2a\Theta (1 - \Theta))^2 \right] \quad \text{(II.5.50)}$$

$$\tau_{\text{A}} = \Gamma_{\text{m}} \Theta (1 - \Theta) / [v_0 (1 - 2a\Theta (1 - \Theta))] \quad \text{(II.5.51)}$$

where Γ_{m} is the possible maximal surface concentration, Θ the degree of coverage, c the bulk concentration of the surfactant, D the diffusion coefficient of the surfactant, a the Frumkin interaction coefficient, v the adsorption rate and v_0 the adsorption exchange rate. This follows the general trend that τ_{D} and τ_{A} increase with increasing values of a. For $\Theta = 0.5$, the degree of coverage at adsorption/desorption potential and $a \to 2$ (two-dimensional condensation of the adsorption layer), it results that $\tau_{\text{D}} \to \infty$ and $\tau_{\text{A}} \to \infty$. From this it follows that condensation of the adsorbed molecules leads to a complete loss of reversibility.

Equations (II.5.46) and (II.5.47) correspond to an equivalent circuit in which the capacity C_{HF} is in parallel with the adsorption impedance Z_{A}. This impedance represents a series of an adsorption resistance (determined by the rate of adsorption) and a Warburg-like complex impedance (corresponding to diffusion of the surfactants) and a pure capacity $C_{\text{LF}} - C_{\text{HF}}$. The electrolyte resistance is already eliminated here. For very high frequencies, it follows that $C_{\text{par}} = C_{\text{HF}}$ and $Y'/\omega = 0$ and, for very low frequencies, results in $C_{\text{par}} = C_{\text{LF}}$ and $Y'/\omega = 0$. Figure II.5.8 shows the complex plane plots of the frequency-normalized admittance for adsorption control by exchange rate only (curve 1), adsorption control by diffusion only (curve 2), and adsorption control by exchange rate and diffusion (curve 3). Investigations of adsorption kinetics of sodium decyl sulphate were performed at the mercury/electrolyte interface using the frequency dependence of the electrode admittance at the potential of the more negative ad/desorption peak [20, 21]. It was concluded that diffusion control is the rate-determining step below and above the critical micelle concentration.

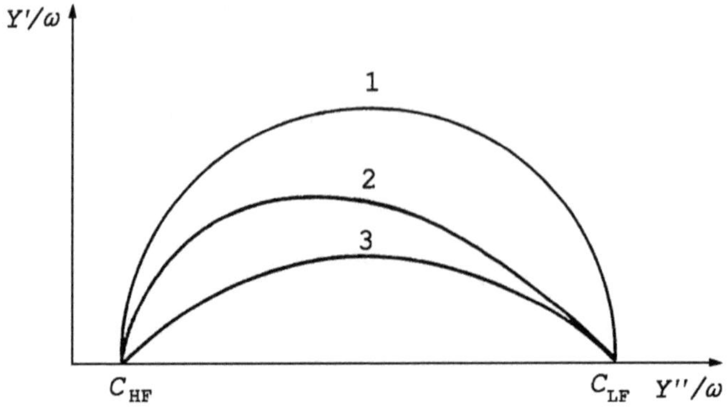

Fig. II.5.8 Complex plane plot of the frequency normalized admittance. *Curve* 1 control by the adsorption exchange rate only; *curve* 2 control by diffusion only; *curve* 3 mixed control by the adsorption exchange rate and diffusion

Lorenz [22] considered the case that a two-dimensional association occurs in the adsorption layer. Here the surfactant monomers are in equilibrium with surfactant clusters of constant size. He derived the corresponding frequency dependence of $\cot\,\delta_{\mathrm{a}} = \omega C_{\mathrm{par}} R_{\mathrm{par}}$. For lower frequencies, a characteristic decrease in $\cot\,\delta_{\mathrm{a}}$ should appear compared to $\cot\,\delta_{\mathrm{a}}$ given by Eqs. (II.5.46) and (II.5.47). Such behaviour was in fact detected for higher concentrations of caproic as well as caprylic acid.

II.5.16 References to Relevant Fields of Applications of EIS

- Corrosion, passive films [23–27]
- Polymer film electrodes [28]
- Characterization of electroactive polymer layers [29]
- Membrane systems [30]
- Solid electrolytes [31]
- Intercalation electrodes [32]
- Fuel cells research and development [33].

References

1. Sluyters-Rehbach M, Sluyters J (1970) Sine wave methods in the study of electrode processes. In: Bard AJ (ed) Electroanalytical chemistry, vol 4. Marcel Dekker, New York, p 1
2. Gabrielli C (1990) Use and applications of electrochemical impedance techniques. Technical Report, Schlumberger Technologies
3. Schöne G, Wiesbeck W, Stoll M, Lorenz W (1987) Ber Bunsenges Phys Chem 91: 496
4. Brug G, van der Eeden A, Sluyters-Rehbach M, Sluyters J (1984) J Electroanal Chem 176: 275
5. Macdonald R (1987) Impedance spectroscopy. Wiley Interscience, New York

6. Marquard E (1963) J Appl Math 11: 431
7. Moré J (1978) Numerical analysis. In: Watson G (ed) Lecture notes in mathematics, vol 630. Springer, Berlin Heidelberg New York, p 105
8. Draper N, Smith H (1967) Applied regression analysis. John Wiley, New York
9. Brett CMA, Brett AM (1993) Electrochemistry – principles, methods, and applications. Oxford University Press, Oxford, pp 224–252
10. Sluyters-Rehbach M, Sluyters J (1984) A.C. techniques. In: Bockris J, Yeager E (eds) Comprehensive treatise of electrochemistry, vol 9. Plenum Press, New York, p 177
11. Vetter K (1952) Z Phys Chem (Leipzig) 199: 300
12. Jehring H, Retter U, Horn E (1983) J Electroanal Chem 149: 153
13. Kahlert H, Retter U, Lohse H, Siegler K, Scholz F (1998) J Phys Chem B 102: 8757
14. De Levie R (1967) Electrochemical response of porous and rough electrodes. In: Delahay P (ed) Advances in electrochemistry and electrochemical engineering, vol 6. Wiley Interscience, New York, p 329
15. Raistrick I (1990) Electrochim Acta 35: 1579
16. Song H, Jung Y, Lee K, Dao L (1999) Electrochim Acta 44: 3513
17. Gomes W, Vanmaekelbergh D (1996) Electrochim Acta 41: 967
18. Lorenz W, Möckel F (1956) Z Electrochem 60: 507
19. Retter U, Jehring H (1973) J Electroanal Chem 46: 375
20. Vollhardt D, Modrow U, Retter U, Jehring H, Siegler K (1981) J Electroanal Chem 125: 149
21. Vollhardt D, Retter U, Szulzewsky K, Jehring H, Lohse H, Siegler K (1981) J Electroanal Chem 125: 157
22. Lorenz W (1958) Z Elektrochem 62: 192
23. Armstrong R, Edmondson K (1973) Electrochim Acta 18: 937
24. Deflorian F, Fedrizzi L, Locaspi A, Bonora P (1993) Electrochim Acta 38: 1945
25. Gabrielli C (1995) Electrochemical impedance spectroscopy: principles, instrumentation, and application. In: Rubinstein I (ed) Physical electrochemistry. Marcel Dekker, New York, p 243
26. Mansfeld F, Lorenz W (1991) Electrochemical impedance spectroscopy (EIS): application in corrosion science and technology. In: Varma R, Selman J (eds) Techniques for characterization of electrodes and electrochemical processes. Wiley Interscience, New York, p 581
27. Mansfeld F, Shih H, Greene H, Tsai C (1993) Analysis of EIS data for common corrosion processes. In: Scully J, Silverman D, Kendig M (eds) Electrochemical impedance: analysis and interpretation. ASTM, Philadelphia, p 37
28. Lang G, Inzelt G (1999) Electrochim Acta 44: 2037
29. Musiani M (1990) Electrochim Acta 35: 1665
30. Buck R (1990) Electrochim Acta 35: 1609
31. Wagner J (1991) Techniques for the study of solid ionic conductors. In: Varma R, Selman J (eds) Techniques for characterization of electrodes and electrochemical processes. Wiley Interscience, New York, p 3
32. Metrot A, Harrach A (1993) Electrochim Acta 38: 2005
33. Selman J, Lin Y (1993) Electrochim Acta 38: 2063

6. [illegible] (19[illegible]) J Appl [illegible]
7. [illegible] Nature [illegible]
[illegible], p 105
8. [illegible] Applied regression analysis. John Wiley, New York
9. [illegible]
[illegible]

Chapter II.6
UV/Vis/NIR Spectroelectrochemistry

Andreas Neudeck, Frank Marken, and Richard G. Compton

II.6.1 Introduction – Why Couple Techniques?

Voltammetric techniques used in electrochemistry monitor the flow of current as a function of potential, time, and mass transport. A huge variety of different experiments are possible, giving information about reaction energies, reaction intermediates, and the kinetics of a process [1–4]. However, additional data are often required and are accessible, in particular, via in situ spectroelectrochemical approaches. By coupling a spectroscopic technique such as UV/Vis/NIR spectroscopy [5, 6] to an electrochemical experiment, a wealth of complementary information as a function of the potential, time, and mass transport is available. In a recently published book dedicated to spectroelectrochemical techniques [7] the diversity of methods and new chemical information obtained is apparent. Both spectroscopic information about short-lived unstable intermediates and spectroscopic information disentangling the composition of complex mixtures of reactants can be obtained. Figure II.6.1 shows a schematic diagram for the case of a computer-controlled potentiostat system connected to a conventional electrochemical cell (working electrode WE, reference electrode RE, counter electrode CE) and simultaneously controlling the emitter and detector of a spectrometer. This kind of experimental arrangement allows the electrochemical and the spectroscopic data to be recorded simultaneously and, therefore, in contrast to the analysis of two independent data sets, direct correlation of data as a function of time and potential is possible.

Using electrochemical techniques it is possible to generate a well-defined amount of intermediates controlled by the charge forced through the working electrode. In this way, interesting intermediates (e.g. radicals or radical ions) can be generated electrochemically in a much more controlled and localised manner compared to, for example, what is possible with conventional photochemical methodology. In order to follow reactions of these intermediates, a wide range of spectroscopic methods are available, which have been successfully coupled to electrochemical systems. A

A. Neudeck (✉)
Textilforschungsinstitut Thüringen-Vogtland e.V., Stellv. Bereichsleiter Forschung und Entwicklung –Vorlaufforschung, Zeulenrodaer Str. 42, 07973 Greiz, Germany
e-mail: andreas.neudeck@t-online.de

F. Scholz (ed.), *Electroanalytical Methods*, 2nd ed.,
DOI 10.1007/978-3-642-02915-8_9,

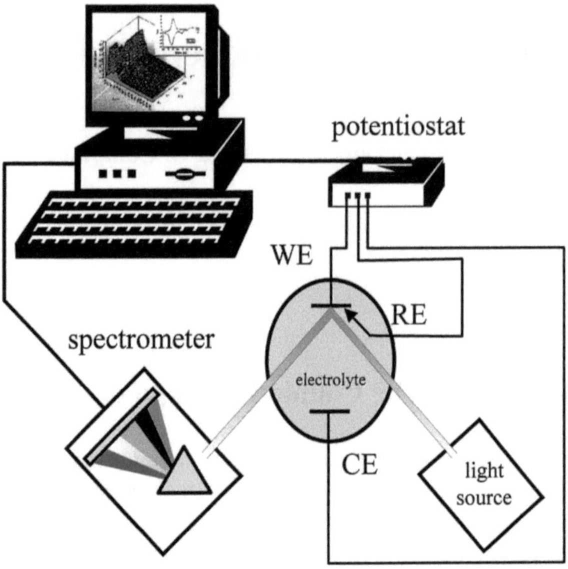

Fig. II.6.1 Schematic diagram of a spectroelectrochemical system with a conventional three-electrode electrochemical cell (*WE* working electrode, *RE* reference electrode, *CE* counter electrode) controlled by a computer-based potentiostat

list of some more commonly used coupled or 'hyphenated' spectroelectrochemical methodologies is given in Table II.6.1.

The techniques listed in Table II.6.1 have been chosen to demonstrate the capability of spectroelectrochemical methods to probe both processes in the solution adjacent to the electrode and heterogeneous processes occurring directly at the surface of the electrode. Given the wide range of existing in situ spectroelectrochemical techniques, Table II.6.1 can only highlight some of the more important methods.

Table II.6.1 Examples for spectroelectrochemical techniques probing both the solution phase in the vicinity of the working electrode (homogeneous) and the electrode surface (heterogeneous), which have been used in situ during the course of voltammetric experiments

Homogeneous Spectroelectrochemical Probes
In Situ Electron Spin Resonance (ESR) [8–11]
Molecules with unpaired electrons may be detected by microwave spectroscopy in the presence of a magnetic field. This technique is of considerable importance for the detection of radical reaction intermediates produced during electrolysis.
In Situ UV/Vis/NIR Optical Spectroelectrochemistry [12, 13]
The light absorption of reaction intermediates generated at the electrode surface can be monitored in transmission or in reflection mode, and this allows sensitive detection and/or quantitative kinetic studies. The sensitivity of this technique can be substantially improved by introducing cavity-ring-down cells [14] or multiple-reflection path cells.
In Situ Raman and Resonance Raman Spectroscopy or Microscopy [15, 16]
Raman allows the intensity and oscillator strength of vibrational bands to be measured with an optical laser system. The technique is particularly powerful in conjunction with confocal microscope.

Table II.6.1 (continued)

In Situ Luminescence Spectroscopy [17, 18]
Compared to the absorption of light in the visible wavelength range, luminescence is a much more sensitive technique due to the absence of background radiation. Even single photons may be counted and analysed in chemiluminescence processes [19].
In Situ Nuclear Magnetic Resonance Spectroscopy (NMR) [20]
NMR is very widely used and a powerful tool for structure determination. Although commonly used for diamagnetic compounds, it may also be applied to paramagnetic materials [21].
In Situ Differential Electrochemical Mass spectrometry (DEMS) [22]
Solution-phase reaction intermediates can be directly injected (within milliseconds) into appropriate mass spectrometers to provide localised solution composition information.
Heterogeneous Spectroelectrochemical Probes
In Situ UV/Vis/NIR Optical Spectroelectrochemistry [23–26]
UV/Vis/NIR spectroscopy may be employed directly for the characterisation of materials formed or deposited on electrode surfaces.
In Situ NMR Spectroscopy [27]
It has been demonstrated that NMR spectroscopy may be employed to study molecules adsorbed onto electrode surfaces during the course of electrochemical measurements. With higher magnetic field strengths and the availability of nano-particle samples this methodology becomes increasingly powerful.
In Situ Surface-Enhanced Raman Scattering (SERS) [28–31]
Raman scattering gives information about molecular or solid-state vibrations. In the presence of some electrode materials (e.g. silver or copper) a considerably enhanced sensitivity for molecules adsorbed at the electrode surface in the form of a monolayer can be achieved. This surface-enhanced Raman scattering effect can be transferred to other electrode materials coated in the form of a very thin layer onto a silver or copper substrate [32]. The recently developed TERS technique (Tip-Enhanced Raman Spectroscopy [33]) provides additional and highly resolved mapping information.
In Situ X-ray Diffraction (XRD) and Synchrotron Techniques [34–36]
XRD techniques are used mainly to study the crystal structure of crystalline solids at the electrode–solution interface. Structural changes, solid-state reactions, precipitation processes, dissolution processes, and intercalation processes can be followed as a function of applied potential and time. For non-crystalline samples, X-ray absorption and EXAFS allow the structure to be studied in more detail.
In Situ Mössbauer Spectroscopy [37]
Mössbauer transitions occur in atomic nuclei and involve absorption or emission of high-energy X-ray photons. Detailed information about the local chemical environment, oxidation state, and coordination symmetry of atoms can be obtained. However, the technique has been used for Fe and Sn only.
In Situ Spectroellipsometry [38] **and Optical Waveguide Techniques** [39]
Ellipsometry allows the optical properties of thin layers at electrode surfaces to be studied as a function of wavelength, time, and potential. Theoretical models allow conclusions about the film thickness and structure. Optical waveguide techniques improve the interaction of the electromagnetic wave with the sample for example in surface plamon resonance processes.
In Situ Non-linear Optical Methods and Second Harmonic Generation [40]
Steady-state electric field inhomogeneity at the electrode–solution interface causes submono-layer amounts of molecules to undergo ‘unusual’ but detectable two-photon transitions, which are ‘forbidden’ by selection rules in the bulk phase. Therefore, this technique allows molecules at a surface to be detected and studied independently and selectively.
In Situ Spectromicroscopy and Scanning Probe Methods [41]
A diverse range of scanning probe methods for use within electrochemical systems is now available.

Electrochemically generated intermediates are usually easier to detect spectroscopically than the substrate material itself. The electrochemical reduction or oxidation of many organic materials yields products, which are often coloured with high absorption coefficients (UV/Vis), or which give a characteristic radical 'signature' in electron spin resonance (ESR) spectroscopy [42]. Based on this highly sensitive detection, spectroelectrochemical detectors have been proposed and employed for analytical applications [43].

II.6.2 Flowing Versus Stagnant Systems – Achieving Spatial, Temporal, and Mechanistic Resolution

II.6.2.1 Steady-State or Transient Techniques

Voltammetric techniques may be broadly divided into *steady-state* techniques, such as channel flow cell [44–46], rotating disc [47, 48], or microelectrode [49] voltammetry at sufficiently low potential scan rate to give a current response *independent* of time, and *transient* techniques, such as cyclic voltammetry or chronoamperometry, giving a current response which is *dependent* on time.

The rotating-disc electrode voltammetry technique is a very commonly employed steady-state method. However, due to the moving parts in the experimental set-up, coupling to in situ spectroelectrochemical techniques has not been very widely used [50]. An elegant and extremely versatile approach to steady-state spectroelectrochemical measurements is based on the channel flow cell technique [51–53]. In this technique (see Fig. II.6.2), a continuous flow of electrolyte solution from a reservoir is allowed to pass through a rectangular duct with the working electrode embedded. The reference and counter electrodes are located outside of the flow cell in the upstream and downstream parts of the system. After setting the potential to a fixed value, downstream detection of the change in UV/Vis absorption, changes in ESR signal intensities, or changes in fluorescence signals may be monitored as a function of the flow rate. Data generated by this experimental technique have been analysed quantitatively with the help of computer models [54], which predict the spectroelectrochemical response based on a given reaction mechanism.

Transient spectroelectrochemical studies are possible with modern and fast spectroscopic probes, e.g. based on diode array spectrometers [55]. Compared to steady-state techniques, the design of transient spectroelectrochemical experiments is more challenging and the data analysis has to take into account the development of diffusion- and convection-driven concentration profiles [56]. However, these experiments open up the possibility to study time-dependent processes such as the development of a diffusion profile. For example, it is known, but usually ignored, that changing the redox state of a molecule changes its diffusivity. It has recently been shown [57] that, for the oxidation of *N,N,N′,N′*-tetra-methylphenylenediamine (TMPD) in water, ethanol, and acetonitrile (Eq. II.6.1), the diffusion coefficients of the reduced form, TMPD, and the diffusion coefficient of the oxidised form,

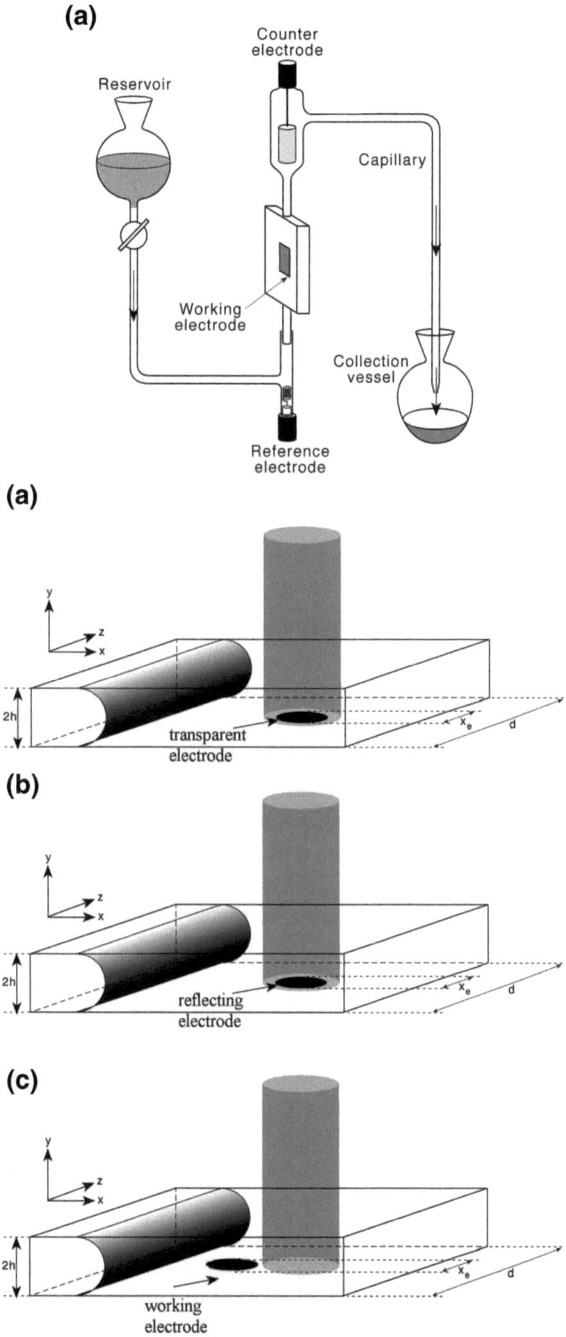

Fig. II.6.2 Channel-flow system (**a**) with gravity-controlled solution flow from a reservoir, passing a reference electrode, flowing through a channel cell and a counter electrode, then passing through a capillary controlling the flow rate, and finally being collected. Rectangular channel cell corresponding to (**a**) with (**b**) transmission detection, (**c**) reflection detection, and (**d**) downstream transmission detection

$TMPD^+$, can be determined separately and simultaneously in a single transient spectroelectrochemical experiment.

$$\mathrm{TMPD} \rightleftarrows \mathrm{TMPD}^+ + \mathrm{e}^- \tag{II.6.1}$$

The experiment was based on a channel flow cell system (see Fig. II.6.2) with UV/Vis detection downstream of the working electrode (Fig. II.6.2d). Switching the electrode potential from the potential region with no faradaic current into a region with diffusion-limited faradaic current allowed the transient change in the UV/Vis absorption to be monitored. The data analysis for this transient UV/Vis response was based on a computer simulation model, which allowed D(TMPD) and $D(TMPD^+)$ to be varied independently. Interestingly, the difference in D(TMPD) and $D(TMPD^+)$ was relatively high in water and ethanol (15% slower diffusion of the radical cation) and considerably lower (5%) in the less polar solvent acetonitrile. This example demonstrates the ability of transient spectroelectrochemical experiments, in conjunction with computer simulation-based data analysis, to unravel even complex processes.

II.6.2.2 Cell Geometry and Experimental Design Considerations

The requirements for a successful spectroelectrochemical experiment may be summarised as (i) a cell with sufficiently good electrical properties (low iR_U drop, homogeneous potential distribution over the entire working electrode surface area, sufficiently high counter electrode area), (ii) a potentiostat system with sufficient output current and output voltage to sustain potential steps and current transient without signal distortion, and (iii) a cell geometry which separates the working electrode from the counter electrode to avoid product generated at the counter electrode interfering with the electrochemical process at the working electrode and with the spectroscopic detection.

For a flow system such as that shown in Fig. II.6.2, the tubing connection between counter and reference electrode has to be kept reasonably short and with a not too narrow bore to avoid the uncompensated solution resistance R_U increasing. Further, even with a powerful potentiostat system, the lack of 'communication' between counter and reference electrode can cause the entire system to oscillate with high frequency (a check with a high-frequency oscilloscope is recommended) depending on the type of electrolyte solution used. These oscillations can be suppressed by improving the high-frequency 'communication' between counter and reference electrode, e.g. by introducing an additional counter electrode upstream of the reference or simply by connecting a small capacitor across reference and counter electrode. A good test for the quality of voltammetric data obtained with a channel-flow system can be based on cyclic voltammetry under stationary (no flow) conditions. Highly reversible electrochemical systems such as the oxidation of ferrocene in acetonitrile (1 mM ferrocene in 0.1 M NBu_4PF_6) can be employed and voltammograms are expected to give well-defined characteristics at sufficiently high scan rate [58]. Any deviation from ideal behaviour is an indication of a cell design or potentiostat problem.

II.6.2.3 Time-Scale Considerations

Long-lived reactive intermediates are readily detected by voltammetric and spectroelectrochemical techniques after they have been generated electrochemically, e.g. by bulk electrolysis. In the simple UV/Vis/NIR spectroelectrochemical cell shown in Fig. II.6.3, the concentration of the electrolysis product will build up in the vicinity of the working electrode over several tens of minutes to give locally a relatively high and detectable concentration of the product. However, a short-lived reaction product or intermediate will be present in very low concentration and go undetected under these conditions.

Increasing the scan rate in cyclic voltammetry allows faster reactions to be studied. At a planar electrode, the diffusion layer grows into the solution phase during the progress of the potential cycle. The thickness of the diffusion layer at the time of the current peak, δ_{peak}, for the case of a reversible cyclic voltammetric response is given approximately by Eq. (II.6.2).

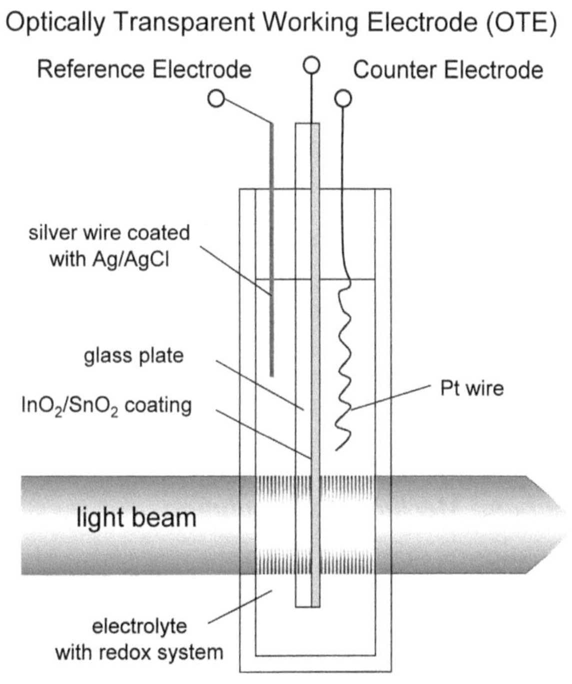

Fig. II.6.3 Simple in situ UV/Vis spectroelectrochemical cell based on a quartz cuvette with an optically transparent working electrode, counter electrode, and reference electrode immersed in solution

$$\delta_{\text{peak}} = \frac{1}{0.446}\sqrt{\frac{RTD}{nFv}} \tag{II.6.2}$$

In this equation 0.446^{-1} is a constant for the case of a reversible diffusion controlled process [59], R is the gas constant 8.31 J K^{-1} mol^{-1}, T is the temperature in Kelvin, D is the diffusion coefficient in m^2 s^{-1}, n denotes the number of electrons transferred per molecule diffusing to the electrode, F is the Faraday constant 96487 C mol^{-1}, and v is the scan rate in Vs^{-1}. The diffusion layer thickness may be compared to the reaction layer thickness, δ_{reaction}, which is given approximately by Eq. (II.6.3).

$$\delta_{\text{reaction}} = \sqrt{\frac{D}{k}} \tag{II.6.3}$$

In this expression D denotes the diffusion coefficient in m^2 s^{-1} and k is the first-order rate constant (s^{-1}) for a chemical reaction step. Similar expressions may be written for other types of chemical processes [60]. The scan rate required for the intermediate to be detected voltammetrically may be estimated based on matching the reaction layer and the diffusion layer (Eq. II.6.4).

$$v = \frac{1}{0.446^2}\frac{kRT}{nF} \tag{II.6.4}$$

For example, an intermediate generated at the electrode surface with $D = 1 \times 10^{-9}$ m^2 s^{-1}, which undergoes a first-order follow-up chemical reaction with $k =$ 100 s^{-1}, requires a scan rate of ca. 13 V s^{-1} to be detectable in cyclic voltammetry. The reaction layer thickness in this case can be estimated as approximately 3 μm. During a fast-scan voltammogram, only a fraction of the charge is flowing compared to the experiment at slower scan rate. Therefore, the sensitivity of the spectroscopic detection is reduced considerably. An upper limit of the absorption of the intermediate generated at the electrode surface can be estimated based on the Beer–Lambert law [61] and by assuming 1/2 δ_{reaction} to be approximately the 'effective' optical path length. By repeating the experiment and by accumulation of the spectroscopic data, a considerable improvement in the signal-to-noise ratio is possible.

II.6.2.4 Spatial and Mechanistic Resolution

Apart from the detection and characterisation of reaction intermediates it is often desirable to gain quantitative mechanistic information in transient or steady-state electrochemical experiments. From voltammetric experiments, it is known that in some cases one type of voltammetric current response may be interpreted by more than one mechanism [62]. Depending on the type of experiment, coupling a spectroscopic method to a voltammetric experiment may provide complementary data, which then allows a definite decision on which reaction pathway is followed.

Therefore, spectroelectrochemical experiments may be used to obtain more than just data to identify the nature of the reaction intermediate. The spatial distribution and concentration gradients of reactants and products give direct insight into the reaction mechanism of a particular electrode process [63]. A spectroelectrochemical technique with high spatial resolution can yield very valuable data. However, the experimental approach shown, for example, in Fig. II.6.2 can be seen to give an absorbance measurement equivalent to the integral value of all the material produced in the diffusion layer without spatial resolution.

A very elegant approach overcoming this problem has been proposed based on a channel flow cell geometry with downstream detection (Fig. II.6.2d). The potential of the electrode is stepped during steady-state flow of the solution across the electrode. A downstream UV/Vis detector system is then employed to measure the time dependence of the concentration profile formation at the electrode surface. A computer program is employed to relate the time-dependent absorbance signal to the concentration profile of reactant and product at the electrode surface. Alternatively, direct measurement of the concentration profiles at the electrode surface has also been reported based on confocal Raman spectroelectrochemistry [16].

II.6.3 UV/Vis/NIR Spectroelectrochemical Techniques

II.6.3.1 Spectroelectrochemistry in Transmission Mode

A very powerful and convenient approach to couple a spectroscopic method with electrochemical experiments is based on the spectroscopy in the ultraviolet (UV), visible (Vis), and near infrared (NIR) range of the electromagnetic spectrum. This can be achieved with UV/Vis/NIR spectroscopy in transmission mode [64] with an optically transparent electrode employed as the working electrode. There are two distinct types of optically transparent electrodes useful in this approach, a partially transparent metal grid or mesh electrode or a glass plate coated with a conductive but optically transparent layer of indium tin oxide (ITO) as optically transparent electrode (OTE) in a glass or quartz cuvette (see Fig. II.6.4). Also, coatings of very thin metal layers (below 100 nm) on glass substrates can be employed as working OTE in spectroelectrochemical studies [65].

A very simple spectroelectrochemical experiment based on a conventional cuvette for UV/Vis spectroscopy with electrodes immersed in the solution phase is shown in Fig. II.6.4. The optically transparent working electrode is located directly in the beam path and additionally a large surface area counter electrode (Pt wire) and a reference electrode (e.g. a coated silver wire) are located in the solution phase above the beam path. After filling the cuvette with electrolyte solution, the reference spectrum can be recorded taking into account the transparency of the OTE and optical properties of the solution. This reference spectrum is later used to subtract the absorption of the cuvette, the electrolyte, and the OTE from experimental data obtained after applying a potential. Next, the redox reagent is added and the first spectrum of the starting material is recorded. The potential is applied

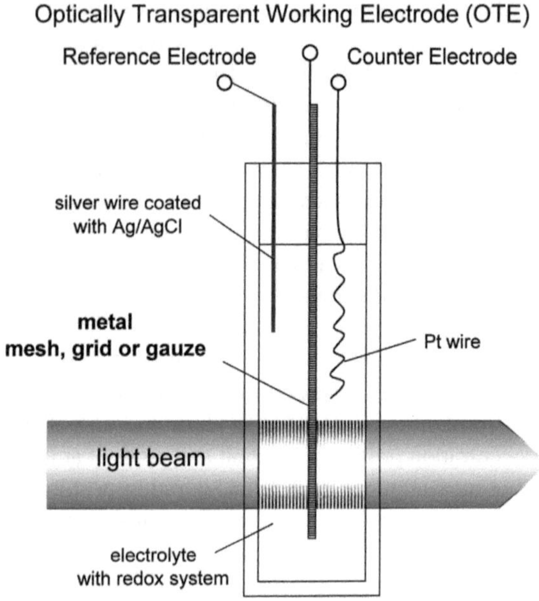

Fig. II.6.4 Simple in situ UV/Vis/NIR spectroelectrochemical cell with an optically transparent electrode prepared from a partly transparent metal mesh, grid, or gauze electrode

to the optically transparent working electrode using a potentiostat system. Initially, a potential in a zero current region is applied and then a step of the potential is applied (chronoamperometry) into a potential region in which the electrochemical conversion proceeds. By recording spectra at regular time intervals, the conversion of the starting material into products during the course of reduction or oxidation at the working electrode can be followed qualitatively (Fig. II.6.4). At the end-point of the electrolysis process in the limited volume of the cuvette, the product can be identified based on the change in absorption. In the case of an unstable intermediate being generated at the OTE, conclusions concerning the presence of intermediates or follow-up reactions may be possible. In the experimental set-up shown in Fig. II.6.4, reaction intermediates with a half-life in the order of minutes may be detected. In a cuvette with a shorter (1 mm) path length the conversion can be driven to completion.

In order to improve the detection of short-lived intermediates, the potential step or chronoamperometric experiment can be replaced by a cyclic voltammetric experiment, which involves applying a triangular potential ramp. With a fast UV/Vis spectrometer, e.g. a diode array system, additional UV/Vis/NIR spectroscopic information as a function of the potential can be recorded simultaneously to the voltammetric data. However, recording cyclic voltammograms with the simple cell shown in Fig. II.6.4 is complicated by the presence of ohmic drop in the solution phase, which is amplified by poor cell design. In this kind of cell, the peak-to-peak separation in cyclic voltammograms of a reversible redox couple may increase by several hundreds of millivolts. Voltammetric data (and simultaneously recorded spectroscopic data) are therefore very difficult to interpret quantitatively.

The size of the working electrode is large, and with the size the high current results in a high ohmic drop distortion. Additionally, resistance of the electrode itself in the case of a glass plate with a very thin conducting layer may further increase the ohmic iR_U drop problem. Considerable improvements can be achieved by minimising the active electrode area of the working electrode. This can be achieved by insulation of the OTE to leave only a small well-defined working electrode area exposed to the solution and by an optimal placement of reference close to the working electrode, as shown in Fig. II.6.5.

As a further requirement, many electrochemical investigations have to be carried out under an inert atmosphere of argon or nitrogen, and sometimes in pre-dried organic solvents. Therefore, a lid with gas inlet and gas outlet is used combined with a septum seal, which allows solvent to be introduced into the cell without it coming into contact with the ambient atmosphere. The insulation of the OTE requires a material resistant to commonly used organic solvents. Commercially available lamination foil, used for protecting documents, has been employed for this purpose and shown to be suitable even in organic solution [66, 67] (see Fig. II.6.5). The laminating foil is applied to the working electrode simply by thermal encapsulation leaving the active electrode area uncovered.

As a further practical tip, for an OTE based on a thin conductive optically transparent layer, the resistance of the working electrode can be reduced if an additional thicker metal layer is coated on the optically transparent film in the region not exposed to the light beam under the insulation layer. If the insulation is prepared through lamination, a thin metal foil can be simply inserted instead of the deposited metal layer. This new cell design (Fig. II.6.5) enables cyclic voltammograms as well as current time curves in the case of a potential step experiment (chronoamperograms) simultaneously with a series of time-resolved spectra to be recorded. This experimental approach allows absorbance–potential or absorbance–time curves to be recorded to analyse the reaction kinetics of the generation of intermediates and the final product up to a timescale of approximately 0.1 s lifetime.

Problems with cells with relatively long path length (1–10 mm) may arise when the starting material itself is strongly absorbing. As soon as a coloured starting material with broad absorption bands is investigated, the spectra of the intermediates and of the final product are superimposed onto the more intense spectrum of the starting material. This is caused by the fact that the overall absorbance is proportional to the thickness of the investigated layer – the optical path lengths through the material. For the intermediate or product the path length corresponds to the diffusion layer, which is in the range 0.01–1 mm. The size of the diffusion layer and reaction layer depends on the timescale of the experiment. For a cuvette with a size of 5 mm the layers of the products may be a factor 10–10000 times thinner than the layer for the starting material. By reducing the size of the cuvette from a thin layer cell down to a capillary slit cell this problem can be solved. Effectively, the optical path length of the cell and the diffusion layer for the electrochemical process can be matched.

Figure II.6.6 shows two possible designs of capillary slit cells. The second type of cell has the added advantage of being suitable also for in situ ESR spectroelectrochemical experiments. Capillary slit cells permit fast electrochemical conversion

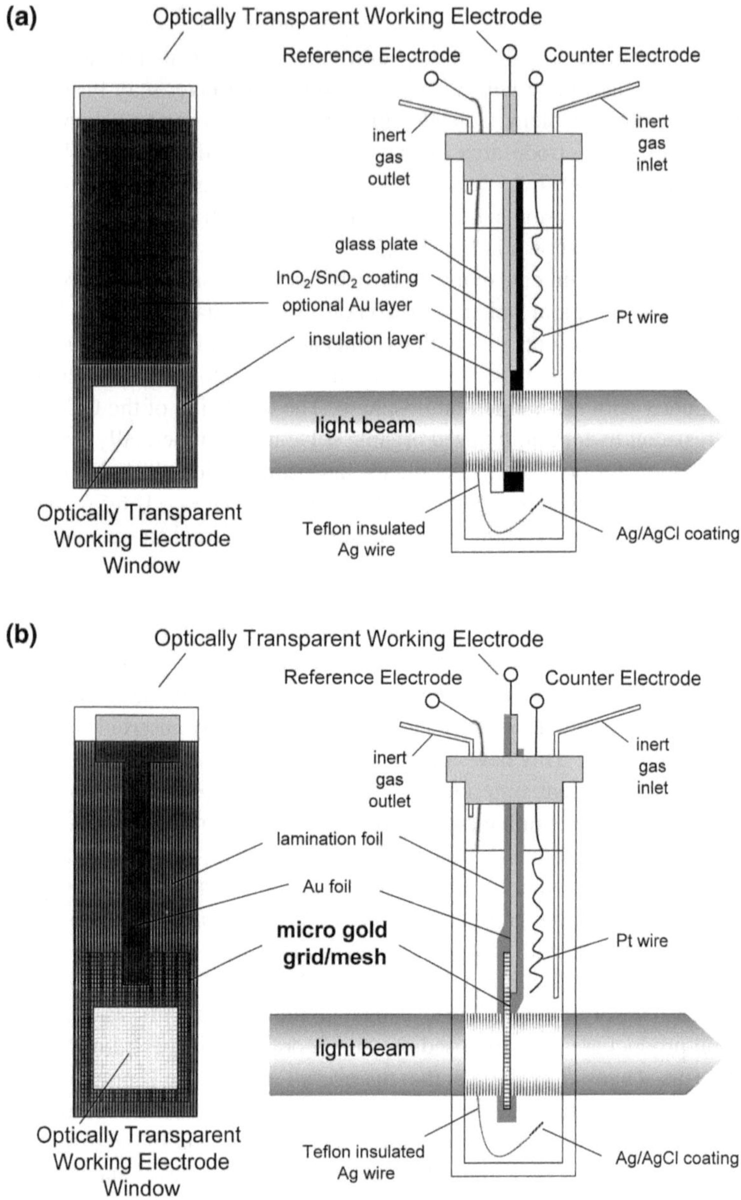

Fig. II.6.5 An improved in situ UV/Vis/NIR spectroelectrochemical cell design with an optically transparent electrode prepared from *top* a glass substrate coated by a thin conductive optically transparent ITO layer and *bottom* a metal mesh, grid, or gauze

of the starting material into unstable radical intermediates, which may be followed in time during follow-up chemical reaction steps. The thickness of the capillary slit, together with the diffusion coefficient D of the reactant, determine the approximate conversion time $\tau_{\text{diffusion}}$ (Eq. II.6.5).

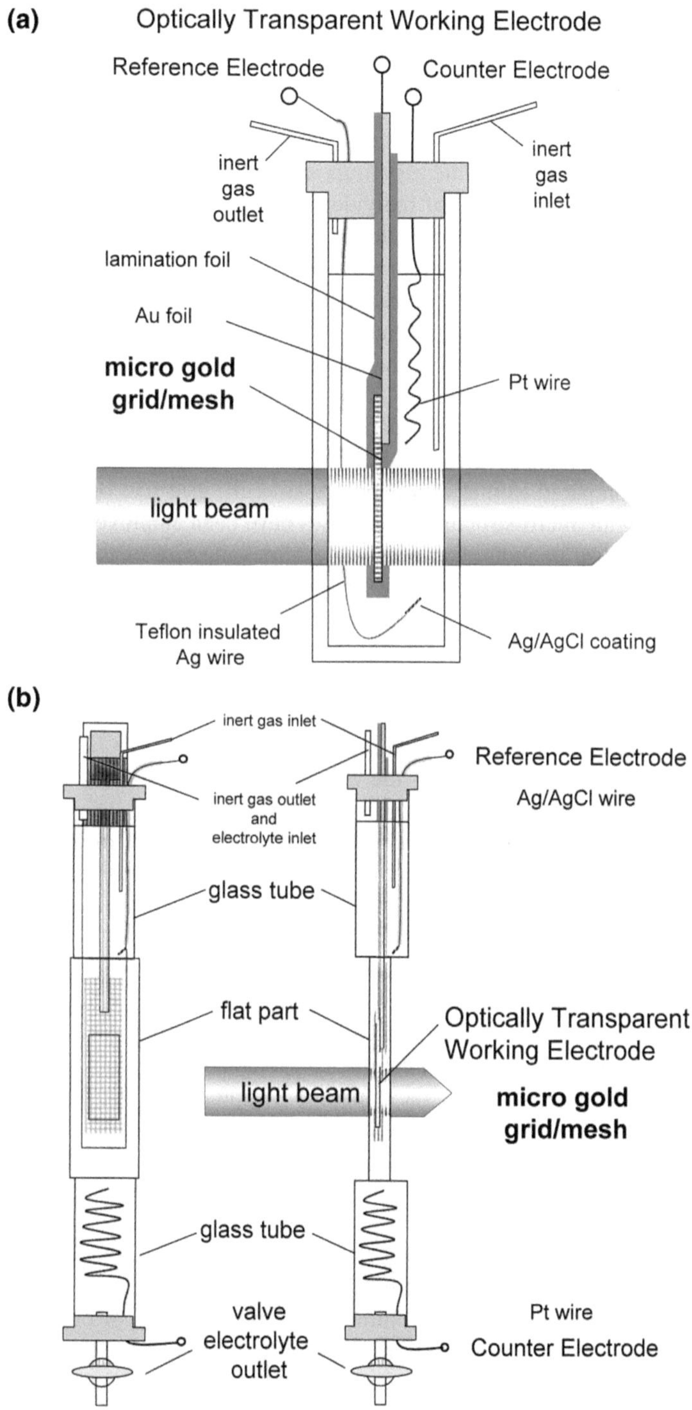

Fig. II.6.6 Capillary slit in situ UV/Vis/NIR spectroelectrochemical cells with an optically transparent electrode prepared from a metal mesh, grid, or gauze in a *top* cuvette cell and in a *bottom* flat cell with outlet allowing the solution to flow through the slit

$$\tau_{\text{diffusion}} = \frac{\delta^2}{D} \tag{II.6.5}$$

Unfortunately, even when capillary slit cells are used in the experiment, the spectra of the different species are still superimposed. Therefore, it is necessary to separate or to 'deconvolute' the superimposed spectra in order to obtain information about the reaction kinetics of individual species. In the literature, techniques have been proposed for the deconvolution of the superimposed spectra [68, 69]. Data processing and deconvolution may be achieved with spreadsheet software on a suitable computer system. As soon as the time dependence of the concentration of each component is known, the absorbance–time curves and the charge–time curves calculated from the current passing the electrode can be used to determine rate or equilibrium constants for the chemical system under study.

Instead of a capillary slit, a thin partly transparent three-dimensional structure may also be employed in spectroelectrochemical studies. To achieve the same level of sensitivity compared to that in a capillary slit cell, grid-like optically transparent electrodes with micro-structured electrodes have been employed [70–72]. In the grid system the size of individual pores determines the size of the diffusion-layer thickness and therefore the timescale of the experiment. A more sophisticated approach has been introduced by replacing the simple grid electrode by a LIGA structure (prepared by the LIGA technique: lithographic galvanic up-forming based on a synchrotron radiation patterned template). A thick honeycombed structure (see Fig. II.6.7) has been used with hexagonal holes of micron dimension. The size of the holes restricts the diffusion-layer thickness into a fast timescale regime, although the optical-path length of the cell is kept relatively long to give strong absorption responses in the spectroscopic measurement.

Figure II.6.8 shows the data obtained from the UV/Vis spectroelectrochemical study of a oligo-thiophene (see structure in Fig. II.6.8) in a LIGA cell [73]. The oligo-thiophene was dissolved in a solution of 0.1 M NBu_4PF_6 in acetonitrile and voltammograms were recorded at two different temperatures, at (A) +20°C and (B) −40°C. What appears to be an electrochemically reversible oxidation process at +1.16 V can be seen to become much more complicated at lower temperature. The analysis of spectroelectrochemical data clearly indicates the presence of two distinct products of which one, the radical cation $A^{\bullet+}$, is detected at ambient temperature (Fig. II.6.8 A). A second product, the dimer A_2^{2+}, is present at 20°C in low concentration. However, upon reducing the temperature in the spectroelectrochemical cell to −40°C, the dimer becomes the dominant product (Fig. II.6.8B) causing a change in both the voltammetric and the UV/Vis spectroelectrochemical response.

The spectroelectrochemical cell based on the LIGA-OTE can be employed in a flow-through-type system. Then, this cell allows the rapid renewal of the solution inside the cell after a spectroelectrochemical experiment. Experiments can be repeated rapidly and with a small volume of sample. A series of experiments can be conducted by varying the conditions without the need to open the cell. The use of optical waveguides to connect the cell with a light source and the spectrometer offers further scope for improvements in the experimental methodology.

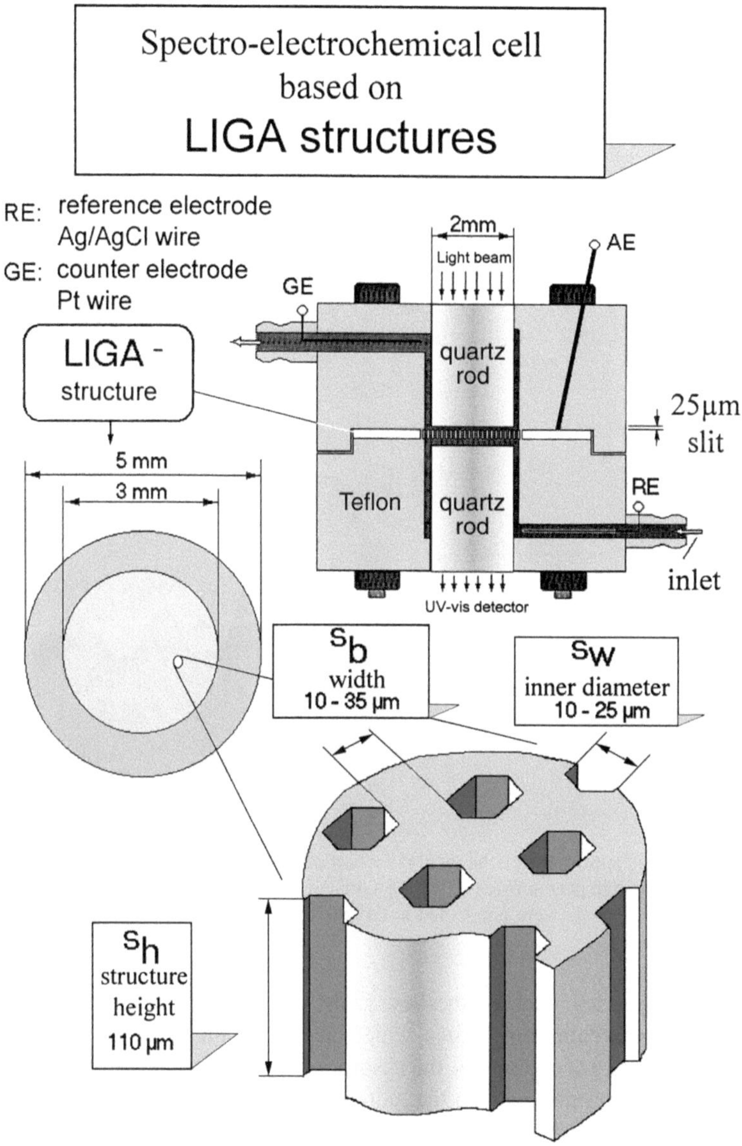

Fig. II.6.7 In situ UV/Vis/NIR spectroelectrochemical cell based on a microstructured optically partly transparent electrode with LIGA design

Conducting ITO-coated glass electrodes generate relatively simple 'planar' concentration profiles. The detected absorption of products generated at this type of electrode is easily related to the overall charge passed through the cell, because the absorption is directly proportional to the concentration integrated over the entire diffusion layer. Compared to optically transparent conductive glass electrodes, partly

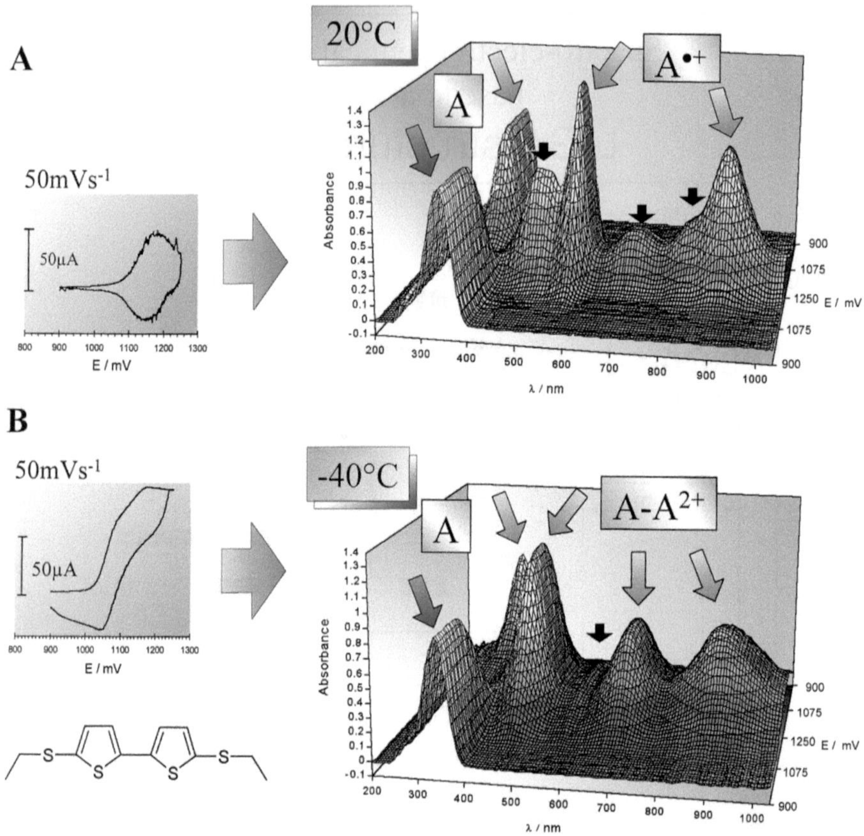

Fig. II.6.8 UV/Vis spectroelectrochemical detection of the oxidation products for the oxidation of an oligo-thiophene in acetonitrile (0.1 M NBu_4PF_6). The voltammetric and in situ spectroelectrochemical measurements were conducted at (**A**) 20°C and (**B**) −40°C in a LIGA cell system

transparent electrodes, such as meshes, grids, gauzes, and micro-structures, give more complex concentration profiles, which change geometry as a function of time. For these latter types of electrodes, the absorption signal detected during electrolysis is *not* necessarily proportional to the amount of the product generated. For quantitative investigations, details of the geometry of the diffusion layer and the time dependence have to be taken into account. For example, locally, a concentration gradient is generated at each wire of a grid electrode orthogonal to the light beam.

The development of the concentration profile for a grid electrode may then be considered to occur in two steps. First, the concentration profile grows into the space between individual wire electrodes perpendicular or orthogonal to the incoming light beam. In this situation, the absorbance detected experimentally is not proportional to the concentration of product generated due to the complex shape of the concentration gradient. Next, the individual concentration gradients merge to a planar diffusion front, which then is proportional to the amount of electrogenerated product similar to the case observed for ITO-coated electrodes. In cyclic voltammetric experiments with in situ spectroelectrochemical detection of the UV/Vis/NIR absorbance one has to therefore carefully consider the grid size and the scan rate to stay in the planar diffusion regime. This then allows the absorbance to be compared to the charge passed determined by integrating the current response.

Based on the development of a diffusion front with dimension $d_{\text{diffusion}}$ (Eq. II.6.6), at very fine meshes, grids, or gauzes with a wire density of 1000 wires per inch or more, the effect of the changing diffusion layer geometry becomes negligible after a time t of ca. 50 ms.

$$\delta_{\text{diffusion}} \approx \sqrt{Dt} \tag{II.6.6}$$

However, diode array spectrometer systems permit spectra to be recorded in about 0.5-ms steps. Therefore, it is necessary to analyse the data carefully with respect to consistency by considering the type of diffusion profile at a given time in the course of the experiment.

A very similar effect of a non-linear concentration to absorbance dependence occurs in grazing angle or long path length cells. In this type of cell (see Fig. II.6.9), the detector beam of the spectrometer passes parallel to the working electrode through the electrochemical cell. The advantage of spectroelectrochemical measurements in transmission through a long path length cell is the high sensitivity, which can be achieved even with a small electrode.

The first long path length cell was described from Niu et al. [74] for analytical applications where high sensitivity is required. In order to overcome the

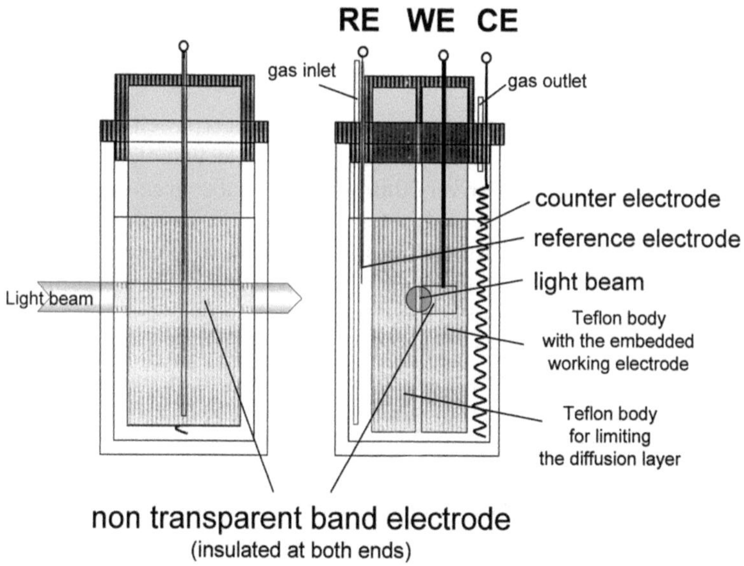

Fig. II.6.9 In situ UV/Vis/NIR spectroelectrochemical 'long path length cell' based on a non-transparent band electrode in a capillary slit and a light beam focused through the diffusion layer

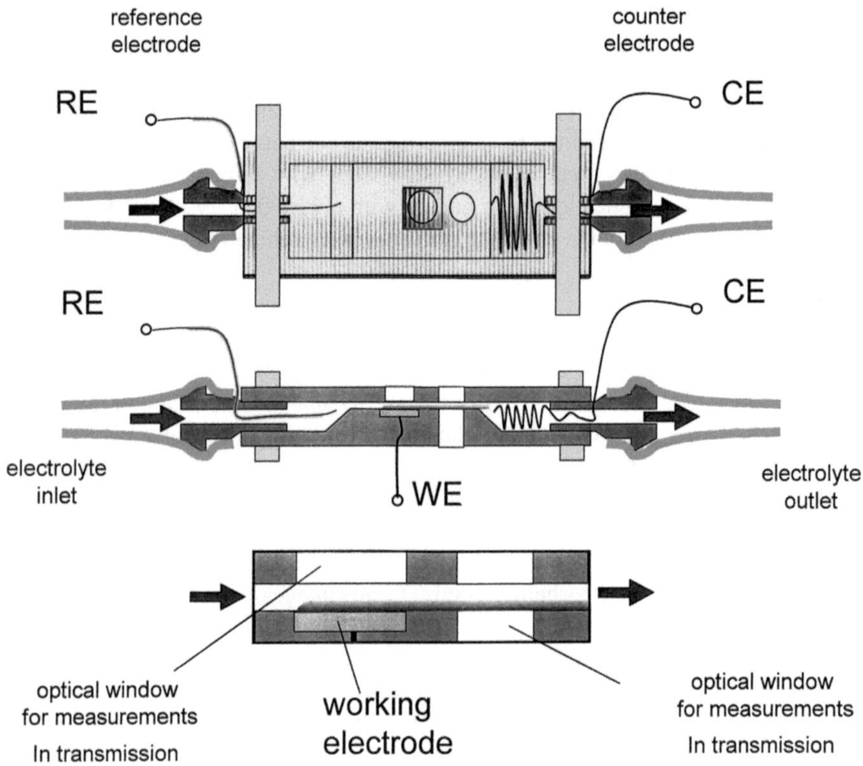

Fig. II.6.10 In situ UV/Vis/NIR spectroelectrochemical flow cell design, which allows spectra to be recorded in transmission and in reflection mode

non-linearity in the absorbance response, a capillary slit type cell is employed with a size similar to the diffusion layer. This allows the detection of all of the product generated electrochemically. For very thin slits, the total conversion of the starting material to the products occurs in only a few milliseconds. This short timescale permits the reaction kinetics of fast reactions to be followed. Instead of using a wall to limit the diffusion-layer thickness, it is also possible to focus the light beam into a very thin layer close to the electrode with a suitable lens system. In this configuration, only the concentration of reagent immediately at the electrode surface is measured in the spectroscopic detection. The current signal from a cyclic voltammogram can be processed by semi-integration [75] to give a signal directly proportional to the surface concentration and is, therefore, ideal for the analysis of the spectroelectrochemical data.

A laser beam focused into the diffusion layer in combination with a diode array detector may even be used to directly image the concentration profile inside the diffusion layer, as described by Posdorfer et al. [76, 77].

The in situ UV/Vis/NIR spectroelectrochemical cells shown in Figs. II.6.6 and II.6.7 permit solution to flow through the cell. However, the flow is only used to renew the solution at the electrode and to allow repeat experiments under identical conditions or on different timescales. Experiments may also be conducted under flowing conditions at a fixed flow rate or as a function of flow rate. In the case of short-lived intermediates being formed a continuous electrolysis in a streaming electrolyte allows the steady-state concentration of intermediates to be measured independent of time. The concentration of the intermediate in the flowing solution might be low, but the concentration is constant over a long time and this permits spectra to be recorded over a long timescale and with high sensitivity. An in situ spectroelectrochemical cell system, which allows spectra to be recorded under flowing solution conditions, is shown in Fig. II.6.10. Both measurements in transmission and measurements in reflection mode are possible and software for modelling concentration profiles and reaction kinetics under flowing conditions has been described.

II.6.3.2 Spectroelectrochemistry in Reflection Mode

In the previous section, it was shown that detailed information about electrochemical processes and the kinetics of follow-up chemical reaction steps can be investigated by UV/Vis/NIR spectroelectrochemical experiments in transmission mode

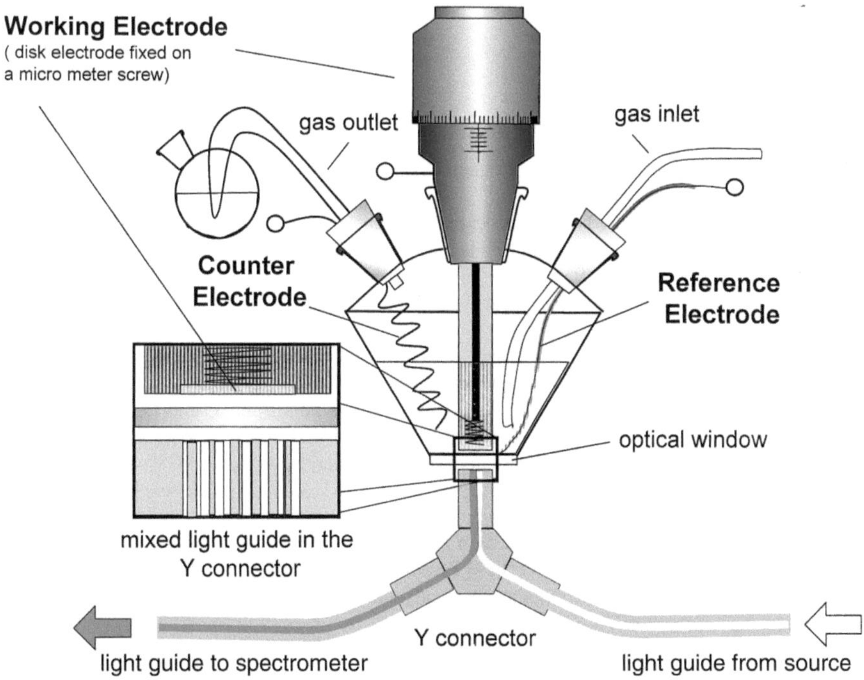

Fig. II.6.11 In situ UV/Vis/NIR spectroelectrochemical cell for measurements in reflection mode at conventional disc electrodes

in the diffusion layer at optically transparent or microstructured non-transparent electrodes. Many metal electrodes show a high reflectivity, and therefore optical spectra may also be recorded under in situ conditions in reflection mode [78, 79]. This approach is essential for the study of adsorbed species, the formation of solid layers at the electrode surface, reactions of solids [80–82], and the redox behaviour of conducting polymer layers [83–85]. Furthermore, in reflection mode, the angle of incidence may be modified and polarised light may be used in ellipsometry studies [86].

By using optical waveguides for illumination and detection, it is possible to record spectra at a disc electrode during the course of cyclic voltammetric measurements (see Fig. II.6.11). Fixing the disc electrode onto a micrometer screw allows the distance between the electrode and the optical window to be varied in a very much easier way compared with the procedure required in in situ spectroelectrochemical cells working in the transmission mode. In this experimental arrangement, the gap between the electrode surface and the wall with the optical waveguide defines the diffusion layer thickness and therefore the type of absorbance response or timescale.

In conclusion, UV/Vis/NIR spectroelectrochemistry in both transmission and reflection mode are extremely useful techniques that yield a wealth of complementary data additional to those obtained in pure electrochemical voltammetric experiments. Especially when based on computer simulation models, this data may be used to unravel the kinetics and thermodynamics of complex electrode processes.

References

1. Bard AJ, Faulkner LR (2001) Electrochemical methods – fundamentals and applications, 2nd edn. John Wiley, New York
2. Dong S, Niu J, Cotton TM (1994) Biochemical spectroscopy. In: Abelson JN, Simon MI (eds) Methods in enzymology. vol 246, Academic Press, Orlando, Florida
3. Broman RF, Heineman WR, Kuwana T (1973) Faraday Discuss Chem Soc 56: 816
4. Aylmer-Kelly AWB, Bewick A, Cantrill PR, Tuxford AM (1973) Faraday Discuss Chem Soc 56: 96
5. Heineman WR, Hawkridge FM, Blount HN (1984) Electroanal Chem 13: 55
6. Kuwana T, Winograd N (1974) Electroanal Chem 7: 74
7. Kaim W, Klein A (2008) Spectroelectrochemistry. RSC Publishing, Cambridge, UK
8. Goldberg IB, McKinney TM (1984) In: Kissinger PT, Heineman WR (eds) Laboratory techniques in electroanalytical chemistry. Marcel Dekker, New York, p 675
9. Compton RG, Waller AM (1988) ESR spectroscopy and electrode processes. In: Gale RJ (ed) Spectroelectrochemistry – theory and practice. Plenum Press, New York, p 349
10. McKinney TM (1977) Electroanal Chem 10: 77
11. Bagchi RN, Bond AM, Scholz F (1989) Electroanalysis 1: 1
12. Bard AJ, Faulkner LR (1980) Electrochemical methods – fundamentals and applications. John Wiley, New York, p 577
13. Hamann CH, Hamnett A, Vielstich W (1998) Electrochemistry. John Wiley, New York, p 247
14. Mazurenka M, Wilkins L, Macpherson JV, Unwin PR, Mackenzie SR (2006) Anal Chem 78: 6833
15. Browne WR, McGarvey JJ (2007) Coord Chem Rev 251: 454

16. Amatore C, Bonhomme F, Bruneel JL, Servant L, Thouin L (2000) J Electroanal Chem 484: 1
17. Compton RG, Fisher AC, Wellington RG (1991) Electroanalysis 3: 27
18. Kanoufi F, Bard AJ (1999) J Phys Chem B 103: 10469
19. Collinson MM, Wightman RM (1995) Science 268: 1883
20. Prenzler PD, Bramley R, Downing SR, Heath GA (2000) Electrochem Commun 2: 516
21. La Mar GN, Horrocks WDW, Holm RH (1973) NMR of paramagnetic molecules: principles and applications. Academic Press, New York
22. Silva-Chong JA, Guillen-Villafuerte O, Rodriguez JL, Pastor E (2008) J Solid State Electrochem 12: 517
23. Kolb DM (1988) UV-Visible reflectance spectroscopy. In: Gale RJ (ed) Spectroelectrochemistry – theory and practice. Plenum Press, New York, p 87
24. Beden B, Lamy C (1988) Infrared reflectance spectroscopy. In: Gale RJ (ed) Spectroelectrochemistry – theory and practice. Plenum Press, New York, p 189
25. Schröder U, Scholz F (2000) Inorg Chem 39: 1006
26. Kavan L, Dunsch L (2007) ChemPhysChem 8: 975
27. Tong YY, Rice C, Wieckowski A, Oldfield E (2000) J Am Chem Soc 122: 1123
28. Pemberton JE (1991) Surface enhanced Raman scattering. In: Abruna HD (ed) Electrochemical interfaces. VCH, Weinheim, p 195
29. Mrozek MF, Weaver MJ (2000) J Am Chem Soc 122: 150
30. Birke RL, Lu T, Lombardi JR (1991) Surface enhanced Raman spectroscopy. In: Varma R, Selman JR (eds) Characterization of electrodes and electrochemical processes. John Wiley, New York, p 211
31. Murgida DH, Hildebrandt P (2008) Chem Soc Rev 37: 937
32. Wasileski SA, Zou SZ, Weaver MJ (2000) Appl Spectrosc 54: 761
33. Bailo E, Deckert V (2008) Chem Soc Rev 37: 921
34. McBreen J (1995) In situ synchrotron techniques in electrochemistry. In: Rubinstein I (ed) Physical electrochemistry. Marcel Dekker, New York, p 339
35. Abruna HD (1991) Electrochemical interfaces. VCH, Weinheim
36. Robinson J (1988) X-ray techniques. In: Gale RJ (ed) Spectroelectrochemistry – theory and practice. Plenum Press, New York, p 9
37. Scherson DA (1991) Mössbauer spectroscopy. In: Abruna HD (ed) Electrochemical interfaces. VCH, Weinheim, p 413
38. Gottesfeld S, Kim YT, Redondo A (1995) Recent applications of ellipsometry and spectroellipsometry in electrochemical systems. In: Rubinstein I (ed) Physical electrochemistry. Marcel Dekker, New York, p 393
39. Grieshaber D, MacKenzie R, Voros J, Reimhult E (2008) Sensors 8: 1400
40. Richmond GL (1991) Investigations of Electrochemical interfaces by nonlinear optical methods. In: Abruna HD (ed) Electrochemical interfaces. VCH, Weinheim, p 267
41. Szunerits S, Pust SE, Wittstock G (2007) Anal Bioanal Chem 389: 1103
42. Drago RS (1977) Physical methods in chemistry. Saunders and Co., London
43. Xie QJ, Nie LH, Yao SZ (1997) Anal Sci 13: 453
44. Cooper JA, Compton RG (1998) Electroanal 10: 141
45. Brett CMA, Brett AMO (1993) Electrochemistry: principles, methods, and applications. Oxford University Press, Oxford
46. Wang ZH, Zhao M, Scherson DA (1994) Anal Chem 66: 4560
47. Rieger PH (1994) Electrochemistry. Chapman & Hall, London
48. Wang ZH, Zhao M, Scherson DA (1994) Anal Chem 66: 1993
49. Montenegro MI, Queiros MA, Daschbach JL (1991) Micro electrodes: theory and applications, Nato ASI Series, vol 197. Kluwer, London
50. Zhao M, Scherson DA (1992) Anal Chem 64: 3064
51. Compton RG, Dryfe RAW (1995) Prog Reaction Kinetics 20: 245
52. Albery WJ, Compton RG, Kerr IS (1980) J Chem Soc Perkin Trans 2: 825
53. Albery WJ, Chadwick AT, Coles BA, Hampson NA (1977) J Electroanal Chem 75: 229

54. Prieto F, Webster RD, Alden JA, Aixill WJ, Waller GA, Compton RG, Rueda M (1997) J Electroanal Chem 437: 183
55. Kress L, Neudeck A, Petr A, Dunsch L (1996) J Electroanal Chem 414: 31
56. Wang RL, Tam KY, Marken F, Compton RG (1997) Electroanalysis 9: 284
57. Wang RL, Tam KY, Compton RG (1997) J Electroanal Chem 434: 105
58. Marken F, Neudeck A, Bond AM, Cyclic Voltammetry this Volume
59. Marken F, Neudeck A, Bond AM, Cyclic Voltammetry this Volume
60. Marken F, Neudeck A, Bond AM,, Cyclic Voltammetry this Volume
61. Atkins PW (2000) Physical chemistry, 6th edn. Oxford Univ Press, Oxford
62. Marken F, Neudeck A, Bond AM, Cyclic Voltammetry this Volume
63. Amatore C, Szunerits S, Thouin L (2000) Electrochem Commun 2: 248
64. (a) Niu J, Dong S (1996) Rev Anal Chem 15: 1; (b) Salbeck J (1993) Anal Chem 65: 2169
65. Porter JD, Heller A, Aspnes DE (1985) Nature 313: 664
66. Neudeck A, Kress L (1997) J Electroanal Chem 437: 141
67. Rapta P, Neudeck A, Petr A, Dunsch L (1998) J Chem Soc Faraday Trans 94: 3625
68. Massart DL, Vandeginste BGM, Deming SN, Michotte Y, Kaufman L (1988) Chemometrics: a textbook. Elsevier, Oxford
69. Keesey RL, Ryan MD (1999) Anal Chem 71: 1744 and references cited therein
70. Neudeck A, Dunsch L (1995) Electrochim Acta 40: 1427
71. Neudeck A, Dunsch L (1995) J Electroanal Chem 386: 135
72. Neudeck A, Dunsch L (1993) Ber Bunsges Phys Chem 97: 407
73. Neudeck A, Audebert P, Guyard L, Dunsch L, Guiriec P, Hapiot P (1999) Acta Chem Scand 53: 867
74. Niu J, Dong S (1995) Electrochim Acta 3: 1059
75. Oldham KB, Myland JC (1994) Fundamentals of electrochemical science. Academic Press, London
76. Posdorfer J, Olbrich-Stock M, Schindler RN (1994) Electrochim Acta 39: 2005
77. Posdorfer J, Olbrich-Stock M, Schindler RN (1994) J Electroanal Chem 368: 173
78. Hansen WN (1973) Internal reflection spectroscopy in electrochemistry. In: Muller RH (ed) Advances in electrochemistry and electrochemical engineering, vol 9. John Wiley, New York
79. McIntyre JDE (1973) Specular reflection spectroscopy of the electrode-solution interphase. In: Muller RH (ed) Advances in electrochemistry and electrochemical engineering, vol 9. John Wiley, New York
80. Ramaraj R, Kabbe C, Scholz F (2000) Electrochem Commun 2: 190
81. Bond AM, Marken F, Hill E, Compton RG, Hügel H (1997) J Chem Soc Perkin Trans 2: 1735
82. Schröder U, Scholz F (1997) J Solid State Electrochem 1: 62
83. Rapta P, Neudeck A, Petr A, Dunsch L (1998) J Chem Soc Faraday Trans 94: 3625
84. Neudeck A, Petr A, Dunsch L (1999) J Phys Chem B 103: 912
85. Aubert PH, Neudeck A, Dunsch L, Audebert P, Maumy M (1999) J Electroanal Chem 470: 77
86. Muller RH (1973) Principles of ellipsometry. In: Muller RH (ed) Advances in electrochemistry and electrochemical engineering, vol 9. John Wiley, New York

Chapter II.7
Stripping Voltammetry

Milivoj Lovrić

II.7.1 Introduction

Electrochemical stripping means the oxidative or reductive removal of atoms, ions, or compounds from an electrode surface (or from the electrode body, as in the case of liquid mercury electrodes with dissolved metals) [1–5]. In general, these atoms, ions, or compounds have been preliminarily immobilized on the surface of an inert electrode (or within it) as the result of a preconcentration step, while the products of the electrochemical stripping will dissolve in the electrolytic solution. Often the product of the electrochemical stripping is identical to the analyte before the preconcentration. However, there are exemptions to these rules. Electroanalytical stripping methods comprise two steps: first, the accumulation of a dissolved analyte onto, or in, the working electrode, and, second, the subsequent stripping of the accumulated substance by a voltammetric [3, 5], potentiometric [6, 7], or coulometric [8] technique. In stripping voltammetry, the condition is that there are two independent linear relationships: the first one between the activity of accumulated substance and the concentration of analyte in the sample, and the second between the maximum stripping current and the accumulated substance activity. Hence, a cumulative linear relationship between the maximum response and the analyte concentration exists. However, the electrode capacity for the analyte accumulation is limited and the condition of linearity is satisfied only well below the electrode saturation. For this reason, stripping voltammetry is used mainly in trace analysis. The limit of detection depends on the factor of proportionality between the activity of the accumulated substance and the bulk concentration of the analyte. This factor is a constant in the case of a chemical accumulation, but for electrochemical accumulation it depends on the electrode potential. The factor of proportionality between the maximum stripping current and the analyte concentration is rarely known exactly. In fact, it is frequently ignored. For the analysis it suffices to establish the linear relationship empirically. The slope of this relationship may vary from one sample to another

M. Lovrić (✉)
Rudjer Boskovic Institute, 10001 Zagreb, Croatia
e-mail: mlovric@irb.hr

F. Scholz (ed.), *Electroanalytical Methods*, 2nd ed.,
DOI 10.1007/978-3-642-02915-8_10,

because of different influences of the matrix. In this case the concentration of the analyte is determined by the method of standard additions [1]. After measuring the response of the sample, the concentration of the analyte is deliberately increased by adding a certain volume of its standard solution. The response is measured again, and this procedure is repeated three or four times. The unknown concentration is determined by extrapolation of the regression line to the concentration axis [9]. However, in many analytical methods, the final measurement is performed in a standard matrix that allows the construction of a calibration plot. Still, the slope of this plot depends on the active area of the working electrode surface. Each solid electrode needs a separate calibration plot, and that plot must be checked from time to time because of possible deterioration of the electrode surface [2].

II.7.2 Overview of Preconcentration Methods

The preconcentration, i.e., accumulation on the electrode surface, can be achieved by the following methods:

(a) Electrodeposition of metal atoms on an inert, solid electrode,
(b) reduction of metal ions on a mercury electrode and dissolution of metal atoms in the mercury,
(c) precipitation of sparingly soluble mercuric, mercurous, or silver salts on the surface of mercury, or silver,
(d) precipitation of metal oxides, hydroxides, or other insoluble salts on noble metal electrode surfaces,
(e) chemical reactions of compounds with an electrode material,
(f) adsorption of electroactive organic substances on the mercury surface,
(g) adsorption of complexes of metal ions with organic ligands,
(h) synergistic adsorption of complexes,
(i) anion-induced adsorption of metal ions,
(j) surface complexation of metal ions,
(k) ion-exchange reactions at surface-modified electrodes, and
(l) extraction by ion transfer across the interface of two immiscible electrolyte solutions.

The accumulation is a dynamic process that may turn into a steady state in stirred solutions. Besides, the activity of accumulated substance is not in a time-independent equilibrium with the activity of analyte in the bulk of the solution. All accumulation methods employ fast reactions, either reversible or irreversible. The fast and reversible processes include adsorption and surface complexation, the majority of ion transfers across liquid/liquid interfaces and some electrode reactions of metal ions on mercury. In the case of a reversible reaction, equilibrium between the activity of accumulated substance and the concentration of analyte at the electrode surface is established. It causes the development of a concentration profile near the electrode and the diffusion of analyte towards its surface. As the activity of the accumulated substance increases, the concentration of the analyte at the electrode surface is augmented and the diffusion flux is diminished. Hence, the equilibrium between the accumulated substance and the bulk concentration of the analyte can be established only after an infinitely long accumulation time (see Eqs. II.7.12, II.7.13, II.7.14 and II.7.30). The reduction of metal ions on mercury electrodes in stirred solutions is in the steady state at high overvoltages. Redox reactions of many metal ions, especially at solid electrodes, and the precipitation of some salts and oxides are kinetically controlled. To increase the rate of these reactions, the accumulation is performed with significant overvoltage. Under this condition the generation of the accumulated material does not influence the concentration of analyte in the diffusion layer, but modifies the characteristics of the electrode surface and may block the charge transfer ultimately. The steady state can be established in the formation of compact metal deposits. At the limit of the electrode capacity the mechanisms of accumulation processes can be complicated by attractive, or repulsive, interactions between the particles of the accumulated matter. Nevertheless, in analytical stripping voltammetry, the accumulation is useful only if these interactions are negligible. Hence, the complexity of the mechanism is not of primary interest. Its investigation serves mainly for the determination of the upper limit of detection of a certain analyte.

II.7.2.1 Metal Deposition on Solid Electrodes

The reduction of metal ions on solid electrodes is a process consisting of three main steps: a deposition of adatoms, a two- and three-dimensional nucleation and a three-dimensional crystal growth [10]. The formation of the first monolayer of metal atoms on the foreign substrate follows a quasi-Nernst equation:

$$\frac{(a_{\mathrm{Me}^{z+}})_{x=0}}{f(\Gamma)} = \exp\left[zF(E - E^{\ominus}_{\mathrm{Me}/\mathrm{Me}^{z+}})/RT\right] \tag{II.7.1}$$

where $f(\Gamma)$ is a certain function of the surface concentration of metal atoms in the first monolayer. It is considered as the activity of a two-dimensional metal phase with respect to the activity of metal atoms in the bulk of the three-dimensional metal phase ($a_{\mathrm{Me}} = 1$). The function $f(\Gamma)$ depends on the interactions between the metal atoms and the substrate, on the lateral interactions between metal atoms in the monolayer, on the crystallographic structure of the substrate and on the number of active sites on the electrode surface. Models of a localized adsorption can describe these relationships. The simplest model is the Fowler–Frumkin isotherm:

$$E = E_{1/2} - \frac{RT}{zF}\left[\ln\frac{\Theta}{1-\Theta} - \omega\,(\Theta - 0.5)\right] \tag{II.7.2}$$

where $\Theta = \Gamma/\Gamma_{max}$ is the degree of surface coverage, Γ_{max} is the maximum surface concentration of metal atoms in the monolayer, ω is the lateral interaction parameter and $E_{1/2}$ is the potential that corresponds to $\Theta = \frac{1}{2}$. The function $f(\Gamma)$ can be calculated by comparing Eqs. (II.7.1) and (II.7.2):

$$f(\Gamma) = \frac{\Theta}{1-\Theta} \cdot \exp\left[zF\left(E_1 - E_{1/2}\right)/RT\right] \cdot \exp\left[-\omega\left(\Theta - 0.5\right)\right] \qquad \text{(II.7.3)}$$

where

$$E_1 = E^{\ominus}_{Me/Me^{z+}} + \frac{RT}{zF} \ln\left(a_{Me^{z+}}\right)_{x=0} \qquad \text{(II.7.4)}$$

is the potential that corresponds to $f(\Gamma) = 1$. The relationship between the activity of metal ions at the electrode surface ($x = 0$) and the activity of adatoms in the first monolayer on the electrode surface is obtained by substituting Eq. (II.7.3) in Eq. (II.7.1):

$$\left(a_{Me^{z+}}\right)_{x=0} = \frac{\Theta}{1-\Theta} \exp\left[-\omega\left(\Theta - 0.5\right]\exp\left[zF\left(E - E^*\right)/RT\right] \qquad \text{(II.7.5)}$$

where

$$E^* = E^{\ominus}_{Me/Me^{z+}} + E_{1/2} - E_1 \qquad \text{(II.7.6)}$$

The difference $E_{1/2} - E_1$ is related to the binding energy of the first monolayer. In cyclic voltammetry this is the difference between the potentials of the most positive peak and the peak corresponding to the three-dimensional metal phase. The latter difference is linearly correlated with the difference of work functions of the deposited metal and the electrode material [11]:

$$\Delta E_p = \alpha \cdot \Delta\Phi \qquad \text{(II.7.7)}$$

where $\Delta\Phi = \Phi_{electrode} - \Phi_{Me}$ and $\alpha = 0.5$ V/eV. The work functions of various metals and carbon are listed in Table II.7.1. If it holds that the difference $E_{1/2} - E_1 > 0$, then the metal is deposited at potentials that are higher (more positive) than the equilibrium potential E_1. This is called an underpotential deposition [13, 14]. Some experimentally studied underpotential deposition systems are compiled in Table II.7.2. If $E_{1/2} - E_1 < 0$, the deposition occurs at potentials lower than E_1. This is an overpotential deposition. The formation of a second monolayer of metal atoms on top of the first one can be considered as an adsorption on a modified substrate. The binding energy of the second layer is not equal to the energy of the first one. The structural and electrochemical characteristics of a metal deposit depend on the thickness of the film. Deposition on foreign substrates becomes identical to the deposition on the native substrate if more than 20 monolayers of metal atoms are deposited.

Table II.7.1 Work functions of some elements [12a]

Element	Φ(eV)	Element	Φ(eV)
Pt	5.65	W	4.55
Ir	5.27	Fe	4.5
Ni	5.15	Hg	4.49
Pd	5.12	Sn	4.42
Au	5.10	Zn	4.33
Co	5.0	Ag	4.26
C	5.0	Pb	4.25
Rh	4.98	Bi	4.22
Os	4.83	Cd	4.22
Ru	4.71	Ga	4.20
Cu	4.65	In	4.12
Sb	4.55	Tl	3.84

Table II.7.2 Some experimentally studied underpotential deposition systems [10]

Substrate	Metals
Au	Ag, Bi, Cd, Cu, Hg, Ni, Pb, Sb, Te, Tl
Pt	Ag, Bi, Cu, Pb, Rh, Tl
Ag	Bi, Cd, Pb, Tl
Rh	Ag, Cu, Pb
Cu	Cd, Pb, Tl
Pd	Cu

The growth mode depends on the binding energy of the first monolayer and the crystallographic misfit between the substrate and the three-dimensional metal deposit [10]. If $E_{1/2} > E_1$ and the crystallographic misfit is negligible, the deposit grows in the layer-by-layer mode. If the misfit is high, islands of three-dimensional crystallites on top of the first monolayer are formed. If $E_{1/2} < E_1$, these islands are formed directly on the unmodified substrate. For all three modes of growth, an overpotential is needed. The growth starts with the association of metal atoms to form clusters. The clusters are stable if they consist of more than a critical number of atoms. The stable cluster is called the nucleus of the new phase. There are instantaneous and progressive nucleations. In the former the nuclei grow, but their number is constant, while in the latter the number of nuclei increases continuously. Both the nucleation and growth are irreversible processes. The mechanisms of metal deposition are beyond the scope of this chapter.

Solid electrodes are used for the analysis of metal ions that cannot be determined on mercury electrodes with reasonable efficiency, or not at all. Some examples are listed in Table II.7.3. They are representatives of two groups of ions. The stripping peak potentials of deposits of the first group are higher, or close to, the stripping potential of mercury. For instance, the concentration of bismuth ions cannot be measured on mercury electrodes below 5×10^{-9} mol/L, but on a glassy carbon electrode, these ions can be determined at the concentration level of 10^{-11} mol/L

Table II.7.3 Accumulation potentials and stripping peak potentials of some metals at graphite electrodes [2]

Ion	Electrolyte	E_{acc} (V vs. SCE)	E_P (V vs. SCE)
Fe^{3+}	0.05 M Na tartarate	−1.6	−0.6
Co^{2+}	0.1 M KSCN	−1.2	−0.5
Ni^{2+}	0.1 M KSCN	−1.2	−0.5
Sb^{3+}	1 M HCl	−0.6	−0.22
Cu^{2+}	1 M KNO_3	−0.6	−0.1
Bi^{3+}	0.1 M HCl	−0.4	−0.08
Hg^{2+}	0.1 M KNO_3	−0.4	+ 0.1
Ag^{+}	1 M KNO_3	−0.4	+ 0.1
Te^{4+}	1 M HCl	−0.6	+ 0.37
Pd^{2+}	1 M HCl + 1 M KCl	−0.6	+ 0.4
Au^{3+}	1 M HCl	−0.2	+ 0.5

[15]. Mercury ions can be detected on a glassy carbon electrode in a concentration as low as 5×10^{-14} mol/L [16]. The atoms of the second group are poorly soluble in mercury (Sb, Co, Ni, Fe). Besides, the reduction of some of these ions and the oxidation of their atoms are both kinetically controlled processes on mercury electrodes [1].

The stripping response of the metal deposit may consist of one or more peaks, depending on the binding energy of the atoms. In analytically useful deposits the variation of this energy is minimal. For instance, if the accumulation is performed by underpotential deposition of adatoms only, a single stripping peak can be obtained [17]. In systems with a significant difference in work functions, such as Bi or Ag on a carbon electrode, the deposition can be controlled by adjusting the accumulation potential. If the difference in work functions is negligible (e.g., Co, Ni, Pd, or Au on carbon), the accumulation can be achieved only by overpotential deposition [18, 19]. It can be facilitated by the presence of some ionic impurities that modify the substrate by their underpotential deposition. An example is the mercury deposition on glassy carbon that is enhanced by traces of thallium ions. However, if two or more metals are accumulated simultaneously, the stripping peaks may appear highly complex and analytically useless [2]. Recently, bismuth film electrodes have been developed for anodic stripping voltammetry of heavy metal ions in environmental and biological samples [20, 21]. These electrodes are useful alternative to mercury film electrodes.

II.7.2.2 Metal Deposition on Mercury Electrodes

Amalgams are metallic systems in which mercury is one of the components. The solubility of the alkali metals, the alkaline earths, the rare earths and Au, Zn, Cd, Ga, In, Tl, Sn, Pb, Bi, Ru, Rh, and Pt in mercury is higher than 0.1 atom % [1]. A reversible redox reaction of an amalgam-forming metal ion on a mercury electrode

Table II.7.4 Solubility and diffusion coefficients of some metals in mercury. Stripping peak potentials and standard rate constants of amalgam electrodes in 0.1 M HCl [1]

Element	Solubility (% w/w)	$D^{Hg} \times 10^5$ (cm^2/s)	E_P (V vs. SCE)	k_s (cm/s)
Zn	1.99	2.4	−0.91	0.004
Cd	5.3	2.0	−0.64	0.6
In	57	1.4	−0.59	0.1
Tl	42.8	1.2	−0.54	0.3
Sn	0.6	1.5	−0.49	0.1
Pb	1.1	2.1	−0.44	0.2
Sb	2.9×10^{-5}	1.5	−0.18	0.05
Cu	3.0×10^{-3}	1.1	−0.16	0.05
Bi	1.1	1.5	−0.06	0.3

$$\mathrm{Me\,(Hg)} \rightleftarrows \mathrm{Me}^{z+} + z\mathrm{e}^- \tag{II.7.8}$$

satisfies the Nernst equation:

$$E = E^{\ominus}_{\mathrm{Me(Hg)}} + \frac{RT}{zF} \ln \frac{a_{\mathrm{Me}^{z+}}}{a_{\mathrm{Me(Hg)}}} \tag{II.7.9}$$

$$E^{\ominus}_{\mathrm{Me(Hg)}} = E^{\ominus}_{\mathrm{Me/Me}^{z+}} + \frac{RT}{zF} \ln a^*_{\mathrm{Me(Hg)}} - E_{\mathrm{s}} \tag{II.7.10}$$

$$E_{\mathrm{s}} = -\frac{\Delta G^{\circ}}{zF} + \frac{RT}{zF} \ln (a^*_{\mathrm{Hg}})^y \tag{II.7.11}$$

where $a^*_{\mathrm{Me(Hg)}}$ and a^*_{Hg} are the metal activity and the mercury activity in the saturated amalgam, ΔG° is the change in the Gibbs free energy of formation of the solid-phase MeHg_y in the amalgam, y is the number of atoms of mercury coordinated to one metal atom in the solid phase, E_{s} is the potential difference of a concentration cell consisting of the pure metal and of its saturated amalgam, immersed in a solution of a salt of the metal, and $E^{\ominus}_{\mathrm{Me/Me}^{z+}}$ is the standard potential of the electrode reaction $\mathrm{Me} \rightleftarrows \mathrm{Me}^{z+} + z\mathrm{e}^-$. When no metal compounds are formed with mercury, the value of E_{s} is close to or equals zero [1]. Compounds of the alkali metals and the alkaline earths with mercury have the best-defined composition, while copper and zinc do not form any intermetallic compounds with mercury. The compositions of many intermetallic compounds are variable. In diluted amalgams, the compounds are dissociated to various degrees.

The accumulation of amalgams can be used in anodic stripping voltammetry if both the reduction of ions and the oxidation of metal atoms occur within the working window of the mercury electrode [22]. Ions that give the best responses are listed in Table II.7.4. In electroanalysis of traces, mercury film electrodes are used [23–25].

When several metals are simultaneously electrodeposited in mercury, intermetallic compounds between them may be formed. In anodic stripping voltammetry the following compounds of copper, zinc, and antimony may influence the measurements: CuZn, CuSn, CuGa, SbZn, SbCd, and SbIn. The values of their solubility products are between 2×10^{-9} (SbZn) and 4×10^{-6} (CuZn) [1].

Considering the low solubility of Sb and Cu in mercury, a concentration of Zn atoms higher than 10^{-4} % w/w may cause the precipitation of these compounds. In general, the formation of intermetallic compounds is suppressed in diluted amalgams.

II.7.2.2.1 Pseudopolarography

The concentration of metal atoms in mercury electrodes depends on the potential (E_{acc}) and the duration (t_{acc}) of accumulation, the bulk concentration of ions ($c_{\mathrm{Me(b)}}$) and the hydrodynamic conditions in the solution [26]. The simplest model considers the reversible electrode reaction on a thin mercury film rotating disk electrode with a fully developed diffusion layer in the solution and uniform distribution of metal atoms in the mercury film:

$$\frac{i}{zFS} = D\left[c_{\mathrm{Me(b)}} - (c_{\mathrm{Me}})_{x=0}\right]/\delta \qquad \text{(II.7.12)}$$

$$c_{\mathrm{Me(Hg)}} = \frac{1}{zFSl}\int_0^t i\mathrm{d}\tau \qquad \text{(II.7.13)}$$

where $(c_{\mathrm{Me}})_{x=0}$ is the concentration of metal ions at the electrode surface, i/S is the current density, S is the electrode surface area, l is the mercury film thickness, D is the diffusion coefficient, and δ is the diffusion layer thickness. The solution of Eqs. (II.7.9), (II.7.12) and (II.7.13) is the dimensionless reduction current $\Phi = \frac{i\delta}{zFSDc_{\mathrm{Me(b)}}}$ and the dimensionless concentration of the metal atoms in mercury film $I = \frac{c_{\mathrm{Me(Hg)}}}{c_{\mathrm{Me(Hg),SS}}}$ [27]:

$$\Phi = \exp\left(-\frac{ut}{t_{\mathrm{acc}}}\right) \qquad \text{(II.7.14)}$$

$$I = \frac{1-\exp(-u)}{u} \qquad \text{(II.7.15)}$$

where $u = \frac{Dt_{\mathrm{acc}}\exp(\varphi_{\mathrm{acc}})}{l\delta}$, $\varphi_{\mathrm{acc}} = zF\left(E_{\mathrm{acc}} - E^{\ominus}_{\mathrm{Me(Hg)}}\right)/RT$, and $c_{\mathrm{Me(Hg),SS}} = \frac{Dt_{\mathrm{acc}}c_{\mathrm{Me(b)}}}{l\delta}$. Equation (II.7.14) shows that the reduction current decreases exponentially with time, but $\lim_{u\to 0}\Phi = 1$. This limit is approached if $E_{\mathrm{acc}} \ll E^{\ominus}_{\mathrm{Me(Hg)}}$. Hence, if the accumulation potential is much lower than the standard potential, a steady state is established and the concentration of metal atoms in the film, $c_{\mathrm{Me(Hg)SS}}$, is linearly proportional to the duration of accumulation t_{acc}. The function I, defined by Eq. (II.7.15), has the form of a polarogram: if $u \to \infty$ then $I = 0$, and if $u \to 0$ then $I = 1$. Its half-wave potential is defined by the condition $I = 0.5$, and hence

$$E_{1/2} = E^{\ominus}_{\mathrm{Me(Hg)}} + \frac{RT}{zF} \ln \frac{1.549\, l\delta}{Dt_{\mathrm{acc}}} \tag{II.7.16}$$

The function I is called the pseudopolarogram [28]. It is constructed by plotting the peak current in anodic stripping voltammetry as a function of the accumulation potential, because the peak current is linearly proportional to the concentration of metal atoms in the mercury electrode [29]. The half-wave potential of a pseudopolarogram depends on the mercury film thickness, the electrode rotation rate and the duration of accumulation. It can be used for the estimation of the optimal accumulation potential. Besides, if metal ions form labile complexes with ligands in the solution, the half-wave potential of a pseudopolarogram depends on the ligand concentration, so that the stability constant of the complexes can be determined from this dependence [30]. If the redox reaction of metal ions is kinetically controlled, the half-wave potential of the pseudopolarogram depends on the rate constant (k_{s}) and the transfer coefficient (α), but it is independent of the duration of accumulation [31]:

$$E_{1/2,\,\mathrm{irr}} = E^{\ominus}_{\mathrm{Me(Hg)}} + \frac{RT}{azF} \ln \frac{k_s \delta}{D} \tag{II.7.17}$$

II.7.2.3 Deposition of Sparingly Soluble Salts on Electrodes

A precipitation of insoluble salts on an electrode surface can be initiated and controlled by varying the concentration of metal ions. The most effective method is electrodeposition of mercuric and mercurous salts induced by an anodic polarization of a mercury drop electrode in a solution of anions [32–35]:

$$2\mathrm{Hg} + 2\mathrm{X}^- \rightleftarrows \mathrm{Hg_2X_2} + 2\mathrm{e}^- \tag{II.7.18}$$

The insoluble salt Hg_2X_2 is precipitated as a submonomolecular layer on the mercury electrode surface. It is assumed that no lateral interactions between the deposited particles exist. At equilibrium, this redox reaction satisfies the Nernst equation [33, 36]:

$$E = E_1^{\ominus} + \frac{RT}{2F} \ln \frac{\Gamma_{\mathrm{Hg_2X_2}}}{(a_{\mathrm{X}^-})^2_{x=0}} \tag{II.7.19}$$

where $E_1^{\ominus} = E^{\ominus}_{\mathrm{2Hg/Hg_2^{2+}}} + \frac{RT}{2F} \ln \kappa_{\Theta}$ and $\kappa_{\Theta} = (a_{\mathrm{Hg_2^{2+}}})_{x=0} (a_{\mathrm{X}^-})^2_{x=0} / \Gamma_{\mathrm{Hg_2X_2}}$ is the equilibrium constant of the reaction $\mathrm{Hg_2^{2+}} + 2\mathrm{X}^- \leftrightarrows \mathrm{Hg_2X_2}$. The surface activity $\Gamma_{\mathrm{Hg_2X_2}}$ is related to the degree of surface coverage Θ by the equation $\Gamma_{\mathrm{Hg_2X_2}} = \Theta \Gamma_{\max}$, where $\Gamma_{\max}$ is the maximum surface concentration of the deposited film with constant, but unknown activity, as appears in the thermodynamics. Hence, the solubility product $K_{\mathrm{s}} = (a_{\mathrm{Hg_2^{2+}}})_{\Theta=1,x=0} (a_{\mathrm{X}^-})^2_{\Theta=1,x=0}$ is related to the equilibrium

Table II.7.5 Solubility product constants of some mercuric and mercurous salts at 25°C [12b]

Salt	K_s
Hg_2Cl_2	1.45×10^{-18}
Hg_2Br_2	6.41×10^{-23}
Hg_2I_2	5.33×10^{-29}
Hg_2CO_3	3.67×10^{-17}
$Hg_2C_2O_4$	1.75×10^{-13}
Hg_2SO_4	7.99×10^{-7}
$Hg_2(SCN)_2$	3.12×10^{-20}
$Hg(OH)_2$	3.13×10^{-26}
HgI_2	2.82×10^{-29}
HgS	6.44×10^{-53}

constant κ_{Θ} by the equation $K_s = \kappa_{\Theta=1}\Gamma_{\max}$. The solubility products of some mercuric and mercurous salts are reported in Table II.7.5.

The redox reaction (Eq. II.7.18) can be considered as an electrosorption reaction in which the ligand X^- is adsorbed to the electrode surface by forming a more or less polarized covalent bond with mercury atoms [37, 38]. When the quantity of adsorbed ligand is less than a monolayer, the adsorbate on the surface behaves as a two-dimensional gas. The adsorption constant of the ligand depends on the electrode potential

$$K_{\text{ads}} = K^{\text{o}}_{\text{ads}} \exp\left[2F\left(E - E_{\text{PZC}}\right)/RT\right] \tag{II.7.20}$$

where E_{PZC} is the potential at which the electrode surface bears no net charge. The adsorption constant $K^{\text{o}}_{\text{ads}}$ is related to equilibrium constant κ_{Θ}:

$$K^{\text{o}}_{\text{ads}} = \kappa_{\Theta}^{-1} \exp\left[2F\left(E_{\text{PZC}} - E^{\ominus}_{2\text{Hg}/\text{Hg}_2^{2+}}\right)/RT\right] \tag{II.7.21}$$

The adsorbate is stripped off by cathodic polarization of the mercury electrode. Apart from inorganic anions [32–35], the method can be used for the accumulation of various sulfur-containing organic molecules such as thiols [32, 36, 39–41], thioethers [32], thiopentone [42], and phenothiazines [43].

Sparingly soluble silver salts (AgCl, AgBr, AgI, AgCNS, Ag_2S, etc.) can be electrodeposited on the surface of silver electrodes [2, 4], but this alternative offers no advantage over the use of mercury drop electrodes. If each adsorptive accumulation is performed on a fresh surface of a new mercury drop, the stripping measurements are more reproducible than measurements on solid electrodes, including mercury film electrodes.

Manganese dioxide can be accumulated on the surface of a platinum disk electrode by oxidation of Mn^{2+} ions at 0.9 V vs. SCE in 0.1 mol/L NH_4Cl [4, 44]. The precipitate is subsequently reduced back to Mn^{2+} at 0.3 V. This reaction is the best example of electroprecipitation on noble metal electrodes [2, 45, 46].

Chemical reactions of compounds with electrode materials have been used for the determination of iodide, chloride, bromide, hydrogen sulfide, and mercury [47–50]. Iodide, chloride, and bromide were oxidized to the free halogens. The halogens were purged by nitrogen from the solution and the gas stream reacted with a silver disk, which later became the electrode for cathodic stripping voltammetry [47, 48]. In the case of sulfide, the hydrogen sulfide reacted with a silver plate [49]. The mercury reacted with a gold plate that later served as the electrode in anodic stripping voltammetry [50].

II.7.2.4 Adsorptive Preconcentration

The adsorption of ions and molecules on the surface of mercury electrodes is a thoroughly investigated phenomenon [51]. Surface-active substances are either electroactive [52] or electroinactive [53]. The former can be analyzed by adsorptive stripping voltammetry [54]. This is the common name for several electroanalytical methods based on the adsorptive accumulation of the reactant and the reduction, or oxidation, of the adsorbate by some voltammetric technique, regardless of the mechanisms of the adsorption and the electrode reaction [55, 56]. Frequently, the product of the electrode reaction remains adsorbed to the electrode surface. Hence, the term stripping should not be taken literally in all cases. Besides, some adsorbates may be formed by electrosorption reactions, so that their reduction includes covalently bound mercury atoms. The boundary between adsorption followed by reduction, on the one hand, and electrosorption, on the other, is not strictly defined. Moreover, it is not uncommon that, upon cathodic polarization, the current response is caused by a catalytic evolution of hydrogen, and not by the reduction of the adsorbate itself [57]. However, what is common to all methods is a linear relationship between the surface concentration of the adsorbate and the concentration of analyte at the electrode surface:

$$\beta \Gamma_{\max} (c_{\mathrm{R}})_{x=0} = \Gamma_{\mathrm{R}} \qquad \text{(II.7.22)}$$

where β is the adsorption constant and $\Gamma_{\max}$ is the maximum surface concentration of the reactant R. Examples of organic substances that can be determined by adsorptive stripping voltammetry are listed in Table II.7.6. The list is certainly not exhaustive.

Complexes of metal ions with surface-active ligands can be adsorbed on the surface of mercury electrodes [54, 58–60]. The adsorption of labile metal complexes with inorganic ligands is called anion-induced adsorption [61–64]. The adsorption is either direct:

$$\mathrm{MeX}_m \rightleftarrows (\mathrm{MeX}_m)_{\mathrm{ads}} \qquad \text{(II.7.23)}$$

or competitive:

$$(\mathrm{MeX}_n)^- + \mathrm{X}^-_{\mathrm{ads}} \rightleftarrows (\mathrm{MeX}_n)^-_{\mathrm{ads}} + \mathrm{X}^- \qquad \text{(II.7.24)}$$

Table II.7.6 Organic compounds determined by adsorptive stripping voltammetry on mercury electrodes

Compound	pH	E_{acc} (V)*	E_P (V)	Footnote
Amethopterine	7	−0.3	−0.72	a
Berberine	2	−0.85	−1.13	b
Bromazepam	5	−0.6	−0.92	c
Buprenorphine	9	0.0	0.32	d
Caprolactam	4.5	0.0	−0.30	e
Carnosine	9.6	0.0	−0.45	f
Dihydrozeatin	4.5	−0.6	−1.4	g
Dopamine	1	−0.2	−0.45	h
Fenitrothion	7	−0.2	−0.48	i
Fenthion	2.6	−0.28	−0.79	j
Fluvoxamine	2	−0.5	−0.75	k
Gestodene	4.5	−0.8	−1.3	l
Lamivudine	3.4	−0.3	−1.7	m
Lormetazepam	3	−0.5	−0.8	n
Methyl parathion	10	−0.3	−0.55	o
Metamitron	2	0.0	−0.5	p
Morphine	10	−0.1	0.25	r
Nitrobenzene	7	−0.20	−0.55	s
Nogalamycin	8.13	0.0	−0.6	t
Noscapine	10	−0.1	0.25	r
Oxybutynin	4	−0.3	−0.55	u
Oxytetracycline	2	−0.5	−1.06	v
Paracetamol	4.7	−0.10	−0.55	w
Probucole	7	−0.5	−0.92	x
Rifamycin SV	3.48	−0.15	−0.9	y
Thyram	10	−0.20	−0.84	z

*V vs SCE; (a) Cataldi TR, Guerrieri A, Palmisano F, Lambonin PG (1988) Analyst 113: 869; (b) Komorsky-Lovrić Š (2000) Electroanalysis 12: 599; (c) Hernandez L, Zapradeil A, Antonio J, Lopez P, Bermejo F (1987) Analyst 112: 1149; (d) Garcia-Fernandez MA, Fernandez-Abedul TM, Costa-Garcia A (2000) Electroanalysis 12: 483; (e) Tocksteinova Z, Kopanica M (1987) Anal Chim Acta 199: 77; (f) Wu XP, Duan JP, Chen HQ, Chen GN (1999) Electroanalysis 11: 641; (g) Blanco MH, Quintana MC, Hernandez L (2000) Electroanalysis 12: 147; (h) Siria JW, Baldwin RP (1980) Anal Lett 13: 577; (i) Galeano-Diaz T, Guiberteau-Cabanillas A, Espinosa-Mansilla A, Lopez-Soto MD (2008) Anal Chim Acta 618: 131; (j) Galeano-Diaz T, Guiberteau-Cabanillas A, Lopez-Soto MD, Ortiz JM (2008) Talanta 76: 809; (k) Berzas Nevado JJ, Rodriguez Flores J, Castaneda Penalvo G (2000) Electroanalysis 12: 1059; (l).Berzas Nevado JJ, Rodriguez Flores J, Castaneda Penalvo G (1999) Electroanalysis 11: 268; (m) Jain R, Jadon N, Radhapyari K (2007) J Coll Interface Sci 313: 254; (n) Zapardiel A, Bermejo E, Perez Lopez JA, Mateo P, Hernandez L (1992) Electroanalysis 4: 811; (o) dos Santos LBO, Masini JC (2008) Anal Chim Acta 606: 209; (p) Lopez de Armentia C, Sampedro C, Goicolea MA, Gomez de Balugera Z, Rodriguez E, Barrio RJ (1999) Electroanalysis 11: 1222; (r) Niazi A, Ghasemi J, Zendehdel M (2007) Talanta 74: 247; (s) Kalvoda R (1984) Anal Chim Acta 162: 197; (t) Ibrahim MS (2000) Anal Chim Acta 409: 105; (u) Jain R, Radhapyari K, Jadon N (2007) J Coll Interface Sci 314: 572; (v) Pinilla GF, Calvo Blazquez L, Garcia-Monco Corra RM, Sanches Misiego A (1988) Fresenius Z Anal Chem 332: 821; (w) Ivaska A, Ryan TH (1981) Collect Czech Chem Commun 46: 187; (x) Mirčeski V, Lovrić M, Jordanoski B (1999) Electroanalysis 11: 660; (y) Asuncion Alonso M, Sanllorente S, Sarabia LA, Arcos MJ (2000) Anal Chim Acta 405: 123; (z) Procopio JR, Escribano MTS, Hernandez LH (1988) Fresenius Z Anal Chem 331: 27

or it follows the surface-complexation mechanism [65]:

$$MeX_m + X^-_{ads} \rightleftarrows (MeX_n)^-_{ads} \qquad (II.7.25)$$

Examples are the adsorptions of $BiCl_4^-$ [66] and $PbBr_2$ complexes [67]. The adsorption of highly stable and inert complexes with surface-active organic ligands containing nitrogen and oxygen as electron donors is utilized for the determination of trace metals [68, 69]. The accumulation is a combination of direct and competitive adsorption and surface complexation. In addition, electrosorption [70]:

$$Me^{m+} + L_{ads} \rightleftarrows (MeL)^{n+}_{ads} + (m-n)\,e^- \qquad (II.7.26)$$

and a synergistic mechanism [71] are possible:

$$MeL_1 + L_{2,ads} \rightleftarrows (L_1MeL_2)_{ads} \qquad (II.7.27)$$

In the first case, an ion Me^{m+} does not form a surface-active complex with L, but the complex is formed between the oxidized, or reduced, ion Me^{n+} and the adsorbed ligand L_{ads}. Synergistic adsorption may occur in the presence of a chelating ligand L_1 that forms an uncharged complex with the metal ion, and the second ligand that is hydrophobic and strongly adsorbed to the mercury surface. The second ligand reacts with the complex MeL_1 and forms the hydrophobic mixed complex L_1MeL_2 at the surface [72]. The phenomenon was analyzed in a solution of uranyl ions, salicylic acid, and tributyl phosphate [73]. Due to the high stability of the complexes, the ligands do not have to be added in great excess. Hence, their adsorption does not prevent the accumulation of complexes. Sulfur-containing ligands are chemisorbed on mercury, while others are adsorbed mainly by π-electron interactions with the electrode surface. In principle, the ligands are electrochemically inactive, but electroactive ligands may also be used if their electrode reactions do not interfere with the stripping reactions of metal ions. Some of the ligands are listed in Table II.7.7.

Table II.7.7 Ligands used to determine trace elements by adsorptive stripping voltammetry [68]

Ligand	Elements
Catechol	Cu, Fe, Ge, Ga, Sb, Sn, U, V
Cupferron	Sm, Tb, Tl
Dimethylglyoxime	Co, Ni, Pd
2,5-Dimercapto-1,3,4-thiadiazole	Al, As, Cd, Ni, Se, Zn
o-Cresolphthalexon	Ce, La, Pr
4-(2-Pyridylazo)resorcinol	Bi, Tl, U
2-(5-Bromo-2-pyridylazo)-5-diethylaminophenol	Bi, Cu, Fe, Nb, Tl, V
8-Hydroxyquinoline	Cd, Cu, Mo, Pb, U, In
Solochrome violet RS	Al, Ba, Ca, Cs, Dy, Fe, Ga, Ho, K, Mg, Mn, Na, Rb, Sr, Tl, V, Y, Yb, Zr
Thiocyanate	Cu, Nb, Tc
Thymolphthalexon	Ba, Ca, Mg, Sr

Table II.7.8 Trace elements determined by adsorptive accumulation of solochrome violet RS complex on a hanging mercury drop electrode

Element	pH	E_{acc} (V)*	$E_{P,\,stripping}$ (V)	Footnote
Al	4.5	−0.45	−0.61	a
Ba	9.5	−0.8	−1.09	b
Ca	9.5	−0.8	−1.05	b
Cs	5	−0.6	−1.03	c
Dy	11.0	−0.75	−0.98	d
Fe	5.1	−0.4	−0.71	e
Ga	4.8	−0.4	−0.51	f
Ho	11.0	−0.75	−1.00	d
K	4.7	−0.6	−0.95	c
Mg	9.5	−0.8	−1.06	b
Mn	11.5	−0.65	−0.81	g
Na	4.7	−0.6	−0.98	c
Rb	4.7	−0.6	1.01	c
Sr	9.5	−0.8	−1.08	b
Tl	5.1	−0.4	−0.92	h
V	6.8	−0.3	−1.05	i
Y	11.0	−0.75	−0.98	d
Yb	11.0	−0.75	−1.00	d
Zr	5	−0.3	−0.45	j

*V vs. SCE; (a) Wang J, Farias PA, Mahmoud JS (1985) Anal Chim Acta 172: 57; (b) Wang J, Farias PAM, Mahmoud JS (1985) J Electroanal Chem 195: 165; (c) Cotzee RC, Albertonni K (1983) Anal Chem 55: 1516; (d) Wang J, Zadeii JM (1986) Talanta 33: 321; (e) Wang J, Tuzhi P, Martinez T (1987) Anal Chim Acta 201: 43; (f) Wang J, Zadeii J (1986) Anal Chim Acta 185: 229; (g) Romanus A, Müller H, Kirsch D (1991) Fresenius Z Anal Chem 340: 363; (h) Wang J, Mahmoud JS (1986) J Electroanal Chem 208: 383; (i) Lenderman B, Monien H, Specker H (1972) Anal Lett 5: 837; (j) Wang J, Grabarić BS (1990) Microchim Acta 100: 31.

Adsorptive stripping voltammetry of metal complexes is complementary to anodic stripping voltammetry. It makes possible the determination of about 40 elements in a great variety of matrices [68]. By changing the conditions in the solution, the same ligand can be used for the determination of many metal ions, as can be seen in Table II.7.8. Several metal ions can be determined simultaneously using the same ligand if their stripping peak potentials differ significantly, as in the case of In and Fe, Zn and Ni, Ti and Fe, U and Ni, Fe and Ga, or Mn and Fe [74].

In the stripping phase the electrode is usually polarized cathodically [75–77]. In the majority of complexes the metal ion is reduced either partly or completely [78, 79], but reduction of the ligand in the adsorbed complex is also possible [59]. Upon partial reduction of the metal ion, the number of ligands in the complex may be changed, but the new complex remains adsorbed at the electrode surface [60, 72]. If the ligand is electroactive, it is essential that the stripping peak potential of the ligand in the adsorbed complex is significantly different from the potential of the free ligand [80]. The stripping response of the adsorbed complex may be enhanced by a catalytic reduction of anions coordinated to the metal ion in the complex [81]. Examples are the reduction of chlorate catalyzed by molybdenum [82] and titanium [83] complexes with mandelic acid, of bromate by iron–triethylamine complex [84], of nitrate by the chromium–dimethylenetriaminepentaacetic acid complex [85] and of nitrite by copper, nickel, and cobalt complexes with dimethylglyoxyme, thiabendazole, and 6-mercaptopurine-9-d-riboside [86, 87]. This type of reactions was reviewed recently [88].

II.7.2.5 Preconcentration by Surface Complexation

Hydrophobic organic ligands added to a carbon-paste electrode, or immobilized on a chemically modified electrode, can be used for the accumulation of metal ions by the surface-complexation mechanism [89–91]. The same goal can be achieved by immobilized ion exchangers such as natural or synthetic zeolites [89], clay minerals [92], silica [93], and ion exchange resins [94]:

$$n\mathrm{HY} + \mathrm{Me}^{n+} \rightleftarrows \mathrm{MeY}_n + n\mathrm{H}^+ \tag{II.7.28}$$

In Eq. (II.7.28) HY is a protonated ligand, or a structural unit of the ion exchanger. If less than 10% of the recipient capacity is consumed, the surface concentration of the bound metal is linearly proportional to the concentration of metal ions at the electrode surface [95]:

$$\Gamma_{\mathrm{MeY}_n} = \lambda \left(c_{\mathrm{Me}^{n+}}\right)_{x=0} \tag{II.7.29}$$

where $\lambda = K\Gamma_{\mathrm{HY}}^{n} c_{\mathrm{H}^+}^{-n}$, K is the equilibrium constant, and Γ_{HY} is the surface concentration of the recipient. If the sample solution is stirred during the accumulation, the rate of increase of Γ_{MeY_n} is equal to the flux of metal ions in the solution:

$$\frac{\mathrm{d}\Gamma_{\mathrm{MeY}_n}}{\mathrm{d}t} = \frac{D}{\delta}\left(c_{\mathrm{Me}^{n+}}^{*} - (c_{\mathrm{Me}^{n+}})_{x=0}\right) \tag{II.7.30}$$

where $c_{\mathrm{Me}^{n+}}^{*}$ is the concentration of metal ions in the bulk of the sample solution. The solution of Eqs. (II.7.29) and (II.7.30) is

$$\Gamma_{\mathrm{MeY}_n} = \lambda c_{\mathrm{Me}^{n+}}^{*}\left[1 - \exp\left(-D\lambda^{-1}\delta^{-1}t\right)\right] \tag{II.7.31}$$

The main advantage of this procedure is that neither a complexing agent nor an electrolyte is added to the sample solution. The metal ions are chemically collected and then transferred into an electrolytic solution for the voltammetric measurements [91].

II.7.3 Stripping Voltammetry at Two Immiscible Liquid Electrolyte Solutions

Finally, the stripping procedure can also be applied to the interface between two immiscible electrolyte solutions [96]. By a proper polarization of the interface, a certain ion can be transferred from the sample solution into a small volume of the second solution. After this accumulation, the ion can be stripped off by linear scan voltammetry, or some other voltammetric technique. The stripping peak current is linearly proportional to the concentration of ions in the second solution and indirectly to the concentration of ions in the sample solution. The method is used for the determination of electroinactive ions, such as perchlorate anion [97]. The principles of the procedure are the same as in the case of faradaic reactions, and the differences arise from the particular properties of phenomena on the interface that are beyond the scope of this chapter.

II.7.4 General Features of Stripping Voltammetry

There is no essential difference between anodic and cathodic stripping voltammetry, in spite of common use of this division. The direction of electrode polarization depends on the properties of the accumulated substance. Anodic polarization is applied to amalgams, metal deposits, and some adsorbed organic compounds that can be oxidized within the window of the working electrode. Cathodic polarization is suitable for oxides, mercuric and mercurous salts, adsorbed metal ion complexes, and reducible organic molecules. Metal ions collected by ion exchangers at a carbon-paste electrode are firstly reduced to atoms by cathodic polarization and then oxidized in an anodic stripping step [91]. The choice of both the voltammetric method and the electrode material is restricted by the properties and the concentration of the analyte. For instance, bismuth or copper ions can be determined by anodic stripping voltammetry on a mercury drop electrode [98], but the oxidation of mercury that may interfere at lower metal-ion concentrations can be avoided by adsorptive accumulation of organic complexes of these ions that is followed by a cathodic stripping determination [99]. This is possible because the complexes are reduced at lower potential than the free metal ions. Ultimately, mercury can be excluded by depositing bismuth or copper on a glassy-carbon electrode [15]. The accumulation procedure and the electrode material are interconnected. The applications of mercury drop, mercury film, noble metals, paraffin-impregnated graphite, glassy-carbon, and carbon-paste electrodes were explained above. Using carbon-paste electrodes, the collection mode can be varied by changing the additives [100, 101]. The working window of these electrodes is wide and their surface is renewable, hence they can be used for the accumulation of both reducible and oxidizable substances. The hydrodynamic conditions depend on the analyte concentration. Usually, the solution is stirred during the accumulation, but not in the stripping phase, so that 5–15 s of a quieting period in between is needed.

Voltammetric techniques that can be applied in the stripping step are staircase, pulse, differential pulse, and square-wave voltammetry. Each of them has been described in detail in previous chapters. Their common characteristic is a bell-shaped form of the response caused by the definite amount of accumulated substance. Staircase voltammetry is provided by computer-controlled instruments as a substitution for the classical linear scan voltammetry [102]. Normal pulse stripping voltammetry is sometimes called reverse pulse voltammetry. Its favorable property is the re-plating of the electroactive substance in between the pulses [103]. Differential pulse voltammetry has the most rigorously discriminating capacitive current, whereas square-wave voltammetry is the fastest stripping technique. All four techniques are insensitive to fast and reversible surface reactions in which both the reactant and product are immobilized on the electrode surface [104, 105]. In all techniques mentioned above, the maximum response, or the peak current, depends linearly on the surface, or volume, concentration of the accumulated substance. The factor of this linear proportionality is the amperometric constant of the voltammetric technique. It determines the sensitivity of the method. The lowest detectable concentration of the analyte depends on the smallest peak current that can be reliably measured and on the efficacy of accumulation. For instance, in linear scan voltammetry of the reversible surface reaction $R_{ads} + ne^- \leftrightarrows P_{ads}$, the peak current is [52]

$$i_p = \frac{n^2 F^2 S v \Gamma_{R,T}}{4RT} \qquad (II.7.32)$$

where $\Gamma_{R,T}$ is the surface concentration of the adsorbed reactant at the end of the accumulation period, v is the scan rate, and S is the electrode surface area. If the minimum peak current that can be measured is 10 nA, the corresponding minimum surface concentration is 1.3×10^{-12} mol/cm^2 (for $n = 2$, $v = 0.1$ V/s, and $S = 0.02$ cm^2). Considering Eq. (II.7.22), together with Eqs. (II.7.29), (II.7.30) and (II.7.31), and assuming that $\beta\Gamma_{max} = 0.1$ cm, the detection limit of this method is 1.3×10^{-8} mol/L. This limit can be achieved after a rather long accumulation from the stirred solution. If the adsorption is ten times weaker ($\beta\Gamma_{max} = 0.01$ cm), the lowest detectable concentration is only 1.3×10^{-7} mol/L.

In anodic stripping voltammetry of amalgams and metal deposits, there is no theoretical limit of detection of metal ions. If the accumulation potential is on the plateau of the pseudopolarogram and the solution is stirred, a steady state is established and the concentration of metal ions is linearly proportional to the duration of the accumulation:

$$c_{Me(Hg)} = \frac{c_{Me(b)} D t_{acc}}{l\delta} \qquad (II.7.33)$$

(for the meaning of symbols, see Eqs. II.7.14–II.7.16). Hence, the minimum concentration of atoms that is needed for the minimum measurable stripping peak can be obtained by adjusting the accumulation time, regardless of the bulk concentration of metal ions. For instance, the determination of mercury ions at the concentration levels of 10^{-11} and 10^{-13} mol/L requires accumulations of 1800 and 2400 s, respectively [106]. However, an infinite accumulation is not possible because of traces of surface-active substances that block the electrode surface. This phenomenon determines the limits of detection in natural solutions.

Surface-active impurities may hinder the stripping measurements [107, 108]. Their influence is highest in the anodic stripping voltammetry of metal ions, in which they may inhibit both the accumulation and stripping phases [109, 110]. The inhibitory effect upon the kinetics of electrode reactions of indium and cadmium is severe, but negligible in the case of the reaction of thallium [111, 112]. The blocking effect exerted in the accumulation phase is observed in the potentiometric stripping analysis, in which the deposit is chemically oxidized [113]. These hindrances can be diminished by covering the electrode surface with porous membranes [114, 115], protective gel layers [116], or polymer films [117]. In adsorptive stripping voltammetry, the foreign surfactants act either competitively [45, 68, 118], or synergistically [66, 71–73, 119, 120].

The application of stripping voltammetry includes the measurements of metal ions and organic compounds in a variety of chemical, environmental, metallurgical, geological, biological, biochemical, pharmaceutical, and clinical materials [2, 121–123]. They are used in routine trace metal analysis of waters [124] and can serve as reliable, sensitive, and precise methods for the verification of results obtained by atomic absorption spectroscopy, or some chromatographic techniques [125].

References

1. Vydra F, Štulik K, Julakova E (1976) Electrochemical stripping analysis. Ellis Horwood, Chichester
2. Brainina K, Neyman E (1993) Electroanalytical stripping methods. John Wiley, New York
3. Dewald HD (1996) Stripping analysis. In: Vanysek P (ed) Modern techniques in electroanalysis. John Wiley, New York, p 151
4. Henze G, Neeb R (1986) Elekrochemische Analytik. Springer, Berlin Heidelberg New York
5. Wang J (1994) Analytical electrochemistry. VCH, Weinheim
6. Jagner D, Graneli A (1976) Anal Chim Acta 83: 19
7. Macdonald DD (1977) Transient techniques in electrochemistry. Plenum Press, New York, p 119
8. Southampton electrochemistry group (1985) Instrumental methods in electrochemistry. John Wiley, New York, p 44
9. Gawrys M, Golimowski J (1999) Electroanalysis 11: 1318
10. Budevski E, Staikov G, Lorenz WJ (1996) Electrochemical phase formation and growth. VCH, Weinheim
11. Kolb DM, Przasnyski M, Gerischer H (1974) J Electroanal Chem 54: 25
12. Weast RC, Lide DR, Astle MJ, Beyer WH (1989) CRC handbook of chemistry and physics, 70th edn. CRC Press, Boca Raton, pp E-93 (a) and B-207 (b)
13. Xing X, Tae Bae I, Scherson DA (1995) Electrochim Acta 40: 29
14. Uchida H, Miura M, Watanabe M (1995) J Electroanal Chem 386: 261
15. Komorsky-Lovrić Š (1988) Anal Chim Acta 204: 161
16. Meyer S, Scholz F, Trittler R (1996) Fresenius J Anal Chem 356: 247
17. Giacomino A, Abollino O, Malandrino M, Mentasti E (2008) Talanta 75: 266
18. Kruusma J, Banks CE, Nei L, Compton RG (2004) Anal Chim Acta 510: 85

19. Spano N, Panzanelli A, Piu PC, Pilo MI, Sanna G, Seeber R, Tapparo A (2005) Anal Chim Acta 553: 201
20. Hwang GH, Han WK, Park JS, Kang SG (2008) Talanta 76: 301
21. Serrano N, Alberich A, Diaz-Cruz JM, Ariño C, Esteban M (2008) Electrochim Acta 53: 6616
22. Štulikova M (1973) J Electroanal Chem 48: 33
23. Malakhova NA, Miroshnikova EG, Stojko NY, Brainina KhZ (2004) Anal Chim Acta 516: 49
24. Kapturski P, Bobrowski A (2007) Electroanalysis 19: 1863
25. Piech R, Baś B, Kubiak WW (2008) J Electroanal Chem 621: 43
26. Branica M, Novak DM, Bubić S (1977) Croat Chem Acta 49: 539
27. Lovrić M (1998) Electroanalysis 10: 1022
28. Kounaves SP (1992) Anal Chem 64: 2998
29. Roe DK, Toni JEA (1965) Anal Chem 37: 1503
30. Komorsky-Lovrić Š, Branica M (1987) J Electroanal Chem 226: 253
31. Branica G, Lovrić M (1997) Electrochim Acta 42: 1247
32. Florence TM (1979) J Electroanal Chem 97: 237
33. Lovrić M, Pižeta I, Komorsky-Lovrić Š (1992) Electroanalysis 4: 327
34. Zelić M, Sipos L, Branica M (1985) Croat Chem Acta 58: 43
35. Piech R, Kubiak WW (2007) Electrochim Acta 53: 584
36. Mirčeski V, Lovrić M (1998) Electroanalysis 10: 976
37. Schultze JW, Koppitz FD (1976) Electrochim Acta 21: 327
38. Szulborska A, Baranski A (1994) J Electroanal Chem 377: 269
39. Forsman V (1983) Anal Chim Acta 146: 71
40. Lopez Fonseca JM, Otero A, Garcia Monteagudo J (1988) Talanta 35: 71
41. von Wandruszka R, Yuan X, Morra MJ (1993) Talanta 40: 37
42. Ali AMM, Farghaly OA, Ghandour MA (2000) Anal Chim Acta 412: 99
43. Kontrec J, Svetličić V (1998) Electrochim Acta 43: 589
44. Jin JY, Xu F, Miwa T (2000) Electroanalysis 12: 610
45. Korolczuk M (2000) Electroanalysis 12: 837
46. Banks CE, Kruusma J, Moore RR, Tomčik P, Peters J, Davis J, Komorsky-Lovrić Š, Compton RG (2005) Talanta 65: 423
47. Scholz F, Nitschke L, Henrion G (1987) J Electroanal Chem 224: 303
48. Nitschke L, Scholz F, Henrion G (1988) Z Chem 28: 452
49. Scholz F, Nitschke L, Henrion G (1987) Z Chem 27: 305
50. Scholz F, Nitschke L, Henrion G (1987) Anal Chim Acta 199: 167
51. Frumkin AN, Damaskin BB (1964) Adsorption of organic compounds at electrodes. In: O'Bockris J, Conway BE (eds) Modern aspects of electrochemistry, vol 3. Butterworths, London, p 149
52. Laviron E (1982) Voltammetric methods for the study of adsorbed species. In: Bard AJ (ed) Electroanalytical chemistry, vol 12. Marcel Dekker, New York, p 53
53. Damaskin BB (1964) Electrochim Acta 9: 231
54. Wang J (1989) Voltammetry after nonelectrolytic preconcentration. In: Bard AJ (ed) Electroanalytical chemistry, vol 16. Marcel Dekker, New York, p 1
55. Kalvoda R (2000) Electroanalysis 12: 1207
56. Fogg AG, Zanoni MVB, Barros AA, Radrigues JA, Birch BJ (2000) Electroanalysis 12: 1227
57. Heyrovsky M (2000) Electroanalysis 12: 935
58. Anson FC, Barclay DJ (1968) Anal Chem 40: 1791
59. van den Berg CMG (1989) Analyst 114: 1527
60. Kalvoda R, Kopanica M (1989) Pure Appl Chem 61: 97
61. Caselli M, Papoff P (1969) J Electroanal Chem 23: 41
62. Zelić M, Lovrić M (1990) Electrochim Acta 35: 1701
63. Zelić M, Branica M (1991) J Electroanal Chem 309: 227

64. Guaus E, Sanz F (1999) Electroanalysis 11: 424
65. Lovrić M (1989) Anal Chim Acta 218: 7
66. Komorsky-Lovrić Š, Branica M (1993) J Electroanal Chem 358: 273
67. Lovrić M, Komorsky-Lovrić Š (1995) Langmuir 11: 1784
68. Paneli MG, Voulgaropoulos A (1993) Electroanalysis 5: 355
69. Cobelo-Garcia A, Santos-Echeandia J, Prego R, Nieto O (2005) Electroanalysis 17: 906
70. Neiman EY, Dracheva LV (1989) Zhur Anal Khim 45: 222
71. Mlakar M, Branica M (1989) Anal Chim Acta 221: 279
72. Mlakar M, Lovrić M, Branica M (1990) Collect Czech Chem Commun 55: 903
73. Mlakar M, Branica M (1989) J Electroanal Chem 257: 269
74. Wang J, Mahmoud J, Zadeii J (1989) Electroanalysis 1: 229
75. Arancibia V, Muñoz C (2007) Talanta 73: 546
76. Tyszczuk K, Korolczuk M (2008) Anal Chim Acta 624: 232
77. Dalvi AA, Satpati AK, Palrecha MM (2008) Talanta 75: 1382
78. Kapturski P, Bobrowski A (2008) J Electroanal Chem 617: 1
79. Segura R, Toral MI, Arancibia V (2008) Talanta 75: 973
80. Croot PL, Johansson M (2000) Electroanalysis 12: 565
81. Turyan YI (1992) J Electroanal Chem 338: 1
82. Pelzer J, Scholz F, Henrion G, Heininger P (1989) Fresenius Z Anal Chem 334: 331
83. Yokoi K, van den Berg CMG (1991) Anal Chim Acta 245: 165
84. Golimowski J (1989) Anal Lett 22: 481
85. Boussemart M, van den Berg CMG, Ghaddaf M (1992) Anal Chim Acta 262: 103
86. Ion A, Banica FG, Luca C (1998) J Electroanal Chem 444: 11
87. Baxter LAM, Bobrowski A, Bond AM, Heath GA, Paul RL, Mrzljak R, Zarebski J (1998) Anal Chem 70: 1312
88. Bobrowski A, Zarebski J (2000) Electroanalysis 12: 1177
89. Labuda J (1992) Select Electrode Rev 14: 33
90. Baldrianova L, Agrafiotou P, Švancara I, Vytřas K, Sotiropoulos S (2008) Electrochem Comm 10: 918
91. Degefa TH, Chandravanshi BS, Alemu H (1999) Electroanalysis 11: 1305
92. Tonle IK, Ngameni E, Walcarius A (2005) Sensors Actuators B 110: 195
93. Walcarius A, Devoy J, Bessiere J (2000) J Solid State Electrochem 4: 330
94. Mikysek T, Švancara I, Vytřas K, Banica FG (2008) Electrochem Comm 10: 242
95. Wang C, Zhu B, Li H (1999) Electroanalysis 11: 183
96. Vanysek P (1996) Liquid–liquid electrochemistry. In: Vanysek P (ed) Modern techniques in electroanalysis. John Wiley, New York, p 337
97. Lu Z, Sun Z, Dong S (1989) Electroanalysis 1: 271
98. Eskilsson H, Jagner D (1982) Anal Chim Acta 138: 27
99. Jin W, Wang J (1994) Electroanalysis 6: 882
100. Švancara I, Matoušek M, Sikora E, Schachl K, Kalcher K, Vytřas K (1997) Electroanalysis 9: 827
101. Matelka R, Vytřas K, Bobrowski A (2000) J Solid State Electrochem 4: 348
102. Saralathan M, Osteryoung RA, Osteryoung JG (1987) J Electroanal Chem 222: 69
103. Mlakar M, Lovrić M (1990) Analyst 115: 45
104. Komorsky-Lovrić Š, Lovrić M (1996) Electroanalysis 8: 958
105. Lovrić M (1987) J Electroanal Chem 218: 77
106. Meyer S, Kubsch G, Lovrić M, Scholz F (1997) Int J Environ Anal Chem 68: 347
107. Ćosović B (1990) Adsorption kinetics of a complex mixture of organic solutes at model and natural phase boundaries. In: Stumm W (ed) Aquatic chemical kinetics. John Wiley, New York, p 291
108. Krznarić D, Plavšić M, Ćosović B (1994) Electroanalysis 6: 131
109. Plavšić M, Ćosović B (1989) Water Res 23: 1545
110. Komorsky-Lovrić Š, Lovrić M, Branica M (1995) Electroanalysis 7: 652

111. Guidelli R, Foresti ML (1977) J Electroanal Chem 77: 73
112. Komorsky-Lovrić Š, Branica M (1994) Fresenius J Anal Chem 349: 633
113. Komorsky-Lovrić Š, Branica M (1993) Anal Chim Acta 276: 361
114. Stewart EE, Smart RB (1984) Anal Chem 56: 1131
115. Aldstadt JH, Dewald HD (1993) Anal Chem 65: 922
116. Keller OC, Buffle J (2000) Anal Chem 72: 943
117. Jin L, Bi N, Ye J, Fang Y (1991) Microchim Acta 103: 115
118. Gromulska A, Stroka J, Galus Z (1999) Electroanalysis 11: 595
119. Harman AR, Baranski AS (1991) Anal Chem 63: 1158
120. Plavšić M, Krznarić D, Ćosović B (1994) Electroanalysis 6: 469
121. Frenzel W, Brätter P (1987) Applications of electroanalytical flow analysis in the trace element determination of biological materials. In: Brätter P, Schramel P (eds) Trace element analytical chemistry in medicine and biology. Walter de Gruyter, Berlin, p 337
122. van den Berg CMG (1999) Determination of trace elements. Analysis by electrochemical methods. In: Grasshoff K, Kremling K, Ehrhardt M (eds) Methods of seawater analysis. Wiley-VCH, Weinheim, p 302
123. Kalvoda R (2000) Crit Rev Anal Chem 30: 31
124. Mart L (1979) Fresenius Z Anal Chem 296: 350
125. Barek J, Mejstřik V, Muck A, Zima J (2000) Crit Rev Anal Chem 30: 37

111. Giudelli R, Foresti ML (1977) J Electroanal Chem 79 [illegible]
112. Komorsky-Lovrić Š [illegible]
113. [illegible]
114. [illegible]
115. [illegible]
116. [illegible]
117. [illegible]
118. [illegible]
119. [illegible]
120. [illegible]
121. [illegible]
122. [illegible]
123. [illegible]
124. [illegible]

Chapter II.8
Electrochemical Studies of Solid Compounds and Materials

Dirk A. Fiedler and Fritz Scholz

II.8.1 Introduction

Electroanalytical techniques are traditionally associated with studies of solutions; however, direct studies of the electrochemistry of solid materials are very tempting because they can give access to a wealth of information, ranging from elemental composition to thermodynamic and kinetic data, from structure–reactivity relations to new synthetic routes.

Methods describing the electrochemical investigation of solid compounds and materials have significantly expanded to new possibilities over the last two decades. This chapter focuses on the use of a fairly new and straightforward method referred to as voltammetry of immobilized microparticles (VIM). Detailed reviews of the method are available elsewhere [1a–c]. Beside applications in fundamental studies, this method proved to be especially valuable for the analysis of solid materials studied in archeometry [1d].

II.8.2 Experimental

II.8.2.1 Electrodes and Electrode Preparation

In studies which focus on the redox properties of immobilized microparticles, electrode preparation and electrolyte composition are to be carefully considered. Almost any kind of solid electrode may be applied in order to investigate the redox properties of immobilized solid microparticles. However, attention must be drawn to possible catalytic properties of the electrode with respect to the reactions to be studied, and reactions of the electrolyte, which may or may not accompany the reactions of the solids. The surface hardness of the particular electrode should be kept in mind as well. For example, a hard electrode will allow soft or flake-like solids to be

D.A. Fiedler (✉)
Cochlear, Limited, Lane Cove, NSW 2066, Australia
e-mail: dirkfiedler1@optusnet.com.au

F. Scholz (ed.), *Electroanalytical Methods*, 2nd ed.,
DOI 10.1007/978-3-642-02915-8_11,

attached while a smooth and soft electrode will preferably be applied in studies of hard materials to obtain a good embedding of the microparticles in the soft surface.

The method involves attaching an ensemble of solid microcrystalline particles to the surface of a suitable electrode. The electrode with the so immobilized solid is then transferred to an electrolyte solution and investigated either purely electrochemically or by additional methods.

Before attaching solid particles to the electrode, and also after the electrochemical measurement, the electrode has to be carefully cleaned. The best cleaning method has to be chosen from applying either inorganic or organic solvents, such as concentrated aqueous acids when studying metal oxides, or acetone or acetonitrile when organometallic compounds are involved. Alternatively, mechanical cleaning with a razor blade or abrasive cloth or powder, or simply polishing the electrode surface on paper may be equally effective. After either cleaning method, thorough rinsing of the electrode surface, possibly with the solvent to be applied in the measurements, is recommended in order to avoid cross contamination. The success of the cleaning must be checked by recording blank voltammograms.

II.8.2.1.1 Carbon Electrodes

Carbon electrodes probably exhibit the most convenient properties in that they combine a large potential window in either aqueous or nonaqueous solutions with almost any desirable size and hardness, not to mention availability and price.

The hardness of an electrode often limits the ability of a solid compound to adhere to an electrode surface. For example, fairly soft, flake-like materials, such as many organic compounds, will preferably adhere to a hard surface, e.g., that of a glassy carbon (GC) electrode, while hard solids such as oxides or minerals will adhere to a soft surface such as that of a paraffin-impregnated graphite electrode (PIGE).

Glassy carbon electrodes come in different hardness, but are generally the hardest nondiamond-like carbon electrodes one can get. Care should be taken when roughening a glassy carbon electrode because additional complications may arise. For example, it has been observed that an electrolyte film can separate the glassy carbon surface from the immobilized particles. This can give rise to special phenomena associated with the electron transfer from the carbon to the sample particles [2].

PIGEs made from porous spectral-quality graphite rods commonly used in spark analysis can be most favorably used for the electrochemical investigation of solid compounds. Typical size for VIM experiments is 50 mm long by 5 mm diameter. Untreated graphite rods are microporous, and the pores would lead to a high background current due to ingress of electrolyte solution into the graphite rods. Of course, such a penetration is also undesirable as it can lead to contamination of the electrodes with solution constituents. Therefore, impregnation of these graphite rods with (chemically rather stable) solid paraffin is suggested. To prepare PIGEs, paraffin with a low melting point between 56 and 70°C is melted in a closed vessel in a water bath. The graphite rods are given to the melt and the vessel is evacuated. The impregnation is performed until no more gas bubbles evolve from the rods.

Usually, this is the case within a time span of 2–4 h. After ambient pressure has been established, the rods are removed from the melt before the paraffin solidifies. The warm electrodes are placed onto filter paper to allow them to cool and dry. The lower end of the electrode is carefully polished on smooth paper. This surface will later accommodate the solid particles to be studied. The electrodes are now ready for use. For easy cleaning, the PIGE is used without an insulation of its cylindrical surface, as will be described below. The upper end of the electrode is connected to the electrochemical instrument with an alligator clip.

Highly oriented pyrolytic graphite (HOPG) can also be used as electrode; one only has to be sure whether the electrode exhibits either a basal plane or a lamellar structure.

The basal plane orientation offers easy cleavage of used graphitic layers, thus allowing generation of a fresh and clean surface simply with the aid of a razor blade or with sticky tape, simply by removing the uppermost surface layers. The electrodes can be made by shrink-fitting a suitable piece of HOPG into an electrode body made from, e.g., Teflon. Contact is usually established with silver-filled epoxy (either commercial quality or acetone-washed silver powder in standard epoxy; filling ratio about 1:1 *v/v*). External contact is easily established by means of a copper wire, which is connected to a 4-mm socket, on the top of the electrode. This avoids chemical interference between the electrolyte solution and electrode materials except where wanted because of the chemical inertness of Teflon. A tight-shrink fit is easily established because of the high thermal coefficient of expansion of Teflon.

The lamellar HOPG orientation is prone to intercalation and will most probably resist proper cleaning. Therefore, use of this material is not generally recommended.

Financially limited or time-pressing studies may take advantage of pencil-lead electrodes, which are equally well suited for VIM studies [3].

II.8.2.1.2 Metal Electrodes

Almost any metal electrode may be applied. One must take care not to operate too close to the limits of the electrochemical window of the used combination of electrode and electrolyte solution in order to avoid catalytic or undesired side effects such as gas evolution reactions. When using thin film electrodes, such as in combined quartz crystal microbalance (QCMB) studies, extreme care must be taken not to scratch the metal surface, usually gold, in order to maintain an electronically conductive path within the electrode. Examples of such experiments are given in the literature [4–7]. Also, when using metal electrodes in aqueous solutions, the background voltammogram should be examined very closely as the formation of surface oxides or the like may be mistaken for signals of the solid under study.

In general, metal electrodes are the second choice after carbon electrodes, and they should only be used when carbon electrodes fail. Platinum disk electrodes have been applied in studies on metals in nonaqueous electrolyte solutions [8]. MnO_2 and manganates were investigated on Pt electrodes in nonaqueous [9] and strongly alkaline aqueous electrolyte solutions [10–12].

II.8.2.1.3 Other Electrodes

For combined studies involving spectroscopic methods, in particular, the possibility of using optically transparent electrodes made from glass coated with indium–tin–oxide (ITO) or similar materials, e.g., SnO_2/F or SnO_2/Sb, should be taken into account (see Chap. Chapter II.6). However, because of their chemical nature, studies of reduction reactions are naturally limited with such electrodes.

In principle, the only limitation of electrode material is electrolysis of either the electrolyte solution or the electrode itself, so the reader is encouraged to try nonconventional or unusual electrodes where applicable. Suitability is quickly established by running a blank voltammogram at the voltammetric sweep rate, which is to be applied later with the solid-covered electrode.

II.8.2.2 Sample Preparation

The compound should be reasonably stable under ambient conditions. The compound or material to be studied should be ground to a fine powder in order to avoid any preferential crystalline orientations on the electrode. Of course, impurities, which may be electrochemically active, should not be introduced by this procedure.

When working with solids that are sensitive to ambient conditions, special precautions have to be established. A simple bucket filled with argon may do the trick, should a glove box not be to hand. This does not work with gaseous nitrogen because N_2 readily mixes with air, while Ar is heavier than air and remains in the bucket provided movement is minimal. Applying liquid nitrogen will provide a steady source of inert gaseous nitrogen. However, it has certain drawbacks in quickly freezing out water from the ambient, which may not be desirable when working with water-sensitive compounds.

Some solids may be easily attached with a cotton bud; however, it will frequently be necessary to gently rub the electrode onto the ground material, which has been placed on a clean filter paper, a clean sheet of blanc writing paper, a glazed tile or a glass plate. According to experience, the most even particle size distribution when attached to an electrode surface can be achieved when using the coarse filter paper method. For very hard materials, electrochemically inert abrasives such as aluminum oxide or diamond powder may be added to the compound under investigation prior to grinding it.

Excessive powder should be either wiped off the electrode body or blown away or removed by carefully tipping the electrode on its side. The electrode can now be placed into the conventional electrochemical cell as the working electrode. Before starting an experiment, it should be ensured that no visible gas bubbles adhere to the electrode surface because they produce entirely undesired distortions of the voltammetric signal to be recorded. As it is very convenient to work with electrodes which do not have an isolation of the shaft, the electrode surface area exposed to the electrolyte solution can be kept constant by first dipping the electrode into solution followed by lifting it up to a position where the electrolyte solution just adheres to the lower circular surface of the electrode rod.

II.8.2.3 Experimental Setup

Generally speaking, the equipment for VIM studies is identical to that for conventional electrochemical studies. (Please refer to part III of this volume for further details on the experimental setup.)

Generally, just as in conventional solution voltammetric experiments, a three-electrode setup is to be preferred in order to avoid excessive voltage drops across the cell and undue potential shifts of the signals.

Furthermore, the electrode should always be immersed carefully into the electrolyte solution. Otherwise, corrosion products of the electrode body or of electrode connectors are almost guaranteed to spoil the intended electrochemical experiment. Corrosion will preferably occur at loci of high current densities, which are typically produced when contacts to the electrode made by alligator clips are immersed into the electrolyte solution. This is also true for the case where entire solid samples need to be contacted in order to undertake electrochemical studies. Therefore, under no circumstances should the contact area be allowed to be wetted by the electrolyte solution. This problem can be circumvented by suitably covering the contact area; in the case of an electrolyte solution with low surface tension, this process can be hard work and may even lead to the construction of a special cell design.

II.8.2.4 Strategy

In order to achieve interpretable results, one should always pay much attention to the reproducibility of results. Also, it is often helpful to compare the results from VIM studies with those from experiments in solution.

Each set of experiments should start and end with a blank voltammogram in order to verify the cleanliness of the electrode itself. Also, the potential window of the used electrode can be determined in this way.

Once solid microparticles of a compound have been immobilized on the surface of a suitable electrode, the electrochemical behavior can be studied. When a voltammetric method is used, the starting potential should be carefully set to a value where no reaction is expected. Alternatively, one may start at the open circuit potential, which either has been predetermined or which is measured by the instrument before commencing the scan. In principle, one can record the oxidation and the reduction response of the compound. In many cases it is useful to perform a pre-electrolysis of the compound, either a reduction or oxidation, followed by recording the response of the reverse processes. Thus metal sulfide or metal oxide particles can be converted to metal particles, which are oxidized in a follow-up scan to yield metal-specific signals, very much like in stripping voltammetry of solutions. Practically, there is no limitation with respect to electrochemical measurements that can be carried out with immobilized particles.

II.8.2.5 What Compounds and Materials Can Be Studied?

There are only two requirements for the compounds and materials to be used by VIM: They should be highly insoluble in the electrolyte solution used and they must possess electroactivity, i.e., the ability to be either oxidized or reduced in the accessible potential window of the experiment. Most importantly, there is no restriction with respect to the electronic conductivity. Even insulators such as white phosphorus can be studied, because the electrochemical reaction which can take place at the three-phase boundary compound–electrode–solution can often deliver sufficient charge to give measurable currents. One can easily distinguish three different kinds of compounds, those which are not electroactive, those which are irreversibly destroyed in the electrochemical reactions, and those which can be reversibly reduced and oxidized. The latter compounds are characterized by possessing the ability to exchange electrons with the electrode and ions with the solution. This ability requires solid compounds that can house ions through features of their crystal structure, e.g., channels or interlayers.

II.8.3 Electrochemical Methods

Before going into detail, we would like to give the reader a feel for the amount of material and the associated faradaic currents one can expect from VIM studies. With electrode diameters in the range of 0.5–5 mm and a particle size of about 10 μm, peak currents are observed between 1 and 500 μA in cyclic voltammetric experiments, at voltammetric scan rates between 1 and 100 mV/s. This can be translated into amounts in the range of about 10 nmol to 5 μmol of solid being attached to an electrode surface. In other words, no special low-current equipment or complicated shielding is required. The only precaution that can often greatly help to achieve the optimum electrochemical window size, besides working with very clean solvents, is to saturate the electrolyte solution with a solvent-saturated inert gas such as nitrogen or argon.

II.8.3.1 Phase Identifications and Quantitative Analysis of Solids

In the simplest application, VIM can be used to identify solid phases. The voltammograms of solid phases are specific fingerprints, even when the assignment of signals is not obvious. Of course, VIM used in this way affords comparison of sample voltammograms to those of well-defined reference phases. Figure II.8.1 shows anodic dissolution voltammograms of complex sulfides, which have been preliminarily reduced to the metals. The voltammograms exhibit one signal for the anodic oxidation of thallium and one for that of tin. However, the shape and signal ratios depend on the composition of the complex thallium–tin sulfide phase. VIM, with the aim to identify phases, has been demonstrated for numerous minerals and pigments [13–16].

The composition of alloys can be very conveniently and rapidly established by VIM, using the technique of differential pulse voltammetry [17]. Also,

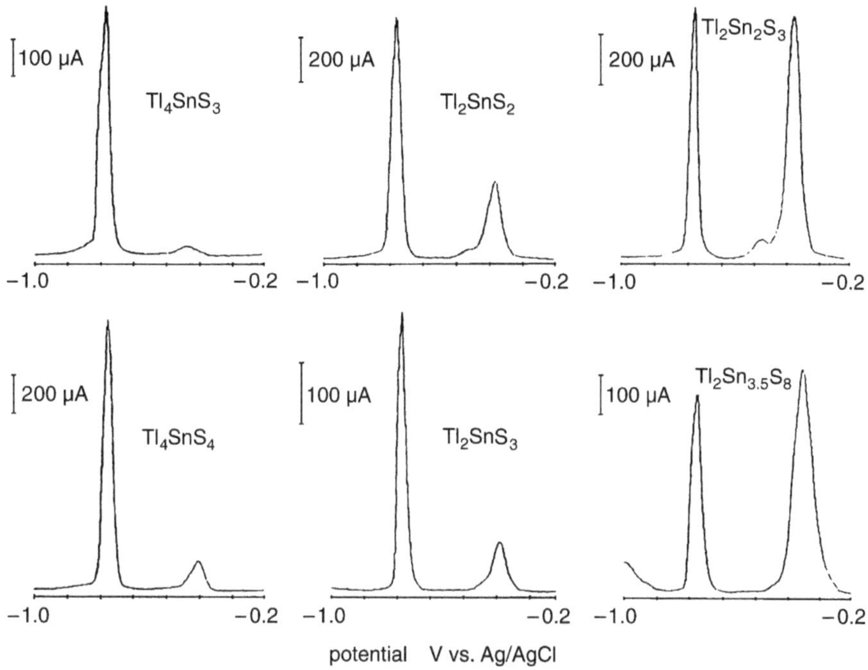

Fig. II.8.1 Anodic differential pulse voltammograms obtained after a preliminary reduction of complex sulfides at −1.0 V vs. Ag/AgCl for 60 s. The electrolyte solution was 1 M HCl with 10^{-4} M $HgCl_2$. The mercury salt was used to simultaneously plate metallic mercury where the tin and thallium metals could dissolve (reprinted with permission from [14])

metal-ion-containing compounds such as sulfides can be analyzed and, in the case of mixtures, the mixing ratio can be established with high accuracy [18]. It should be kept in mind that only electrochemically active ions can be analyzed directly. However, this depends on the electrolyte used, and, by choosing an appropriate solvent, one can, in principle, reduce virtually any metal ion.

Solid solutions are of great importance in various fields of science and technology. Analysis of solid solutions should answer two basic questions: (1) Does the solid solution consist of a single phase or is it a (mechanical) mixture of two different phases? and (2) In the case of a single-phase solid solution, what is its composition? These questions can be answered more or less successfully with the help of many analytical techniques, e.g., X-ray diffraction or infrared spectroscopy. However, the great advantage of electrochemical investigations of immobilized microparticles is that these answers can be found with very minute sample amounts, even with single microparticles. Figure II.8.2 compares schematically the electrochemical responses of two compounds, A and B, to that of a solid solution of A and B. The process involved may be either oxidation or reduction, it may be reduction leading to irreversible dissolution, or it may be a reversible insertion electrochemical reaction. Mixed-phase thermodynamics requires the solid solution to exhibit a single voltammetric signal whereas the mixture of the single compounds will yield

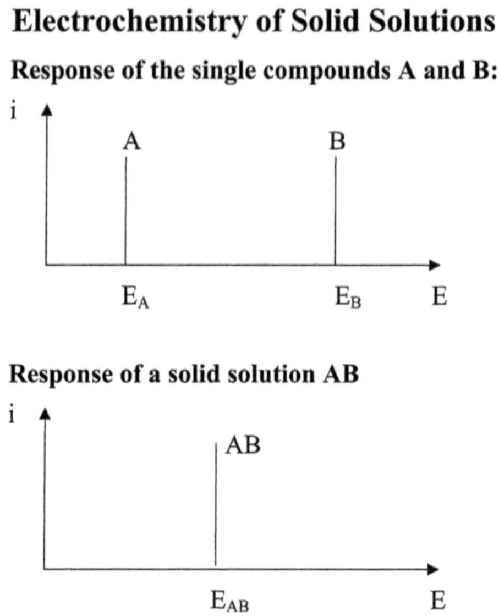

Fig. II.8.2 Schematic situation of the electrochemical signals of two compounds A and B forming solid solutions AB

two such signals. When the entire compositional series of solutions is studied, and assuming strict thermodynamic control of the electrochemical response (application of small voltammetric scan rates), the response will shift continuously from that of compound A to that of compound B. The shift will be nonlinear for two reasons, mixing entropy and possible nonideality of the system. Mixing entropy leads to a very small deviation from linearity; e.g., for a 1:1 solid solution of A and B and a one-electron transfer reaction, the deviation from linearity is approximately 0.018 V.

Figure II.8.3 gives two examples, one for the dependence of a voltammetric peak potential of copper reduction on the composition of copper sulfide–selenide solid solutions, and one for the dependence of the formal potentials (mid-peak potentials from cyclic voltammetry) of mixed iron–copper hexacyanoferrates on the composition of these compounds. The copper sulfide–selenides behave in very nonideal fashion, whereas the solid solution hexacyanoferrates give, within the limit of experimental errors, an almost linear dependence. These and other examples [18–21] are well suited to show that by voltammetric measurements of immobilized microparticles it is extremely facile to answer the two questions, is it a solid solution or not, and what is its composition.

II.8.3.2 Studies of the Electrochemical Behavior of Solid Materials

In the case of battery materials, the interparticle diffusion effects within a normally starved electrolyte cell design force standard voltammetric or coulometric experiments to be extremely time consuming. However, when working with small

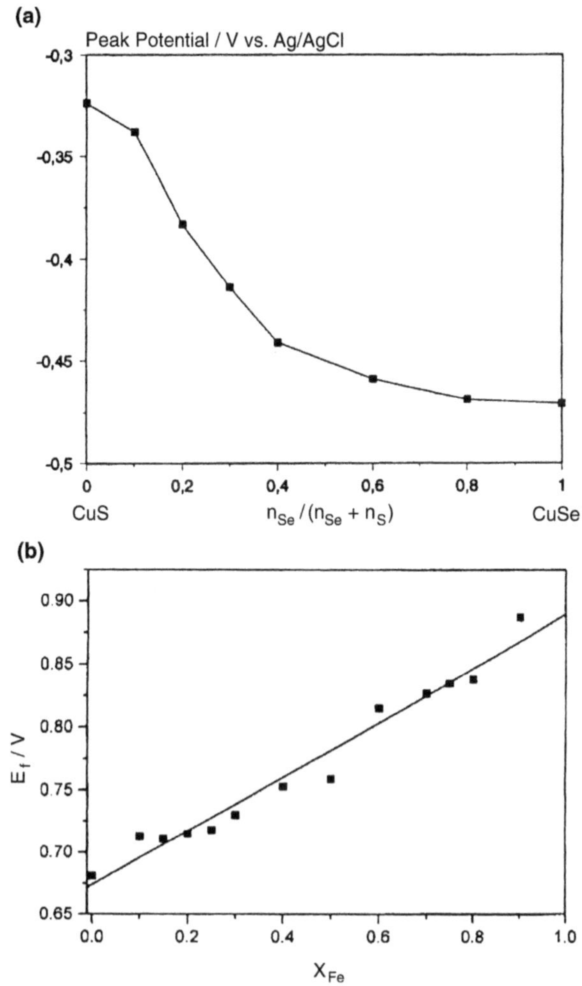

Fig. II.8.3 (**a**) Plot of peak potentials of different mixed CuS–CuSe solid solutions vs. mole fraction $n_{Se}/(n_{Se}+n_S)$ (reprinted with permission from [18]). (**b**) Formal potentials $E_f = (E_{p,\,anodic}+E_{p,\,cathodic})/2$ of the low-spin iron system of the mixed iron–copper hexacyanoferrates as derived from the cyclic voltammograms (scan rate 0.05 V/s) vs. mole fraction of high-spin iron x_{Fe} (reprinted with permission from [21])

samples of either the active material or the corresponding electrode mixture, the effect of interparticle diffusion can be safely excluded. Hence, voltammetric scan rates may be increased from some μV/s to some mV/s, thus reducing the experimental timescale from many hours to several minutes [9, 10]. Furthermore, it has been shown that VIM provides essentially the same information as can be deduced from the response of whole batteries when discussing the repetitive charge/discharge or "cycling" behavior of electrodes [10].

VIM also allows the observation of reaction products that are unstable in solution when working under conditions that allow for electrochemical reaction of the compound while at the same time preventing macroscopic dissolution processes. For example, the rapid isomerization of *cis*-$[Cr(CO)_2(dpe)_2]^+$ (dpe: $Ph_2PCH_2CH_2PPh_2$) to the *trans*$^+$ form in solution was elegantly avoided in the solid state, and so the proposed square scheme involving the *cis*/*cis*$^+$ and the *trans*/*trans*$^+$ couples was nicely proved [22].

VIM is suitable when the kinetics of electrochemical dissolution of particles is of interest because information on size and shape of the particles is desired [23–26].

Theoretical aspects of VIM have been addressed in several papers. These studies concern the geometry of the progressing electrochemical reaction [27–29], the influence of the conductivity of solutions [30], and also the question whether electrical insulators may yield measurable currents [31]. The interested reader is referred to these publications.

II.8.4 Combined Methods

II.8.4.1 Ex Situ Methods

In principle, any desired ex situ combination of VIM with other analytical methods can be applied, keeping in mind the necessary stability over a rather long timescale and the absolute amount one needs in order to gain significant signal-to-noise (S/N) ratios with such methods. Otherwise, in situ detection of products on a shorter timescale may be facilitated by suitable combination as exemplified further below. Of course, significant stability of the product of the VIM experiment is presumed in ex situ combinations. For example, one can conveniently probe the presence or absence of counterions after a VIM experiment by EDX [4, 22, 32–36] or the crystallographic properties by X-ray spectroscopy [6].

Ex situ specular reflectance IR spectroscopy was successfully applied to identify *cis*-$[Cr(CO)_2(dpe)_2]^+$, which is stable on a long timescale only in the solid-state phase, but very rapidly isomerizes to the corresponding *trans*-isomer in solution [37].

Electrochemical analysis of dissolved solid products generated by VIM [34] was elegantly applied by complete electrolysis of the solid and, after having removed and dried the electrode, subsequent dissolution of the solid product in a drop of an organic solvent, and final detection of the now dissolved product with a microelectrode dipped into the drop, using the VIM electrode as counter electrode.

Determination of metals in minerals or oxides in general is another example of VIM followed by an electrochemical measurement. The mineral is first reduced to generate metallic phases on the electrode. Subsequently, the metallic deposits are stripped in the very same electrochemical cell at appropriate potentials. Compositions can be calculated from resulting peak integrals [13, 38].

II.8.4.2 In Situ Methods

Combined VIM and X-ray Diffraction [39]. The crystallographic phase of (solid) products generated immediately by VIM experiments can be studied when performing both experiments simultaneously. The reduction of PbO and Pb(OH)Cl served as an example of this combination, showing strong evidence for a topotactic electrochemical solid-state reaction.

Combined VIM and EPR Spectroscopy [40]. This combination allows for the in situ detection of electrochemically generated radicals. Studying the solid compounds TCNQ (7,7,8,8-tetracyanoquinodimethane) and organometallic trans-$Cr(CO)_2(dpe)_2$ [dpe: 1,2-bis(diphenylphosphino)ethane], it was shown that the radicals generated by either electrochemical reduction (TCNQ) or oxidation [*trans*-$Cr(CO)_2(dpe)_2$] of these solids are also solids. Dissolution and thus distinctively different appearance of the corresponding EPR spectra could be forced by suitable tuning of the electrolyte solution composition.

Combined VIM and Electrochemistry [40]. Rather unexpectedly, solid compounds immobilized to an electrode surface adhere very well to such surfaces even when rotating such an electrode at a speed of up to 10^4 min^{-1}. Reaction products, when soluble, can conveniently be detected with the help of a ring disk electrode.

Combined VIM and Light Microscopy [41, 42]. Observation of color changes of immobilized microparticles is greatly assisted by combination of VIM with light microscopy, as demonstrated with electrochromic octacyanomolybdates and -tungstates. By refining the instrumentation, diffuse reflection can be measured and thus quantified.

Combined VIM and UV/Vis spectroscopy of indigo immobilized on an ITO electrode has been described [43].

Combined VIM and quartz crystal microbalance experiments allowed the determination of mass increases during electrochemical redox reactions, indicating ingress of ions or molecules from the electrolyte solution [4–7].

Combined VIM and Calorimetry [44, 45]. With the help of thermistors and especially adopted electrodes it is possible to measure the heat effects of electrochemical conversions of solid microparticles.

Combined VIM and AFM [46–51]. In situ Atomic Force Microscopy (AFM) is an elegant method to follow the course of electrochemical transformations of solid phases. The combined use of electrochemical measurements and AFM allows detailed analysis of the reaction scenarios.

References

1. (a) Scholz F, Schröder U, Gulaboski R (2005) Electrochemistry of immobilized particles and droplets. Springer, Berlin; (b) Scholz F, Meyer B (1998): Voltammetry of solid microparticles immobilized on electrode surfaces. In: Bard AJ, Rubinstein I (eds) Electroanalytical chemistry. A series of advances, vol 20. Dekker, New York, pp 1–86; (c) Hermes M, Scholz F (2009) Solid state electrochemical reactions of electroactive micro- and nano-particles in a liquid electrolyte environment. In: Handbook of solid state electrochemistry. V Kharton

(edt), VCH-Wiley, Weinheim; (d) Doménech-Carbó A, Doménech-Carbó MT, Costa V (2009) Electrochemical methods in archeometry, conservation and restoration. In: Monographs in Electrochemistry, F. Scholz (edt) Springer, Berlin
2. Lange B, Lovric M, Scholz F (1996) J Electroanal Chem 418: 21
3. Blum D, Leyffer W, Holze R (1996) Electroanalysis 8: 296
4. Dostal A, Meyer B, Scholz F, Schröder U, Bond AM, Marken F, Shaw SJ (1995) J Phys Chem 99: 2096
5. Shaw SJ, Marken F, Bond AM (1996) J Electroanal Chem 404: 227
6. Bond AM, Fletcher S, Marken F, Shaw SJ, Symons PG (1996) J Chem Soc Faraday Trans 92: 3925
7. Shaw SJ, Marken F, Bond AM (1996) Electroanalysis 8: 732
8. Bond AM, Bobrowski A, Scholz F (1991) J Chem Soc Dalton Trans 411
9. Fiedler DA, Besenhard JO, Fooken M (1997) J Power Sources 69: 157
10. Fiedler DA (1998) J Solid State Electrochem 2: 315
11. Fiedler DA, Albering JH, Besenhard JO (1998) J Solid State Electrochem 2: 413
12. Jantscher W, Binder L, Fiedler DA, Andreaus R, Kordesch K (1999) J Power Sources 79: 9
13. Meyer B (1995) PhD thesis, Humboldt University Berlin, Berlin
14. Zhang S, Meyer B, Moh G, Scholz F (1995) Electroanalysis 7: 319
15. Doménech-Carbó A, Domenéch-Carbó MT, Gimeno-Adelantado JV, Moya-Moreno M, Bosch-Reig F (2000) Electroanalysis 12: 120
16. Doménech-Carbó A, Doménech-Carbó MT, Moya-Moreno M, Gimeno-Adelantado JV, Bosch-Reig F (2000) Anal Chim Acta 407: 275
17. Scholz F, Nitschke L, Henrion G (1989) Naturwissenschaften 76: 71
18. Meyer B, Zhang S, Scholz F (1996) Fresenius J Anal Chem 356: 267
19. Bond AM, Scholz F (1991) Langmuir 7: 3197
20. Reddy, SJ, Dostal A, Scholz F (1996) J Electroanal Chem 403: 209
21. Schwudke D, Stößer R, Scholz F (2000) Electrochem Commun 2: 301
22. Bond AM, Colton R, Daniels F, Fernando DR, Marken F, Nagaosa Y, Van Steveninck RFM, Walter JN (1993) J Am Chem Soc 115: 9556
23. Grygar T, Subrt J, Bohacek J (1995) Collect Czech Chem Commun 60: 950
24. Grygar T (1995) Collect Czech Chem Commun 60: 1261
25. Grygar T (1996) Collect Czech Chem Commun 61: 93
26. Grygar T (1996) J Electroanal Chem 405: 117
27. Lovric M, Scholz F (1997) J Solid State Electrochem 1: 108
28. Lovric M, Scholz F (1999) J Solid State Electrochem 3: 172
29. Schröder U, Oldham KB, Myland JC, Mahon PJ, Scholz F (2000) J Solid State Electrochem 4: 314
30. Lovric M, Hermes M, Scholz F (1998) J Solid State Electrochem 2: 401
31. Oldham KB (1998) J Solid State Electrochem 2: 367
32. Downard AJ, Bond AM, Hanton LR, Heath GA (1995) Inorg Chem 34: 6387
33. Dostal A, Kauschka G, Reddy SJ, Scholz F (1996) J Electroanal Chem 406: 155
34. Bond AM, Cooper JB, Marken F, Way DM (1995) J Electroanal Chem 396: 407
35. Bond AM, Marken F (1994) J Electroanal Chem 372: 125
36. Bond AM, Colton R, Mahon PJ, Tan WT (1997) J Solid State Electrochem 1: 53
37. Bond AM, Colton R, Marken F, Walter JN (1994) Organometallics 13: 5122
38. Scholz F, Nitschke L, Henrion G, Damaschun F (1989) Naturwissenschaften 76: 167
39. Meyer B, Ziemer B, Scholz F (1995) J Electroanal Chem 392: 79
40. Bond AM, Fiedler DA (1997) J Electrochem Soc 144: 1566
41. Schröder U, Meyer B, Scholz F (1996) Fresenius J Anal Chem 356: 295
42. Schröder U, Scholz F (1997) J Solid State Electrochem 1: 62
43. Bond AM, Marken F, Hill E, Compton RG, Hügel H (1997) J Chem Soc Perkin Trans 2: 1735
44. Bárcena Soto M, Kubsch G, Scholz F (2002) J Electroanal Chem 528: 18
45. Bárcena Soto M, Scholz F (2002) J Electroanal Chem 528: 27

46. Suarez MF, Marken F, Compton RG, Bond AM, Miao WJ, Raston CL (1999) J Phys Chem 103: 5637
47. Hasse U, Scholz F (2001) Electrochem Commun 3: 429
48. Hasse U, Wagner K, Scholz F (2004) J Solid State Electrochem 8: 842
49. Hasse U, Scholz F (2004) Electrochem Commun 6: 409
50. Hasse U, Fletcher S, Scholz F (2006) J Solid State Electrochem 10: 833
51. Hasse U, Scholz F (2006) Electrochem Commun 8: 1005

Chapter II.9
Potentiometry

Heike Kahlert

II.9.1 Introduction

In potentiometry, the potential of a suitable indicator electrode is measured versus a reference electrode, i.e. an electrode with a constant potential. Whereas the indicator electrode is in direct contact with the analyte solution, the reference electrode is usually separated from the analyte solution by a salt bridge of various forms. The electrode potential of the indicator electrode is normally directly proportional to the logarithm of the activity of the analyte in the solution. Potentiometric methods have been and are still frequently used to indicate the end point of titrations. This use has been known since the end of the nineteenth century. Direct potentiometric determinations using ion-selective electrodes have been mainly developed in the second half of the twentieth century.

It is an attractive feature of potentiometry that the equipment is rather inexpensive and simple: one needs a reference electrode, an indicator electrode and a voltage-measuring instrument with high input impedance. The potential measurement has to be accomplished with as low a current as possible because otherwise the potential of both electrodes would change and falsify the result. In the past, a widespread method was the use of the so-called Poggendorf compensation circuit. In most cases today, amplifier circuits with an input impedance up to 10^{12} Ω are used. The key element for potentiometry is the indicator electrode. Currently, ion-selective electrodes are commercially available for more than 20 different ions and almost all kinds of titrations (acid–base, redox, precipitation and complex titrations) can be indicated. In the following, some indicator electrodes and the origin of the electrode potentials will be described.

II.9.2 Cell Voltage

The Galvani potential difference of a single electrode is not directly measurable, because it is not possible to connect the two phases of an electrode with a measuring

H. Kahlert (✉)
Institut für Biochemie, Universität Greifswald, 17487 Greifswald, Germany
e-mail: hkahlert@uni-greifswald.de

F. Scholz (ed.), *Electroanalytical Methods*, 2nd ed.,
DOI 10.1007/978-3-642-02915-8_12,

system without creating new phase boundaries with additional electrochemical equilibria and thus additional Galvani potential differences. Galvanic cells consist of electrically connected electrodes. The cell reaction is the sum of the single electrode reactions, and the cell voltage E is the sum of the overall potential drops at the phase boundaries and within the phases. For example, when two electrically connected metal electrodes are immersed into the same electrolyte solution (Fig. II.9.1), at each phase boundary metal/electrolyte solution an equilibrium Galvani potential difference $\Delta\phi$ is established, the difference of which is E if identical metals are attached to both electrodes, which is fulfilled by using, e.g. copper wires for connecting the electrodes with the voltage meter:

$$E = \Delta\phi^{\mathrm{I}} - \Delta\phi^{\mathrm{II}} \tag{II.9.1}$$

When one of these Galvani potential differences, e.g. $\Delta\phi^{\mathrm{II}}$ is kept constant, this electrode can be used as a reference electrode. Then, the relative electrode potential of an electrode (indicated here as X) is the cell voltage of a galvanic cell, which consists of the electrode and the reference electrode (R).

$$E_{\mathrm{R}}(\mathrm{X}) = \Delta\phi(\mathrm{X}) - \Delta\phi(\mathrm{R}) \tag{II.9.2}$$

By international agreement, the standard hydrogen electrode (SHE) is used as the basic reference electrode to *report* electrode potentials. Its potential is defined to be zero at any temperature (zero point of potential scale). Relative electrode potentials against the standard hydrogen electrode are named electrode potentials $E(\mathrm{X})$. The electrode potential has a positive value when the standard hydrogen electrode is the anode of the galvanic cell, and it has a negative value when the SHE is the cathode of such a cell. The standard hydrogen electrode, in principle, is very useful for potential measurements, because the equilibrium potential is established very fast and reproducible. However, the handling of this electrode is difficult, especially with respect to a well-defined hydrogen ion activity.

For practical purposes, an important conclusion is that the difference of the relative potentials of two electrodes (indicated as X and Y) measured against the same reference electrode is equal to the cell potential of a cell consisting of these two electrodes X and Y:

$$E_{\mathrm{R}}(\mathrm{X}) - E_{\mathrm{R}}(\mathrm{Y}) = E(\mathrm{X},\mathrm{Y}) \tag{II.9.3}$$

Hence, any reference electrode can be used instead of the SHE, provided its potential is stable and well defined against the standard hydrogen electrode. These requirements are fulfilled by electrodes of the second kind, which are discussed in Chap. III.2.

II.9.3 Indicator Electrodes and Their Potentials

II.9.3.1 Redox Electrodes

When both the oxidised and the reduced forms of a redox couple are dissolved in a solution, an inert metal can attain a potential, which only depends on the ratio

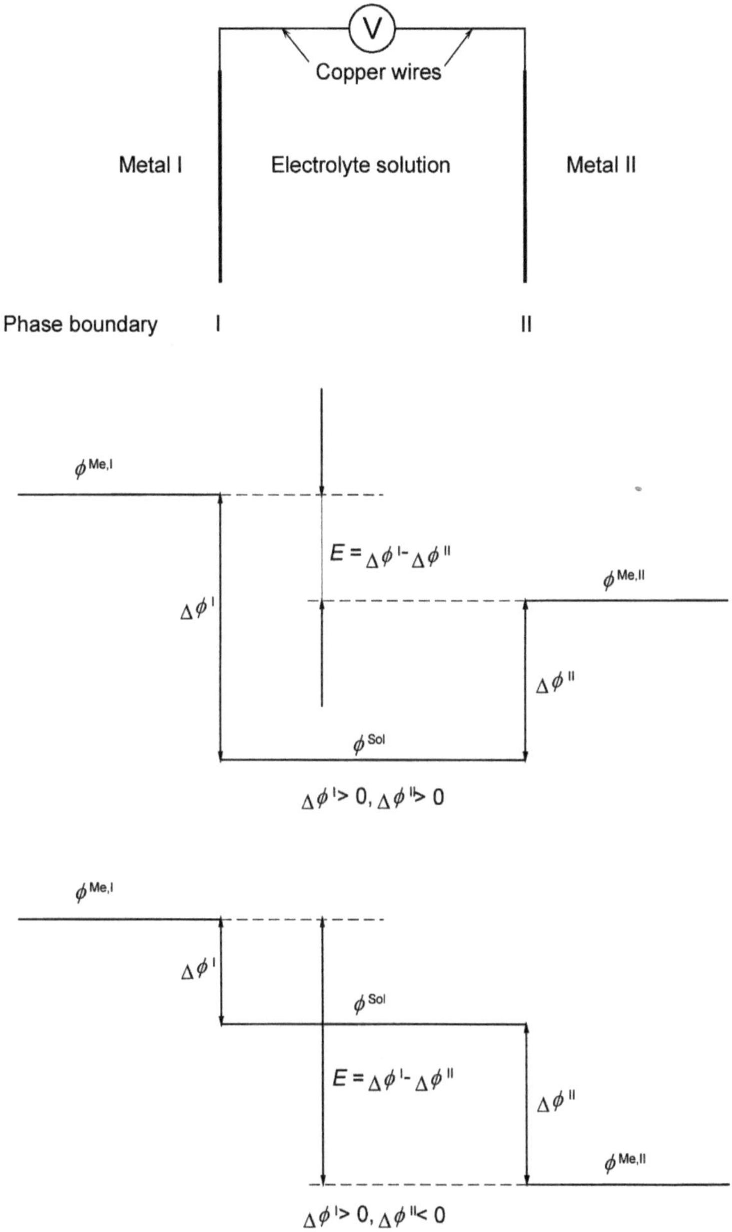

Fig. II.9.1 Cell voltage E as the sum of the Galvani potential differences for two metal electrodes in the same electrolyte solution. Superscript Sol indicates the solution and Me indicates the metal

of the activities of these two species. The inert electrode operates during the measurement as an electron source or drain (Fig. II.9.2a). As an example for a simple redox electrode with a homogeneous ionic equilibrium, in a solution containing Ce^{3+}, the potential of a platinum electrode depends on the activity ratio of Ce^{3+}

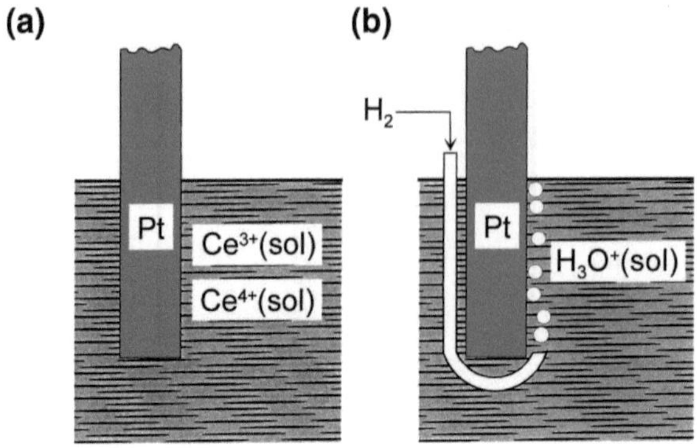

Fig. II.9.2a-b. **(a)** Schematic description of a simple redox electrode. Sol indicates the solution. **(b)** Schematic description of a hydrogen electrode. Sol indicates the solution

and Ce^{4+}:

$$Ce^{3+} \rightleftarrows Ce^{4+} + e^- \tag{II.9.4}$$

In the case of an electrochemical equilibrium, the Nernst equation follows:

$$\Delta\phi = \Delta\phi^{\ominus} + \frac{RT}{F}\ln\frac{a_{Ce^{4+}}}{a_{Ce^{3+}}} \tag{II.9.5}$$

and the electrode potential is

$$E\left(Ce^{3+},Ce^{4+}\right) = E_c^{\ominus'}\left(Ce^{3+},Ce^{4+}\right) + \frac{RT}{F}\ln\frac{a_{Ce^{4+}}}{a_{Ce^{3+}}} \tag{II.9.6}$$

Cf. Chap. I.2 to see the meaning of the formal potential $E_c^{\ominus'}\left(Ce^{3+},Ce^{4+}\right)$. Such inert indicator electrodes are very suitable for the indication of redox titrations.

Another example of a redox electrode with a homogeneous redox equilibrium is the so-called quinhydrone electrode for pH measurements reported by Biilmann in 1921 [1]. Quinhydrone is a charge transfer complex consisting of quinone and hydroquinone in a 1:1 ratio. If this compound is placed into an aqueous solution, the chemical equilibrium appears as follows [2]:

$$\text{Quinhydrone} \rightleftarrows \text{Hydroquinone} + \text{Quinone} \tag{II.9.7}$$

At a platinum wire immersed into this aqueous solution, an electrochemical equilibrium is established according to the electron transfer reaction:

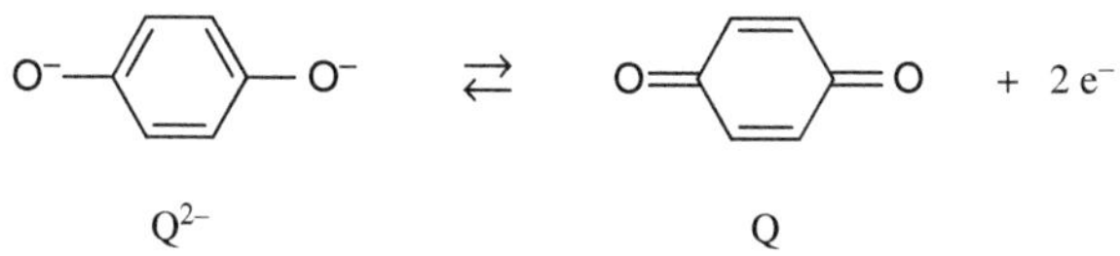

(II.9.8)

The Galvani potential difference can be formulated according to the Nernst equation [3]:

$$\Delta\phi = \Delta\phi^{\ominus} + \frac{RT}{2F}\ln\frac{a_{\mathrm{Q}}}{a_{\mathrm{Q}^{2-}}} \tag{II.9.9}$$

Hydroquinone is a dibasic acid with acid constants $K_{a,1}$ and $K_{a,2}$:

(II.9.10)

$$K_{a,1} = \frac{a_{\mathrm{H_3O^+}} \cdot a_{\mathrm{HQ^-}}}{a_{\mathrm{H_2Q}}} = 1.41 \cdot 10^{-10} \tag{II.9.11}$$

(II.9.12)

$$K_{a,2} = \frac{a_{\mathrm{H_3O^+}} \cdot a_{\mathrm{Q^{2-}}}}{a_{\mathrm{HQ^-}}} = 3.98 \cdot 10^{-12} \tag{II.9.13}$$

The equation of mass balance for hydroquinone is (a_{Red} = analytical concentration of hydroquinone):

$$a_{\mathrm{Red}} = a_{\mathrm{H_2Q}} + a_{\mathrm{HQ^-}} + a_{\mathrm{Q^{2-}}} = a_{\mathrm{Q^{2-}}}\left(\frac{a^2_{\mathrm{H_3O^+}}}{K_{a,1}\cdot K_{a,2}} + \frac{a_{\mathrm{H_3O^+}}}{K_{a,2}} + 1\right) \tag{II.9.14}$$

$a_{\mathrm{Q^{2-}}}$ is now found from (II.9.14):

$$a_{\mathrm{Q^{2-}}} = \frac{a_{\mathrm{Red}} \cdot K_{a,1} \cdot K_{a,2}}{a^2_{\mathrm{H_3O^+}} + a_{\mathrm{H_3O^+}} \cdot K_{a,1} + K_{a,1} \cdot K_{a,2}} \tag{II.9.15}$$

and, hence, the Galvani potential difference is

$$\Delta\phi = \Delta\phi^{\ominus} - \frac{RT}{2F}\ln K_{a,1}K_{a,2} + \frac{RT}{2F}\ln\frac{a_{\mathrm{Q}}}{a_{\mathrm{Red}}} + \frac{RT}{2F}\ln\left(a^2_{\mathrm{H_3O^+}} + K_{a,1}a_{\mathrm{H_3O^+}} + K_{a,1}K_{a,2}\right) \tag{II.9.16}$$

In solutions with pH < 9.5, $a^2_{\mathrm{H_3O^+}} >> K_{a,1}K_{a,2} + K_{a,1}a_{\mathrm{H_3O^+}}$, and, in the absence of compounds which would react only with the species of quinhydrone, the ratio

$\frac{a_Q}{a_{Red}}$ is practically unity, and therefore the quinhydrone electrode is easily applicable as a pH sensor. The electrode potential can be simply formulated as

$$E(25°C) = E_c^{\ominus\prime}(\text{quinhydrone}, H_3O^+) - 0.059\ [V] \times pH \quad (II.9.17)$$

($E_c^{\ominus\prime}$(quinhydrone,H_3O^+) is the formal potential of the quinhydrone electrode at pH < 9.5. At 25°C the formal potential is 0.6995 V vs. SHE). The situation at a redox electrode is more complex when a solid, an electrolyte and a gas are involved in the overall electrode reaction. The well-known hydrogen electrode is an example for such gas electrodes [4]. The theory of the hydrogen electrode is given in Sect. III.2.2. Of course, the hydrogen electrode can be used as a pH-sensitive electrode as its potential depends on pH as follows:

$$E(H_2, H_3O^+, 25°C, 1\ \text{atm}) = -0.059\ [V] \times pH \quad (II.9.18)$$

A schematic construction of a hydrogen electrode is given in Fig. II.9.2b.

II.9.3.2 Metal Electrodes or Electrodes of the First Kind

Electrodes of the first kind consist of a metal wire, which is in contact with a solution containing the cations of this metal. In contrast to redox electrodes, ions as charged particles are exchanged across the phase boundary, i.e. the electrode reaction can be regarded as an ion-transfer reaction. An example of this electrode is the silver electrode, a silver wire in a silver-ion-containing solution (Fig. II.9.3a). With this electrode all reactions can be monitored, which are accompanied by a change of the activity of silver ions:

$$Ag^0 \rightleftarrows Ag^+ + e^- \quad (II.9.19)$$

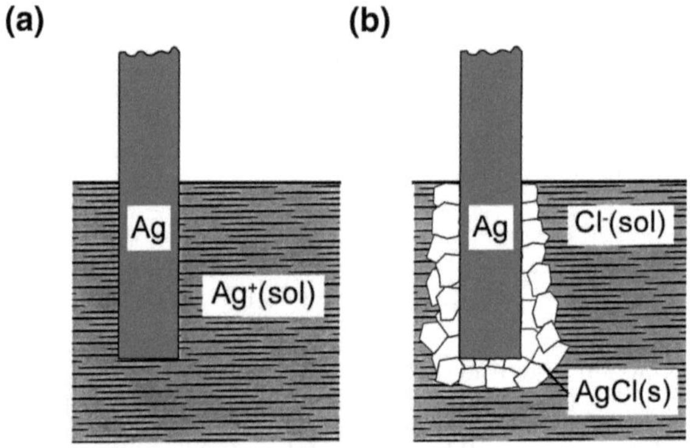

Fig. II.9.3a-b **(a)** Schematic description of the silver electrode as an example for an electrode of the first kind. Sol indicates the solution. **(b)** Schematic description of the silver-silver chloride electrode as an example for an electrode of the second kind. Sol indicates the solution and s indicates the solid salt

In electrochemical equilibrium, the following relation holds true for the electrochemical potentials:

$$0=\tilde{\mu}_{\mathrm{Me}^{z+}}^{\text{solution}}-\left(\tilde{\mu}_{\mathrm{Me}}^{\text{metal}}-\tilde{\mu}_{\mathrm{e}^-}^{\text{metal}}\right) \tag{II.9.20}$$

For neutral reaction participants, the electrochemical potential is equal to the chemical potential $\tilde{\mu}_{\mathrm{Me}}^{\text{metal}}=\mu_{\mathrm{Me}}^{\text{metal}}$. The activity of the electrons in the metal phase is not an independent variable, and therefore the chemical potential of electrons is *not* divided into a standard and an activity term:

$$\begin{aligned}0=\ &\mu_{\mathrm{Me}^{z+}}^{\ominus,\text{solution}}+RT\ln a_{\mathrm{Me}^{z+}}+z_{\mathrm{Me}^{z+}}F\phi^{\text{solution}}\\ &-\left(\mu_{\mathrm{Me}}^{\ominus,\text{metal}}+RT\ln a_{\mathrm{Me}}-\mu_{\mathrm{e}^-}^{\text{metal}}-z_{\mathrm{e}^-}F\phi^{\text{metal}}\right)\end{aligned} \tag{II.9.21}$$

Thus, the Galvani potential difference at the electrode–solution interface is

$$\Delta\phi=\phi^{\text{metal}}-\phi^{\text{solution}}=\frac{\mu_{\mathrm{Me}^{z+}}^{\ominus,\text{solution}}-\mu_{\mathrm{Me}}^{\ominus,\text{metal}}+\mu_{\mathrm{e}^-}^{\text{metal}}}{z_{\mathrm{Me}^{z+}}F}+\frac{RT}{z_{\mathrm{Me}^{z+}}F}\ln\frac{a_{\mathrm{Me}^{z+}}}{a_{\mathrm{Me}}} \tag{II.9.22}$$

Since the metal is a pure phase it possesses unit activity. Hence the electrode potential is written as follows:

$$E=E^{\ominus}(\mathrm{Me},\mathrm{Me}^{z+})+\frac{RT}{z_{\mathrm{Me}^{z+}}F}\ln a_{\mathrm{Me}^{z+}} \tag{II.9.23}$$

Electrodes of the first kind are applicable for the indication of precipitation and complex titrations in particular.

II.9.3.3 Electrodes of the Second Kind

An electrode of the second kind results, when a simple metal electrode is coupled to a precipitation equilibrium of a metal salt. The silver/silver chloride electrode is an example of this kind of electrodes (Fig. II.9.3b). Hence, the electrode reaction consists of the ion-transfer reaction and the precipitation of silver chloride according to

$$\text{Ion-transfer reaction: } \mathrm{Ag}^0 \rightleftarrows \mathrm{Ag}^+ + \mathrm{e}^- \tag{II.9.24a}$$

$$\text{Precipitation: } \mathrm{Ag}^+ + \mathrm{Cl}^- \rightleftarrows \mathrm{AgCl}_{\downarrow} \tag{II.9.24b}$$

$$\text{Netreaction: } \mathrm{Ag}^0 + \mathrm{Cl}^- \rightleftarrows \mathrm{AgCl}_{\downarrow} + \mathrm{e}^- \tag{II.9.24}$$

Here, the activity of the metal ions in the solution depends on the solubility equilibrium, and the activity of the silver ions in the solution can be described by the solubility product K_s according to

$$a_{\mathrm{Ag}^+} = \frac{K_\mathrm{S}}{a_{\mathrm{Cl}^-}} \tag{II.9.25}$$

This results in the following Nernst equation, so that the electrode potential is proportional to the logarithm of the activity of the chloride in the electrolyte solution:

$$E = E^{\ominus}(\mathrm{Ag,Ag^+}) + \frac{RT}{F}\ln a_{\mathrm{Ag}^+} = E^{\ominus}(\mathrm{Ag,Ag^+}) + \frac{RT}{F}\ln K_\mathrm{s} - \frac{RT}{F}\ln a_{\mathrm{Cl}^-} \tag{II.9.26}$$

or

$$E = E_\mathrm{c}^{\ominus\prime}(\mathrm{Ag,AgCl}) - \frac{RT}{F}\ln a_{\mathrm{Cl}^-} \tag{II.9.27}$$

Note that the formal potential $E_\mathrm{c}^{\ominus\prime}$ (Ag,AgCl) includes the solubility product.

As indicator electrodes, electrodes of the second kind are useful for the measurement of anions such as chloride, sulphate and so on. Because this type of electrode shows some outstanding characteristics, they are the most frequently used reference electrodes (cf. Chap. III.2).

II.9.3.4 Membrane Electrodes

When ions are transferred between two phases, say a solution and a membrane, this gives rise to a Galvani potential difference (see also Sect. I.2.5, where the ion transfer between two immiscible electrolyte solutions is considered). This happens also when both phases are ion conductors and electrons are not involved. The term membrane denotes a thin plate separating two liquid phases. Membranes are divided into three groups: 1. *Large meshed membranes.* Rapid mixing of different electrolyte solutions is delayed, and a diffusion potential arises due to the different diffusion coefficients of different ions (see also Chap. III.3, Eq. III.3.33). 2. *Close meshed membranes.* Only ions or molecules up to a certain size can pass the membrane. They work similarly to those which are thick with respect to adjacent mixed phases but they are able to host certain ions or molecules and, at least in principle, may transport them from the side of higher electrochemical potential to the side of lower electrochemical potential. Such membranes are named *semipermeable.* Here, an electric potential difference between the two solutions exists, which is called *Donnan potential.* 3. *Thick membranes.* Two Galvani potential differences can occur at the two interfaces, and diffusion potentials may build up in the membrane. Among the thick membranes, those of glass are most important (the glass membrane is geometrically thin; however, in the sense of the nomenclature given here it is thick).

II.9.3.4.1 Semipermeable Membrane Without Inner Diffusion Potential

The membrane is permeable for the ionic species K and the solvent, i.e. water molecules. When an uncharged membrane is placed between two solutions containing two different activities $a_{\mathrm{K}}^{\mathrm{I}}$ and $a_{\mathrm{K}}^{\mathrm{II}}$ of species K, then a phase transfer of charge carriers occurs. The direction of this transfer depends on the gradient of the electrochemical potential. This results in a charging of the phase boundary and creation of an electric field. The initially favoured ion transfer will be slowed down and in the end a further net transfer will be stopped because of electrostatic repulsion forces, and the forth and back transfer of ions will cancel. In electrochemical equilibrium both reactions have the same rate, and the potential difference is constant. Assuming that (i) no temperature or pressure gradient exists across the membrane, (ii) the solvent in both solutions is the same, e.g. water and (iii) no diffusion potential within the membrane occurs, then the electrochemical potentials in the two phases are equal in case of electrochemical equilibrium:

$$\tilde{\mu}_{\mathrm{K}}^{\mathrm{I}} = \tilde{\mu}_{\mathrm{K}}^{\mathrm{II}} \tag{II.9.28}$$

$$\mu_{\mathrm{K}}^{\ominus,\mathrm{I}} + RT \ln a_{\mathrm{K}}^{\mathrm{I}} + z_{\mathrm{K}} F \phi^{\mathrm{I}} = \mu_{\mathrm{K}}^{\ominus,\mathrm{II}} + RT \ln a_{\mathrm{K}}^{\mathrm{II}} + z_{\mathrm{K}} F \phi^{\mathrm{II}} \tag{II.9.29}$$

Under these conditions the chemical standard potentials in both phases are equal, so that one gets for the equilibrium Galvani potential difference:

$$\Delta\phi^{\mathrm{I,II}} = \frac{RT}{z_{\mathrm{K}} F} \ln \frac{a_{\mathrm{K}}^{\mathrm{II}}}{a_{\mathrm{K}}^{\mathrm{I}}} \tag{II.9.30}$$

This difference between the Galvani potential differences of two volume phases is the so-called Donnan potential (Donnan in 1911). For more detailed information and special cases of membrane potentials, cf. Koryta [5]. If such a semipermeable membrane is permeable only for one ionic species, the Donnan potential can be used for the determination of the activity of this ionic species in an analyte solution. This is what ion-selective membrane electrodes are used for. Figure II.9.4 depicts schematically such an electrode [6]. The membrane separates two aqueous electrolyte solutions. Solution I is the so-called inner solution with a well-defined activity of the analyte $a_{\mathrm{K}}^{\mathrm{I}}$. Solution II is the so-called outer solution with the unknown activity of the analyte $a_{\mathrm{K}}^{\mathrm{II}}$. Two equal reference electrodes, for instance, silver/silver chloride electrodes, are immersed in solutions I and II and are connected via a voltmeter with high input impedance. Then for the overall cell potential it follows that

$$E = const. + \frac{RT}{z_{\mathrm{K}} F} \ln a_{\mathrm{K}}^{\mathrm{II}} \tag{II.9.31}$$

Ion-selective membrane electrodes are in frequent use in analytical measurements. For further information, cf. for instance [7–14].

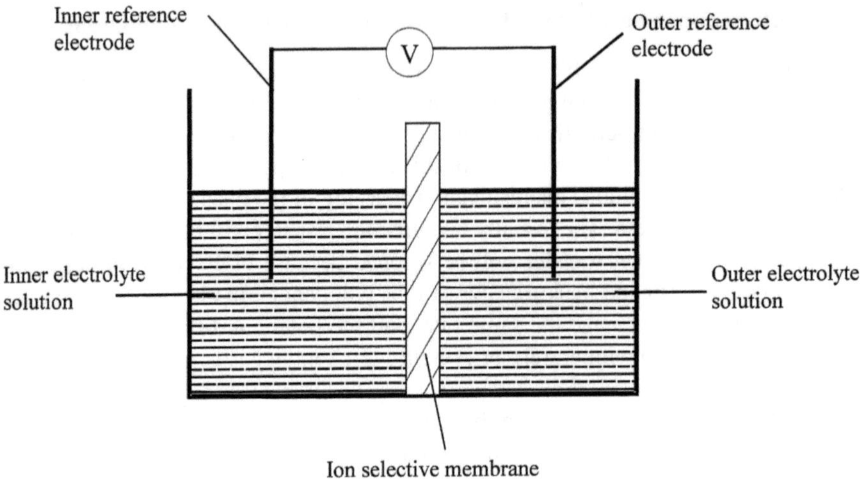

Fig. II.9.4 Schematic description of an ion selective membrane electrode

II.9.3.4.2 Potential Difference at a Glass Electrode

The glass electrode is by far the most frequently used ion-selective electrode. It is applied in almost all laboratories for the determination of pH values. Depending on the composition of the glass, some other ions are also detectable. However, in most textbooks, the theoretical description of the potential-determining electrochemical processes does not take into account important recent findings. For instance, in the past, the potential difference at a glass electrode was described either as an effect of the adsorption of hydronium ions, as a diffusion potential within the glass membrane, as a Donnan potential or as a potential difference at phase boundaries (see Fig. II.9.5). The results are the same, but, obviously, the theory of phase boundary potentials gives the right explanation (cf. Schwabe and Suschke [15]). A diffusion of hydronium ions through the glass membrane seems very improbable and indeed has never been observed. The charge transport inside the membrane occurs on account of the alkali ions inside the glass structure. A thermodynamic description of the phase boundary potential, the so-called "ion exchange theory" was developed by Nicolsky in 1937 and was further developed by others [16–17]. The mechanism found and experimentally verified by Baucke, the so-called dissociation mechanism, gives a deeper insight [18–19]. This includes thermodynamic and kinetic concepts. For an understanding of the function of glass electrodes it is necessary to know that glass material is made of an irregular silicate network with fixed anionic groups and mobile cations. These mobile cations can respond to concentration and potential gradients and can leave the network by replacement by other cations. The anionic groups, which are located immediately at the glass surface, belong to the glass and, on the other hand, are involved in equilibria with ions in the adjacent solution. This produces a phase boundary potential. The most important equilibrium is the dissociation of acidic silanol groups according to

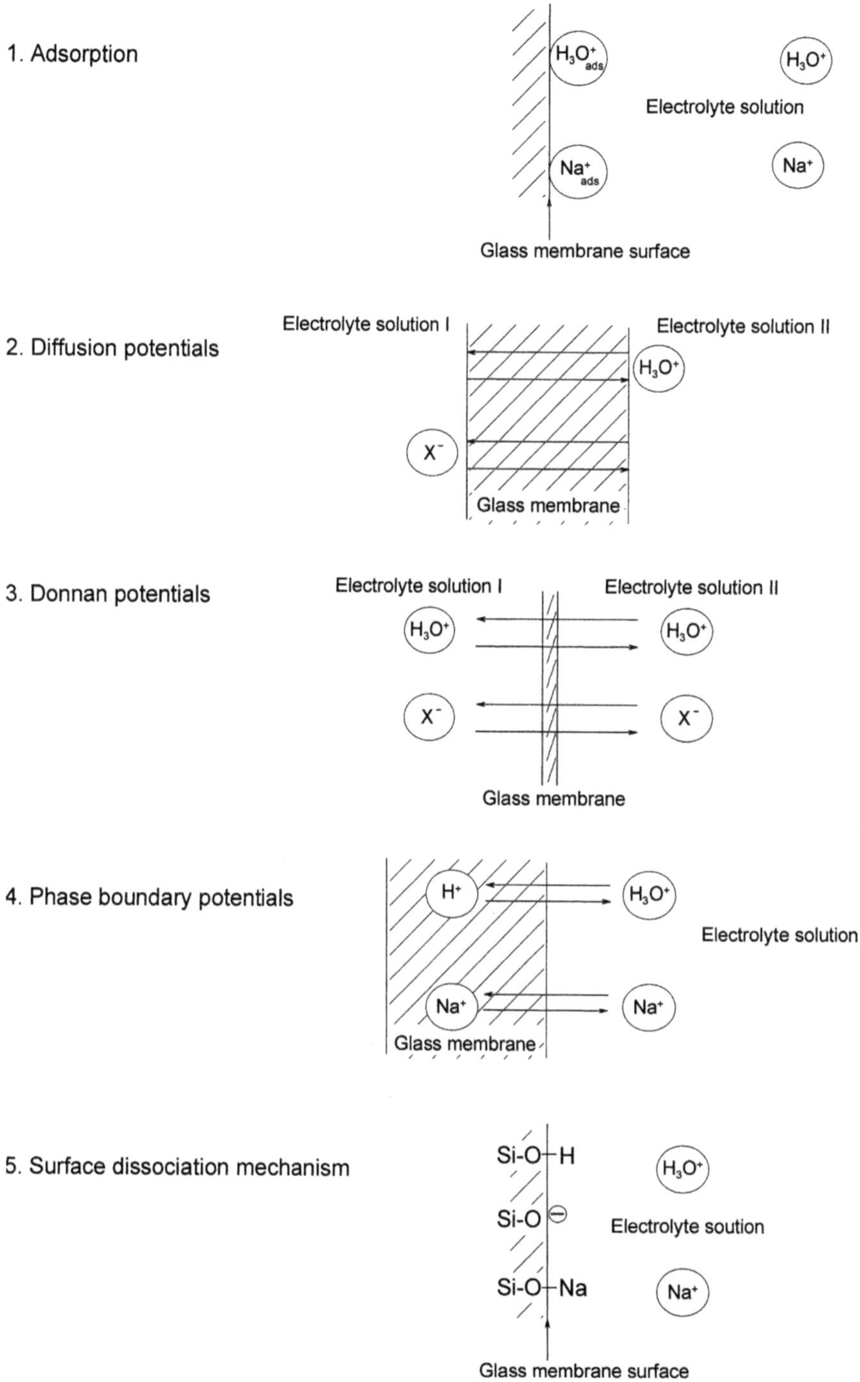

Fig. II.9.5 Different theories of the formation of the potential difference at a glass membrane

$$\equiv \mathrm{SiOH_{(s)}} + \mathrm{H_2O_{(sol.)}} \rightleftarrows \equiv \mathrm{SiO^-_{(s)}} + \mathrm{H_3O^+_{(sol.)}} \tag{II.9.32}$$

(s indicates solid and sol indicates solution)

The electrochemical dissociation equilibrium, Eq. (II.9.32), is characterised by a dissociation constant [20]:

$$K_{\mathrm{d,a}} = \frac{a_{\mathrm{H_3O^+}} c'_{\mathrm{SiO^-}}}{c'_{\mathrm{SiOH}}} \exp\left(-F\Delta\phi/RT\right) \tag{II.9.33}$$

where $c'_{\mathrm{SiO^-}}$ and c'_{SiOH} are the surface concentrations. Thus, one obtains directly the Galvani potential difference at the glass–solution phase boundary as a function of the hydrogen ion activity in the solution:

$$\Delta\phi = -\frac{RT}{F} \ln K_{\mathrm{d,a}} + \frac{RT}{F} \ln \frac{a_{\mathrm{H_3O^+}} c'_{\mathrm{SiO^-}}}{c'_{\mathrm{SiOH}}} \tag{II.9.34}$$

In case of high activities of hydroxyl ions (pH > 7) it must be taken into consideration that the protons from the acidic silanol groups also can react with hydroxyl ions, so that the dissociation equilibrium has to be formulated as

$$\equiv \mathrm{SiOH_{(s)}} + \mathrm{OH^-_{(sol.)}} \rightleftarrows \equiv \mathrm{SiO^-_{(s)}} + \mathrm{H_2O_{(sol.)}} \tag{II.9.35}$$

The formulation of the dissociation constant for this process in alkaline solutions, $K_{\mathrm{d,b}}$, yields a Galvani potential difference that is a function of the hydroxyl ions in the solution:

$$\Delta\phi = -\frac{RT}{F} \ln K_{\mathrm{d,b}} + \frac{RT}{F} \ln \frac{c'_{\mathrm{SiO^-}}}{a_{\mathrm{OH^-}} c'_{\mathrm{SiOH}}} = -\frac{RT}{F} \ln K_{\mathrm{d,b}} K_{\mathrm{W}} + \frac{RT}{F} \ln \frac{a_{\mathrm{H_3O^+}} c'_{\mathrm{SiO^-}}}{c'_{\mathrm{SiOH}}} \tag{II.9.36}$$

The dissociation constants in acidic solutions, $K_{\mathrm{d,a}}$, and the dissociation constant in alkaline solutions, $K_{\mathrm{d,b}}$, are connected via K_{W} $\left(K_{\mathrm{W}} = \frac{K_{\mathrm{d,a}}}{K_{\mathrm{d,b}}}\right)$. Hence, Eq. (II.9.36) effectively results in Eq. (II.9.34):

$$\Delta\phi = -\frac{RT}{F} \ln K_{\mathrm{d,a}} + \frac{RT}{F} \ln \frac{a_{\mathrm{H_3O^+}} c'_{\mathrm{SiO^-}}}{c'_{\mathrm{SiOH}}}$$

This means, that glass electrodes show an application range in both, acidic and alkaline solutions with one and the same function, but one has to remember that, above pH 7 hydroxyl ions are involved in the chemical reaction. Baucke has shown that the formation of a Galvani potential difference can also be derived from a kinetic point of view. Figure II.9.6, in principle, shows the proton exchange reaction between the surface of the glass and the solution. Both the backward and the forward reactions are second-order reactions. The exchange current can be expressed as follows:

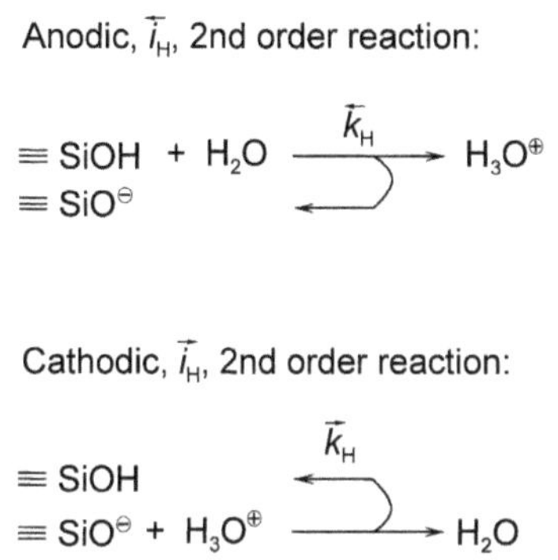

Fig. II.9.6 Anodic and cathodic exchange currents of protons between the functional groups at a glass membrane surface and an electrolyte solution

$$\begin{aligned} i_{0,\mathrm{H}} &= \left|\vec{i}_{\mathrm{H}}\right| = F\left|\vec{k}_{\mathrm{H}}\right| a_{\mathrm{H_3O^+}} c'_{\mathrm{SiO^-}} \exp\left(-\beta_{\mathrm{H}} F\Delta\phi/RT\right) \\ &= \overleftarrow{i}_{\mathrm{H}} = F\,\overleftarrow{k}_{\mathrm{H}} a_{\mathrm{H_2O}} c'_{\mathrm{SiOH}} \exp\left[(1-\beta_{\mathrm{H}})\,F\Delta\phi/RT\right] \end{aligned} \quad \text{(II.9.37)}$$

This follows an expression for the Galvani potential difference, which is in agreement with Eq. II.9.34, but the dissociation constant is expressed by the ratio of rate constants of the forward and backward reactions:

$$\Delta\phi = -\frac{RT}{F}\ln\frac{\overleftarrow{k}_{\mathrm{H}}}{\vec{k}_{\mathrm{H}}} + \frac{RT}{F}\ln\frac{a_{\mathrm{H_3O^+}} c'_{\mathrm{SiO^-}}}{c'_{\mathrm{SiOH}}} \quad \text{(II.9.38)}$$

The current density of protons is of $\mathrm{mA/cm^2}$ [18], which guarantees the stability of the electrode potential and a quick response.

Today glass electrodes are fabricated in a great variety of shapes and sizes depending on the demands. For routine work, the glass electrode with a bulb-shaped membrane is usually applied (see Fig. II.9.7). For the purpose of measuring the Galvani potential differences at both sides of the glass membrane, it is necessary to add supplementary phases. With two additional electrodes (the best is to use reference electrodes), each positioned on one side of the glass membrane, the following scheme of a so-called symmetrical glass electrode cell is obtained:

I	II	III	IV	V	VI	VII	I‘
Ag	AgCl (s), KCl (3 M)	H_3O^+ (analyte solution)	outer glass surface	glass membrane	inner glass surface	AgCl (s), KCl (3 M), H_3O^+ (inner solution)	Ag

The cell potential difference is the sum of the Galvani potential differences at all interfaces. For the interfaces III/IV and VI/VII Eq. (II.9.34) holds true. By using

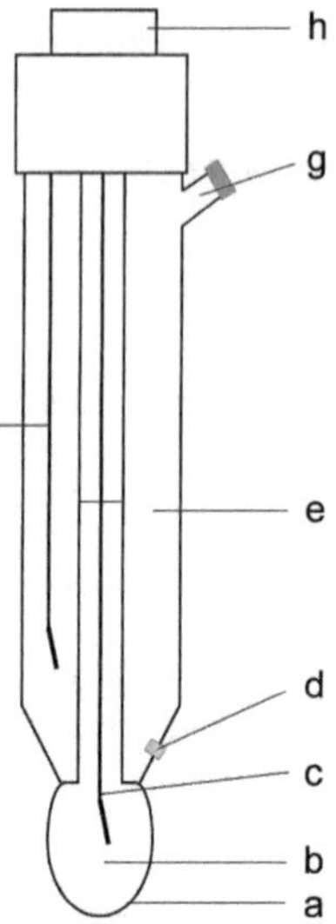

Fig. II.9.7 Schematic description of a glass electrode of bulb shape as a combination electrode (Einstabmesskette). (**a**) Bulb-shaped glass membrane, (**b**) inner buffer solution, (**c**) inner reference system, (**d**) diaphragm, (**e**) outer reference electrolyte solution, (**f**) outer reference system, (**g**) refilling neck, (**h**) head of the combination electrode

two identical reference electrodes, the Galvani potential differences at I/II and I'/VII cancel, and one gets for the cell potential:

$$E = \Delta\phi^{\mathrm{III,IV}} - \Delta\phi^{\mathrm{VI,VII}} = \frac{RT}{F}\ln\frac{a^{\mathrm{III}}_{\mathrm{H_3O^+}}c'^{\mathrm{IV}}_{\mathrm{SiO^-}}c'^{\mathrm{VI}}_{\mathrm{SiOH}}}{a^{\mathrm{VII}}_{\mathrm{H_3O^+}}c'^{\mathrm{IV}}_{\mathrm{SiOH}}c'^{\mathrm{VI}}_{\mathrm{SiO^-}}} \qquad \text{(II.9.39)}$$

At an ideal glass membrane, the surface concentrations of silanol and siloxy groups on both sides of the membrane are equal, and, if the inner solution is a buffer with a constant and well-known pH, the potential of the glass electrode depends directly on the pH of the analyte solution (outside the membrane):

$$E(25°\mathrm{C}) = 0.059[\mathrm{V}] \times \left(\mathrm{pH}^{\mathrm{VII}} - \mathrm{pH}^{\mathrm{III}}\right) \qquad \text{(II.9.40)}$$

Normally, the assumption that inner and outer surfaces of the glass membrane have the same composition does not hold true, especially because the inner buffer solution is kept inside the electrode right from the production of the electrode and always stays there, whereas the outer surface is influenced by different solutions over the time of use. This gives rise to the so-called asymmetry potential. Further, one will usually find a sub-Nernstian slope, which is less than 0.059 V (at 25°C). (For the thermodynamic explanation, see [19]). Therefore an empirical equation of the following kind has to be used:

$$E = E'_{\mathrm{N}}\left(\mathrm{pH}'_0 - \mathrm{pH}\right) \qquad \text{(II.9.41)}$$

with E_N' being the practical slope and pH_0' the real zero point of the electrode. This means that for practical measurements with glass electrodes, a calibration of the electrode is necessary. Because the practical slope and the real zero point are not constant values with respect to temperature and time, this calibration has to be repeated from time to time. For further information, cf. [2, 4, 7, 11, 14, 19] and the literature cited therein.

II.9.3.4.3 Solid-Contact Inner Reference Systems

The development of solid-contact ion-selective membrane electrodes (SC-ISE) aims at electrodes that overcome certain problems of ISE with inner filling aqueous solutions (e.g. drying out of the inner filling solution, temperature and pressure range, miniaturization etc.). The simplest design is realized in the so-called coated-wire electrode (molecular dispersion, or solution, of a salt or a complex in a polymeric matrix painted on an electron conductor like platinum wire). However, well-defined phase boundary potentials between the electronic conductor and the ionically conduction ion-selective membrane, which are the requirement for adequate long-term potential stability and reproducible potentials, are difficult to establish in such a simple arrangement. The most promising arrangement for solid-contact inner reference systems is to use and intermediate layer with both redox and ion exchange properties between the electron conductor and the ion selective membrane. A large variety of ion-to-electron transducers have been suggested in the literature (c.f. [25]). Amongst them conducting polymers (PPy, PEDOT) are considered as the most attractive [26]. More recently, macroporous carbon or single-walled carbon nanotubes have been investigated with respect to their signal transducing properties [27].

II.9.4 Interferences and Detection Limits in Potentiometric Measurements

In the foregoing discussion it was assumed that there is only one electrode reaction determining the potential of the electrode. It was also assumed that the concentration of the analyte is high enough and that the exchange currents are sufficiently high to establish the electrode potential, i.e. to assure electrochemical equilibrium (no net current) and hence an electrode potential following the Nernst equation. In reality, there are always two or more ions that can be exchanged at an interface. Thus, protons and also sodium ions can be transferred from the solution to the $\equiv SiO^-$ group on the glass surface. In both cases a negative charge is annihilated on the glass, i.e. compensated by the positive charge of the attached cation. It depends on the magnitude of the dissociation constant K_d for protons and sodium ions according to Eq. (II.9.33), and on the activities of protons and sodium ions in the solution, to what extent a mutual interference will be observed. The influence of such interfering ions on the electrode potential is expressed in a modified Nernst equation, the Nicolsky–Eisenman equation:

$$E = E_c^{\ominus\prime} + \frac{RT}{z_M F} \ln \left[a_M + \sum_I K_{M,I} \cdot a_I^{z_M/z_I} \right] \qquad \text{(II.9.42)}$$

with M being the measured ion and I the interfering ions. $K_{M,I}$ is the selectivity coefficient for the measurement of M in the presence of I. The smaller its value, the better the selectivity. In case of good ion-selective electrodes it has a value about 10^{-4}. For the determination of such selectivity coefficients different methods have been suggested, c.f. Eisenman et al., Light et al. and Rechnitz et al. [7]. The method of determining the selectivity coefficient by fixed interfering ion activity developed by Moody and Thomas [7] is also recommended by IUPAC [21, 22]. Here, the electrode potential is measured in a solution with a fixed activity of the interfering ion I and varying activities of measured ion M. In the range of higher activities of M the potential is determined by this measured ion M. At lower activities, the interference of I appears gradually (see Fig. II.9.8) and, at a certain activity of M, the electrode potential becomes constant and it is only determined by the interfering ion. By extrapolating both parts of this curve as it is shown in Fig. II.9.8, the selectivity coefficient can be calculated as follows:

$$K_{M,I} = \frac{a_{M,A}}{a_I^{z_M/z_I}} \qquad \text{(II.9.43)}$$

Equation (II.9.42) was derived on a purely thermodynamic basis. The interference can be visualised also in terms of electrode kinetics. Although this is not necessary because the ion transfer is usually reversible, it helps to understand

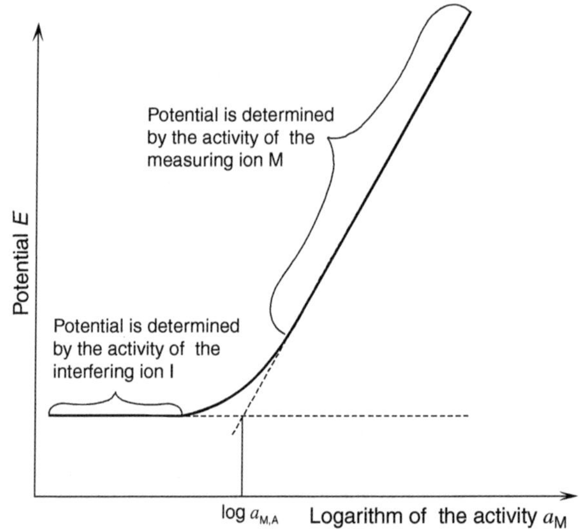

Fig. II.9.8 Graphic description of the evaluation of the selectivity coefficient with the help of the fixed interfering ion activity method

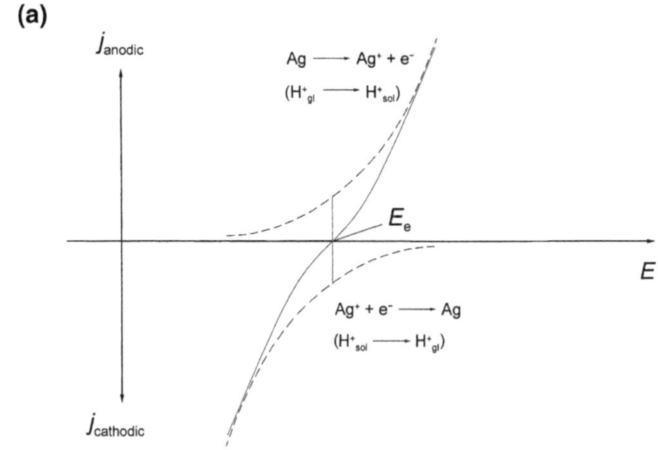

Fig. II.9.9a Schematic presentation of anodic and cathodic partial current densities (*dashed lines*), net current densities (*thin full line*) and the position of the equilibrium electrode potential E_e for a silver electrode and a glass electrode (one glass–solution interface). Subscript gl indicates the surface of a glass membrane and sol indicates the electrolyte solution

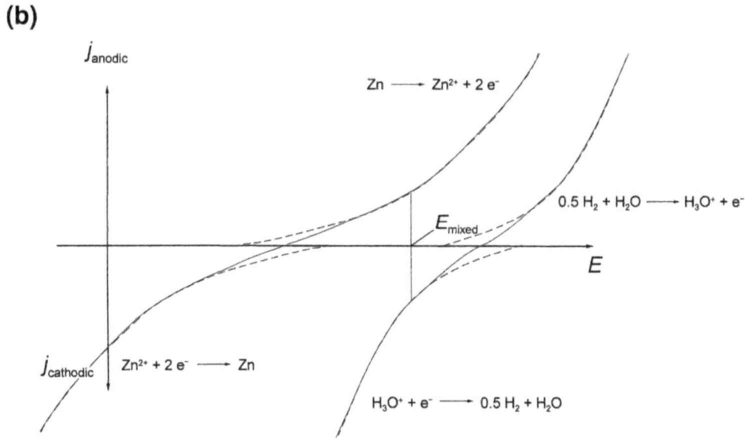

Fig. II.9.9b Schematic presentation of anodic and cathodic current densities (*dashed lines*), net current densities of the single reactions (*thin full lines*) and the position of the mixed potential E_{mixed} at a zinc electrode in an acidic solution

the situation. For this purpose the reader has to remember the origin of *mixed potentials* in case of electrodes with electron exchange (see also the footnote on Table I.2.1, Chap. I.2). Figure II.9.9a depicts the exchange current densities and the position of the equilibrium potential at which the anodic and cathodic exchange current densities have the same magnitude for both cases: ion and/or electron transfer at the electrode surface. Figure II.9.9b depicts the situation when a piece of zinc is dipped into an acidic aqueous solution, where the concentration of zinc ions shall

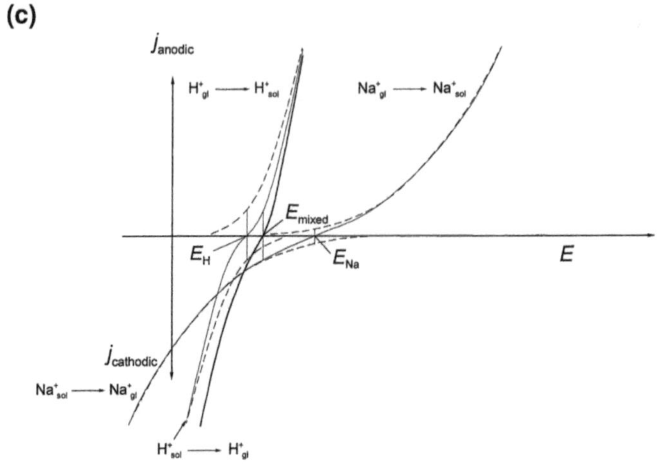

Fig. II.9.9c Schematic presentation of anodic and cathodic current densities (*dashed lines*), net current densities of the single reactions (*thin full lines*), overall net current densities (*thick full line*) and the position of the mixed potential E_{mixed} at a glass electrode. Subscript gl indicates the surface of a glass membrane and sol indicates the electrolyte solution. E_{H} and E_{Na} denotes the equilibrium potentials of this glass electrode in pure solutions of hydrogen or potassium ions

be small, and where also the concentration (fugacity) of hydrogen gas is very small. The latter two facts and the additional fact that the Zn/Zn^{2+} electrode is fairly irreversible, i.e. there occurs an overvoltage for the reduction of zinc ions, lead to the situation that a potential E_{mixed} exists, at which the exchange current densities of proton reduction and zinc oxidation are equal [23]. This is the potential the piece of zinc will attain when no current from an outer source passes through the zinc. The reversible zinc potential means that the reduction of zinc ions and the oxidation of zinc metal occur at the same rate. The electrode always acquires that potential for which the anodic exchange current density occurs at the most negative potential and the cathodic exchange current density at the most positive potential. This means that those processes which most easily occur will determine the potential. In other terms, at the mixed potential the exchange current densities of the two underlying processes have the largest magnitude. The potential is called mixed since the anodic and cathodic exchange currents belong to two different systems, i.e. the hydrogen and the zinc electrodes. The reason for its existence is both the irreversibility of one system, and the low activity of zinc ions and hydrogen gas in the solution.

In the case of membrane electrodes, as mentioned earlier, the irreversibility of ion transfer is not a matter of concern because it is rare. However, extreme concentration differences between two interfering ions may result in the formation of a mixed potential. Figure II.9.9c shows this for the competition of protons and sodium ions at a glass surface. At very high sodium concentration and rather lower hydronium activity (high pH), the electrode potential may be a mixed potential because of a high rate of sodium ion transfer from solution to the glass surface and a small rate of proton transfer from the glass surface to the solution. Of course, in the case of some membranes, the kinetics of ion transfer may also differ for different ions, i.e. the irreversibility of one or the other system can be the reason for the occurrence of mixed potentials.

An electrode attains a potential where the anodic and cathodic exchange currents are equal. This holds true for electron- and ion-exchange reactions. Therefore, the reactants need to be present in such concentration that their exchange currents are sufficiently high. Generally this affords concentration above about 10^{-6} mol L^{-1}. Hence, this is the detection limit for almost all ion-sensitive electrodes. One could think that this is not true for the glass electrode, because measurements almost up to pH 14 are possible, where the concentration of hydronium ions approaches 10^{-14} mol L^{-1}. However this is a fallacy, since, above pH 7, the race is taken over by the hydroxide ions (see Eqs. II.9.35 and II.9.36). It is the autoprotolysis of water and the ionic product of water that form the basis of the extremely wide applicability of the glass electrode almost from pH 0 to pH 14. Similar cases can be realised for sulphide-ion-selective electrodes based on silver sulphide, when the concentration of sulphide ions in the solution is controlled by a precipitation equilibrium with silver ions in the solution. In such cases one can calculate (!) a free sulphide ion concentration of 10^{-46} mol L^{-1} in a 10^{-2} mol L^{-1} silver solution. Indeed, plotting the electrode potential versus log sulphide concentration gives a straight line from -2 to -46 (see p. 209 in [7])! To assume that a concentration of 10^{-46} mol L^{-1} of sulphide ion in, let us say a 100 mL beaker, is certainly nonsense, because this concentration means that it needs 10^{23} L of solution to have one (!) free sulphide ion! All water on earth amounts to about 1.38×10^{18} L. In fact it is again that the race of sulphide ions, which act at high concentration, is taken over by the race of silver ions, when these are at high concentration. Therefore, always treat with caution any potentiometric ion selective measurements that are reported at levels much below 10^{-6} mol L^{-1}. There are certainly rare cases where this detection limit can be lowered by one to two orders of magnitude, but that is not the rule.

An important characteristic of a potentiometric indicator electrode is the response time. The response time is the time required for the establishment of an equilibrium electrode potential. According to an IUPAC recommendation [24], the response time is the time interval from the moment of inserting the potentiometric set-up into the test solution to the moment when the potential deviates from the equilibrium potential by 1 mV. This time interval can span from milliseconds to minutes or hours and depends on many conditions, e.g. concentration of the measuring ion (small exchange currents), speed of stirring, temperature, history and pretreatment of the indicator electrode and so on.

References

1. Biilmann E (1921) Ann Chim 15: 109
2. Galster H (1990) pH-Messung. VCH, Weinheim
3. Koryta J, Dvorak J, Bohackova V (1975) Lehrbuch der Elektrochemie. Springer Verlag, Wien

4. Schwabe K (1976) pH-Meßtechnik. Verlag Theodor Steinkopf, Dresden
5. Koryta J (1972) Anal Chim Acta 61: 329
6. Simon W, Wuhrmann HR, Vašák M, Pioda LAR, Dohner R, Štefanac Z, (1970) Angew Chem 82: 433
7. Yu TR, Ji GL (1993) Electrochemical methods in soil and water research. Pergamon Press, Oxford New York Seoul Tokyo
8. Cammann K, Lemke U, Rohen A, Sander J, Wilken H, Winter B (1991) Angew Chem 103: 519
9. Meyerhoff ME, Fraticelli YM (1982) Anal Chem 54: 27R
10. Fricke GH (1980) Anal Chem 52: 275R
11. Freiser H (1980) Ion-selective electrodes in analytical chemistry. Plenum Press, New York
12. Havas J (1985) Ion- and molecule-selective electrodes in biological systems. Akadémiai Kiadó, Budapest
13. Camman K (2001) Instrumentelle Analytische Chemie. Spektrum Akademischer Verlag, Heidelberg Berlin
14. Diamond D (1998) Principles of chemical and biological sensors. John Wiley & Sons, New York Chichester Weinheim Brisbane Singapore Toronto
15. Schwabe K, Suschke HD (1964) Angew Chem 76: 39
16. Nicolsky BP (1937) Acta Physicochim USSR 7: 597
17. Eisenman G, Rudin DO, Casby JO (1957) Science 126: 831
18. Baucke FGK (1985) J Non-Crys Sol 73: 215
19. Bach H, Baucke FGK, Krause D (2001) Electrochemistry of glasses and glass melts, including glass electrodes. Springer Verlag, Berlin Heidelberg New York Barcelona Hong Kong London Milan Paris Singapore Tokyo
20. Baucke FGK (1974) J Non-Cryst Solids 14: 13
21. IUPAC (1979) Pure Appl Chem 51: 1913
22. IUPAC (1981) Pure Appl Chem 53: 1907
23. Vetter KJ (1961) Elektrochemische Kinetik. Springer Verlag, Berlin Göttingen Heidelberg
24. IUPAC (1976) Pure Appl Chem 48: 127
25. Lindner E, Gyurcsányi RE (2009) J Solid State Electrochem 13: 51
26. Bobacka J (2006) Electroanalysis 18: 7
27. Crespo GA, Macho S, Rius FX (2008) Anal Chem 80: 1316

Chapter II.10
Electrochemical Quartz Crystal Nanobalance

György Inzelt

II.10.1 Introduction

The method of piezoelectric microgravimetry (nanogravimetry) using an electrochemical quartz crystal microbalance (EQCM) or nanobalance (EQCN) can be considered as a novel and much more sensitive version of electrogravimetry. The EQCN technique has become a widely used technique in several areas of electrochemistry, electroanalytical chemistry, bioelectrochemistry, etc. [1–10]. Obviously, mass changes occurring during adsorption, sorption, electrosorption, electrodeposition, or spontaneous deposition can be followed, which is very helpful for the elucidation of reaction mechanism via identification of the species accumulated on the surface. These investigations include metal and alloy deposition, underpotential deposition, electroplating, synthesis of conducting polymers by electropolymerization, adsorption of biologically active materials, and analytical determination of small ions and biomolecules. Of course, the opposite processes, i.e., spontaneous dissolution, electrodissolution, corrosion, can also be studied. Electrochemical oscillations, in which the formation and oxidation of chemisorbed molecular fragments play a determining role, have been studied, too. The majority of the investigations have been devoted to ion and solvent transport associated with the redox transformations of electrochemically active polymers. Similar studies have been carried out regarding polynuclear surface layers such as metal hexacyanometalates as well as inorganic and organic microcrystals of different compositions.

The quartz crystal nanobalance (QCN) can be combined with practically any electrochemical methods, such as cyclic voltammetry, chronoamperometry, chronocoulometry, potentiostatic, galvanostatic, rotating disc electrode [11], or potentiometric measurements. The EQCN can be further combined with other techniques, e.g., with UV–Vis spectroscopy [12], probe beam deflection (PBD) [13], radiotracer [14], atomic force microscopy (AFM) [15], and scanning electrochemical microscopy (SECM) [16]. The concept and the instrumentation of

G. Inzelt (✉)
Department of Physical Chemistry, Eötvös Loránd University, 1518 Budapest 112, Hungary
e-mail: inzeltgy@chem.elte.hu

F. Scholz (ed.), *Electroanalytical Methods*, 2nd ed.,
DOI 10.1007/978-3-642-02915-8_13, © Springer-Verlag Berlin Heidelberg 2010

ac electrogravimetry have also been developed [17]. By the help of the combined technique, the overall mass effect, that can be measured by EQCN, can be deconvoluted, i.e., the transport and sorption of different ionic and neutral species can be separated. QCN can also be used without the combination with electrochemical techniques in gas sensors, immunosensors, DNA biosensors, drug analysis, etc., by coating the crystal with a suitable reacting layer, e.g., antibodies in immunosensors [6, 9].

II.10.2 Theory and Basic Principles of Operation

II.10.2.1 Piezoelectricity, Converse Piezoelectricity, Piezoelectric Materials, and Oscillators

Piezoelectricity is a property of certain classes of material, including natural crystals of quartz. In piezoelectric materials, mechanical strain is generated by application of an electric field or, conversely, by application of mechanical stress an electrical polarization is generated. Piezoelectric materials are found in 20 out of 21 crystal space groups with no symmetry centers. Piezoelectric materials may be organic or inorganic solids. They may contain intrinsic electric fields, especially if unequally charged substances are present in their structures. Piezoelectric materials should be specified, although α-quartz is most commonly used for presently constructed sensors, because the desired features can be achieved through various cuts of its native crystal. These crystals exhibit small temperature dependence of the resonant frequency for room temperature operation, and small dependence of the resonant frequency on built-in stress from applied contacts or surface films.

Most commonly used plate oscillators are cut to exhibit longitudinal (thickness expander) modes or thickness shear modes. For quartz, cuts across the *X*-axis, *Y*-axis or the so-called rotated Y-cuts are used. Various designations of cuts appear in the literature: X-cut (thickness expander), AC, AT (thickness shear), DT (plate shear), BT, ST (horizontal shear), SC (stress compensated), etc. The symbol T specifies a cut with minimal temperature dependence of the resonant frequency. Within these general cuts, there are subgroups, e.g., planar, plano-convex, and others. These cuts should be specified, as their properties are very important for the piezoelectric sensor performance. Electronic surface films and electrical contacts are usually accomplished by vacuum deposition, sputtering, or electroplating of metal films on both crystal faces. The central part of the crystal between the oscillator electrodes is driven into oscillations of the largest amplitude, i.e., the differential mass sensitivity reaches maximum at the center of the plate. The amplitudes of the oscillations decrease with the increase of the distance from the near-resonant regions and certain frequencies are severely damped. Finally, at the edges, the oscillations are not permitted. Usually, the two excitation electrodes have slightly different sizes in order to decrease the capacitance effect. The smaller electrode determines the 'piezoactive area.'

II.10.2.2 The Quartz Crystal Nanobalance

Quartz crystal nanobalance is a thickness-shear-mode acoustic wave mass-sensitive detector based on the effect of an attached foreign mass on the resonant frequency of an oscillating quartz crystal. The QCN responds to any interfacial mass change. The response of QCN is also extremely sensitive to the mass (density) and viscoelastic changes at the solid–solution interface [1–8].

The quartz crystal nanobalance is a *piezoelectric* sensing device that consists of an oscillator circuit and a crystal, which is incorporated into the feedback loop of the circuit. The crystal is the frequency-determining element, as its quality factor (Q) – which is the measure of freedom from energy loss or effectiveness of energy storage – is very high, i.e., Q is in the range 10^6–10^8 in air or vacuum and 1000–3000 in contact with a liquid phase. The piezoelectric crystals are patterned with two excitation electrodes (electronic surface films) on their opposite sides. Due to the *converse piezoelectricity* phenomenon, when alternating voltage is applied to the attached electrodes mechanical oscillations occur within the crystal lattice. These oscillations are stable only at the natural resonant frequency of the crystal. The crystal cut determines the mode of oscillations. Shear vibrations are generated if one large crystal face moves parallel with respect to the underlying planes as in QCNs with AT-cut α-quartz crystals. This crystal wafer is prepared by cutting the quartz at approximately 35.17° from its Z-axis. A typical crystal plate is a cylindrical disk of a diameter 10 mm and thickness about 0.7–0.1 mm for resonant operation in the 2–15-MHz frequency range. The fundamental resonant frequency (f_o) shifts when a thin film deposited on the surface of the quartz crystal. Under the assumption that the density and the shear modulus of the film are the same as those of quartz as well as the film is uniform (constant density and thickness) and covering the acoustically active area of the whole crystal, the Sauerbrey equation [18] describes the relationship between the resonant frequency shift (Δf) and the added mass (Δm):

$$\Delta f = f_c - f_o = -\frac{2 f_o^2 \,\Delta m}{A\,(\rho_q \mu_q)^{1/2}} = -C_f\,\Delta m \qquad \text{(II.10.1)}$$

where f_c is the resonant frequency of the composite oscillator formed from the crystal and the film at the surface, A is the acoustically active surface area, $\rho_q = 2.648$ g cm^{-3} and $\mu_q = 2.947 \times 10^{10}$ N m^{-2} are the density and the shear modulus of quartz, respectively, Δm is the change of the surface mass density, and C_f is the integral mass sensitivity.

Since

$$f_o = \frac{\upsilon_q}{2L_q} \qquad \text{(II.10.2)}$$

where $\upsilon_q = \left(\frac{\mu_q}{\rho_q}\right)^{1/2}$ is the wave speed for quartz ($\nu_q = 3336$ m s^{-1}) and L_q is the thickness of the quartz, in this case the only difference between the film-coated and uncoated quartz is the added thickness (ΔL_f). The film should not be too thick, and the mass loading should be lower than 2% of the mass of the quartz plate, i.e., typically lower than 20 μg cm^{-2}. There may be non-mass-related frequency changes due to stress (e.g., caused by an improper mounting) and temperature effects [1–8].

The values of the integral mass sensitivity for the most frequently used $f_{\mathrm{o}} = 5$ MHz and 10 MHz crystals are $C_{\mathrm{f}} = 5.66 \times 10^7$ and 2.264×10^8 Hz cm^2 g^{-1}, respectively. It follows that by using a crystal with $A = 0.3$ cm^2 surface area 1 Hz change – which can be measured easily and accurately – corresponds to 6 ng and 1.4 ng, respectively. The uniform distribution of the mass over the active area of the quartz plate is of importance since the differential mass sensitivity (c_f) varies across this area. For typically used round-shaped crystals [1, 6, 8, 19]:

$$C_{\mathrm{f}} = 2\pi \int_0^r c_{\mathrm{f}}(r)\, r\mathrm{d}r \qquad \text{(II.10.3)}$$

The resonant frequency also depends on the density and the viscosity of the contacting media. The frequency shift for measurement in air vs. vacuum is smaller than 10 Hz. However, the frequency shift from air to water or aqueous solution for a 10-MHz crystal is about 4–12 kHz. The magnitude of this shift can be calculated by using the following relationship:

$$\Delta f = -f_{\mathrm{o}}^{3/2} \left(\frac{\rho_{\mathrm{L}} \eta_{\mathrm{L}}}{\pi \rho_{\mathrm{q}} \mu_{\mathrm{q}}} \right)^{1/2} \qquad \text{(II.10.4)}$$

where ρ_{L} and η_{L} are the density and viscosity of the liquid contacting one side of the crystal.

The additional surface mass change can be measured and calculated by using the Sauerbrey equation even in the case when the crystal is in contact with a liquid. It makes the in-situ detection of the surface mass changes possible in the course of electrochemical experiments by using an electrochemical quartz crystal nanobalance. A lack of conformance with the Sauerbrey equation may be traced back to high-mass loading, surface roughness, surface stress, interfacial slippage, nonuniform mass distribution, and viscoelastic effects. In the latter case the film does not behave as a rigid layer, i.e., the deformation is not purely elastic. It happens when the film is thick and its morphology changes due to, e.g., its swelling.

The frequency can be measured by the help of a frequency counter with an accuracy of 0.1 Hz and sampling time 0.1–1 s.

II.10.2.3 The Electrochemical Quartz Crystal Nanobalance (EQCN)

Electrochemical quartz crystal microbalance (EQCM) or nanobalance (EQCN) is a combination of a piezoelectric sensor, i.e., a quartz crystal nanobalance (QCN) and an electrochemical cell containing the sample electrolyte solution, the reference electrode, as well as other electrodes as required, driving oscillator, amplifiers, and readout units [1–7]. In most electrochemical experiments, the piezoelectric crystal,

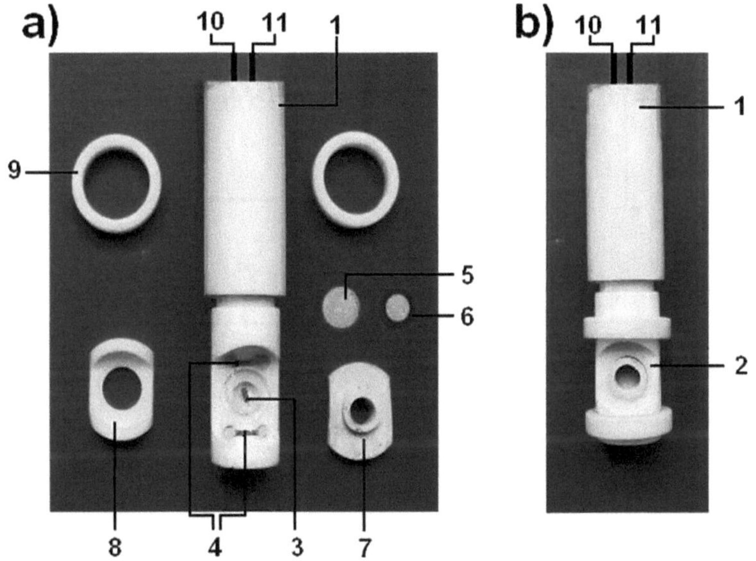

Fig. II.10.1 The electrochemical quartz crystal nanobalance developed in the author's laboratory. (**a**) Picture showing the dismantled device, (**b**) the nanobalance assembly as used as a working electrode in a usual three-electrode electrochemical cell: *1* crystal oscillator electronics inside a Teflon® of Kynar® cylinder, *2* the crystal holder, *3* gold connection to the gold (platinum) electrode of the crystal's back side (not exposed to the electrolyte), *4* gold connections to the gold (platinum) electrode of the crystal's front side exposed to the electrolyte, which serves both as a part of oscillator circuit and as the working electrode, *5* front face of the quartz crystal coated with gold (or Pt), *6* back face of the quartz crystal coated with gold (or Pt), *7* silicon gum with gold wire connections to *4*, *8* upper part of the holder, *9* Teflon O-rings fixing, *10* connection to the frequency counter, *11* connection to the potentiostat

usually a quartz crystal plate, with attached electronic and occasionally selective surface films [1–10] is clamped between two O-rings or is mounted in other suitable holders. Generally, only one crystal face and one oscillator electrode in the electrochemical cell, are in contact with the sample solution (Figs. II.10.1 and II.10.2). The *ac* excitation voltage can be easily separated electronically from the *dc* potential that is applied to carry out electrochemical investigations. In this way, the surface mass change (Δm) occurs during any electrochemical experiments, e.g., galvanostatic, potentiostatic, cyclic voltammetry, can be followed by detecting the frequency changes (Δf).

The electrochemical data, e.g., the charge (Q) consumed can be correlated with the frequency (mass) changes, and – if the requirements of the accurate QCM measurements, e.g., uniform distribution of the added mass, elastic behavior, are met as well as the charge efficiency is 100% – the molar mass (M) of the depositing or exchanged species can be calculated in a simple manner:

$$M = \frac{nFA\Delta f}{C_f Q} \qquad \text{(II.10.5)}$$

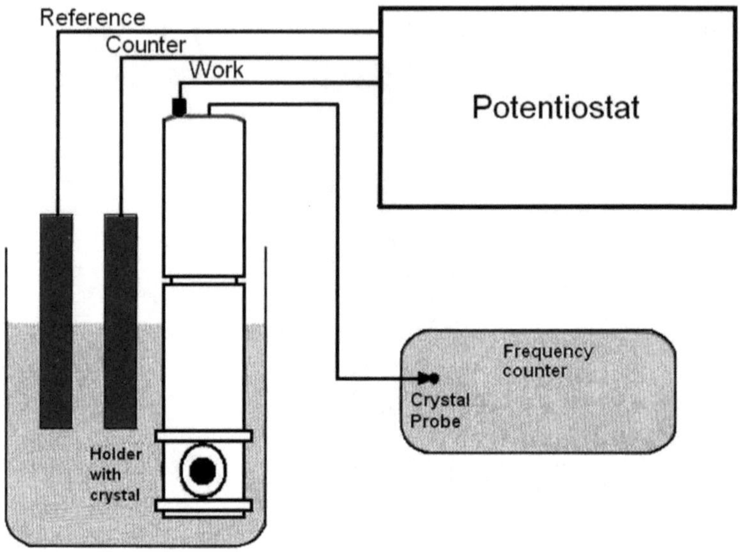

Fig. II.10.2 Basic EQCN setup

where A is the surface area, C_f is the integral mass sensitivity, F is the Faraday constant, and n is the number of electrons transferred. For cyclic voltammetric experiment, the corresponding relationship between the current (I) and the frequency change is as follows:

$$I = \frac{nFv}{C_f M} \frac{\mathrm{d}(\Delta f)}{\mathrm{d}E} \tag{II.10.6}$$

where v is the scan rate, $\mathrm{d}(\Delta f)/\mathrm{d}E$ is the derivative of Δf with respect to potential.

It is useful to calibrate the EQCM, i.e., to determine C_f, e.g., by electrodeposition and electrodissolution of silver. The rigidity layer behavior can be tested by depositing films of different thicknesses. Usually relatively thin films (10 nm – some hundreds nm) show rigid layer behavior. The deviation from the linearity regarding the Δm vs. Q function is related to the appearance of the viscoelastic effect. By the help of impedance measurements the viscoelastic characteristics of the surface film can also be tested [4, 5, 6, 7, 10]. In the absence of any deposition the change of the density and viscosity in the double layer or in the diffusion layer may cause 0.1–10 Hz frequency change. It may interfere with the effect caused by the deposition of monolayers or submonolayers. In some cases other effects, e.g., stress, porosity, pressure, and temperature, should also be considered.

II.10.3 Applications of EQCN: Selected Examples

In order to illustrate the usefulness and the capability of the EQCN method selected examples from different areas of electroanalytical chemistry will be shown below.

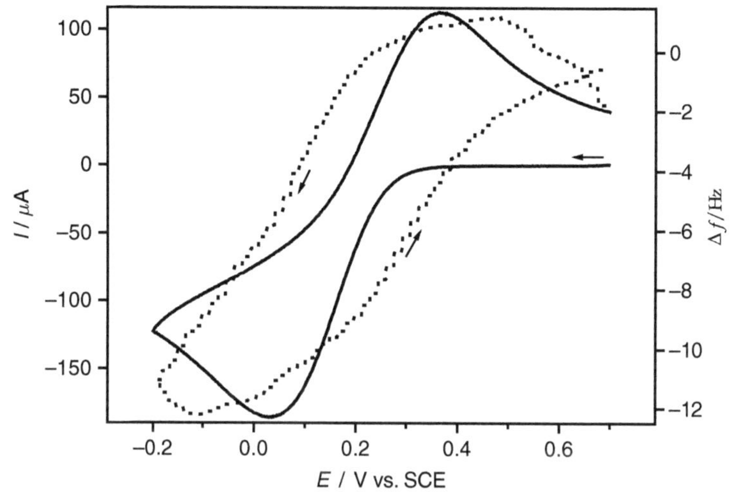

Fig. II.10.3 Cyclic voltammogram (*continuous line*) and the simultaneously obtained EQCM frequency curve (*dotted line*) for a gold electrode ($f_0 = 10$ MHz) in contact with a solution of 0.5 mol dm^{-3} K_2SO_4 and 0.01 mol dm^{-3} $K_3[Fe(CN)_6]$. Scan rate: 10 mV s^{-1}

II.10.3.1 Redox Transformations of $[Fe(CN)_6]^{3-}$ /$[Fe(CN)_6]^{4-}$ System on Gold

The redox pair of $[Fe(CN)_6]^{3-}$/$[Fe(CN)_6]^{4-}$ is frequently used as a test system, e.g., in cyclic voltammetry (see Sect. II.1.5.3) or rotating disk electrode studies, assuming an uncomplicated electrode reaction. However, as seen in Fig. II.10.3, the characteristic cyclic voltammetric curves of $[Fe(CN)_6]^{3-}$ ions are accompanied with a frequency decrease during reduction and an increase during oxidation.

This phenomenon can be assigned to the depletion layer effect [20] and the existence of a persistent monolayer of $[Fe(CN)_6]^{3-}$ /$[Fe(CN)_6]^{4-}$ and K^+ ion exchange [21, 22].

II.10.3.2 Underpotential Deposition of Hydrogen and Oxygen on Pt and Rh Surfaces and the Effect of the Electrochemical Double Layer

By the help of the piezoelectric nanogravimetry, the oxide formation and reduction on noble metal electrodes can well be followed as seen in Figs. II.10.4 and II.10.5.

The smaller changes in the double-layer region due to the ionic adsorption and the changes of the electrolyte concentration near the metal surface can also be detected. The mass change observed in the region of the oxidative desorption or the reductive adsorption of hydrogen both for Pt and Rh [23] is just the opposite that could be expected. Instead of the slight mass decrease during the oxidation and the

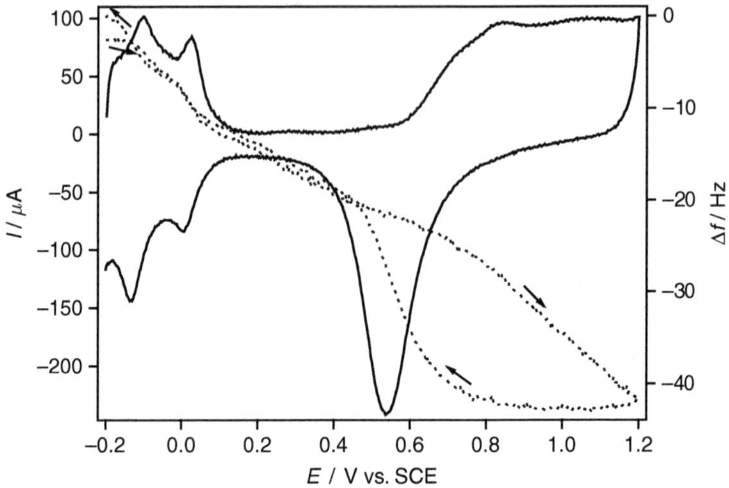

Fig. II.10.4 Cyclic voltammogram (*continuous line*) and the simultaneously obtained EQCM frequency curve (*dotted line*) for a platinum electrode ($f_o = 5$ MHz) in contact with 1 mol dm^{-3} H_2SO_4 solution. Scan rate: 50 mV s^{-1}

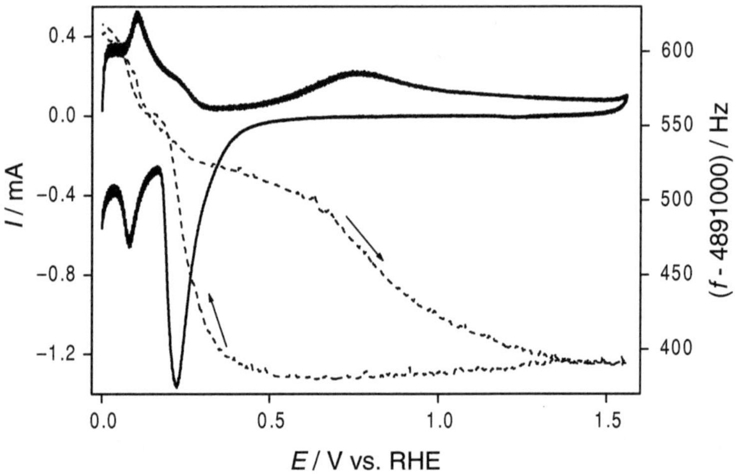

Fig. II.10.5 Cyclic voltammogram (*continuous line*) and the simultaneously obtained EQCM frequency curve (*dotted line*) for a rhodium electrode in contact with 1 mol dm^{-3} H_2SO_4 solution. Scan rate: 5 mV s^{-1}

mass increase during the reduction – which effect would be hardly observable due to the low molar mass of hydrogen – mass increase and mass decrease, respectively, can be detected which effects reveal that the desorption of adsorbed hydrogen is accompanied with the sorption of anions and vice versa. A careful inspection of the curves displayed in Fig. II.10.4 also attests that a slight dissolution of platinum even in sulfate media occurs during the reduction of the oxide layer. It can be explained by the reorganization of the PtOH layer to OHPt type layer, and assuming that some surface Pt atoms become detached from the metal surface together with the oxygen atoms (OH groups).

II.10.3.3 Polymer Film Electrodes: Electrodeposition and Ionic Exchange Processes

The formation of conducting polymers by electropolymerization can be monitored by EQCN as seen in Fig. II.10.6. From the frequency change and the charge consumed, the efficiency can be calculated, but other important conclusions can also be drawn regarding the mechanism of the polymerization. For instance, the polymerization reaction proceeds during the cathodic half-scan, and it is the very

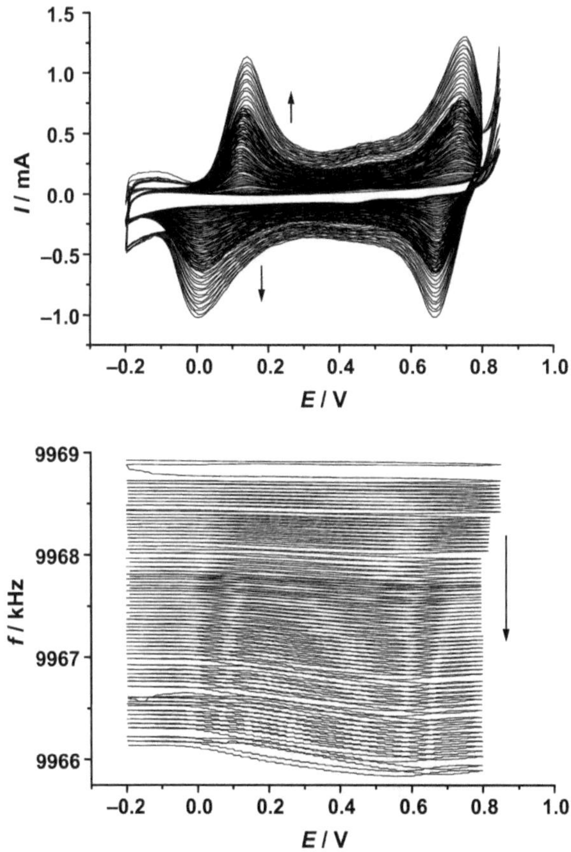

Fig. II.10.6 The cyclic voltammograms and the simultaneously detected EQCM frequency changes during the electropolymerization of aniline at a platinum electrode. Sweep rate: 100 mV s^{-1}. Solution composition: 0.2 mol dm^{-3} aniline in 1 mol dm^{-3} $HClO_4$

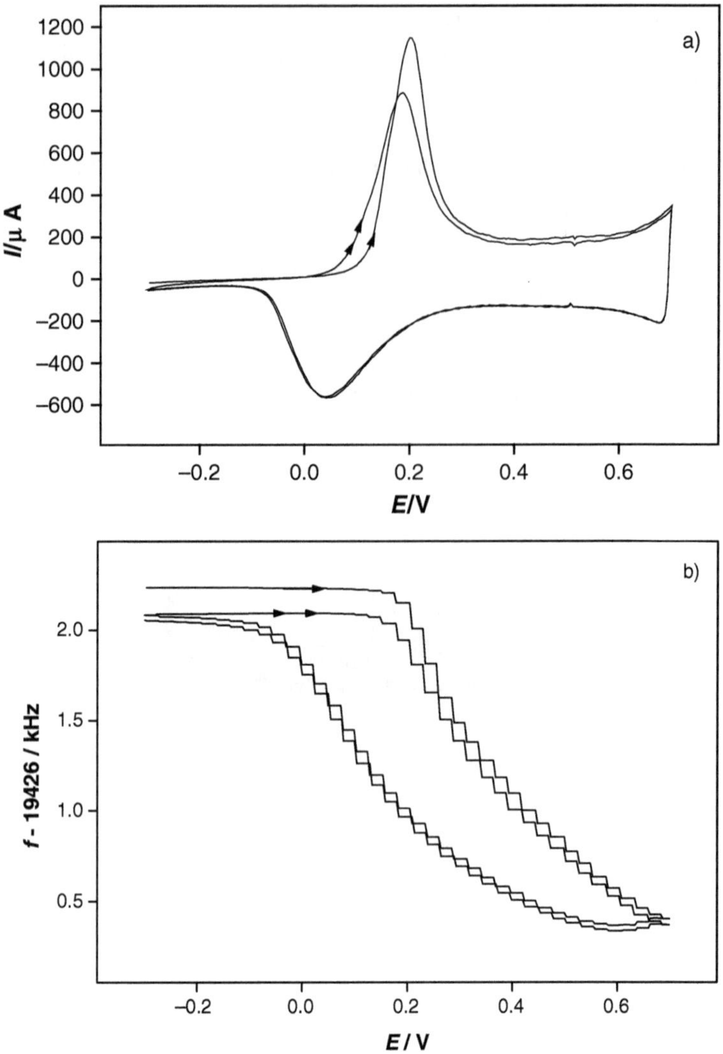

Fig. II.10.7 (**a**) Cyclic voltammograms (two cycles) and (**b**) the simultaneously detected EQCM frequency changes for a polyaniline film ($L = 2.9$ μm) in contact with 1 M H_2SO_4. Sweep rate: 100 mVs^{-1}

reason why the electropolymerization is more effective by using cyclic voltammetry in comparison with potentiostatic method.

The simultaneous measurements of the current–potential and frequency–potential curves are of the utmost importance to gain a better understanding of the ionic and solvent sorption processes occurring in the surface films of conducting and redox polymers, and help to establish an adequate reaction mechanism [24]. For instance, a closer inspection of the curves displayed in Fig. II.10.7 reveals that during the early phase of oxidation of polyaniline protons (hydronium ions) leave the film, indicating a partial protonation of the polyaniline amine groups in the reduced form. The incorporation of anions starts when a substantial portion of polyaniline film has already been oxidized. The mass increase induced by the incorporation of anions and solvent molecules also continues in the 'capacitive' region.

II.10.3.4 Microcrystals

The ingress and egress of ions and solvent molecules accompanying the redox reactions of solid compounds and materials attached to electrode surface (see Chapter II.8) can also be studied by EQCN method. If the charged (salt) form of the compound is soluble in the given solvent (water), a stable response can be achieved by decreasing the water activity, e.g., by using electrolyte of high concentration [25]. 7,7',8,8'-Tetracyanoquinodimethane (TCNQ) is insoluble in water, while its lithium salt, $Li^{+}TCNQ^{-}$ – that forms during an one-electron reduction – is highly soluble. As seen in Fig. II.10.8 stable voltammetric and EQCN frequency response can be detected even at very low scan rate. Both the cyclic voltammograms and the chronoamperometric responses (Fig. II.10.9) show the typical pattern of the solid–solid phase transformation under the control of nucleation and growth kinetics. It has been found that the surface structural changes are accompanied with extensive hydration or dehydration, and the anomalously large frequency transients is due to the occurrence of strain at sites of lattice mismatch.

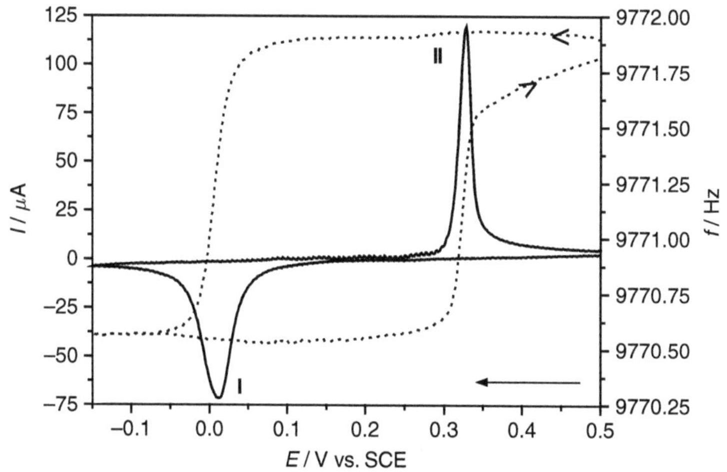

Fig. II.10.8 Cyclic voltammogram (—) and the EQCM frequency response (······) of TCNQ microcrystals attached to a gold electrode (f_o = 10 MHz) in contact with 12 moldm^{-3} LiCl solution. Scan rate: 1 mV s^{-1}

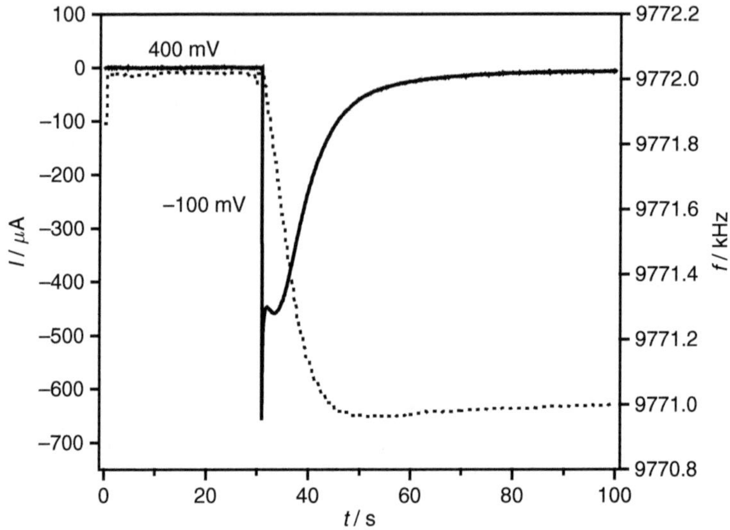

Fig. II.10.9 Chronoamperometric response (—) and the simultaneously obtained EQCM frequency changes (······) for the reduction of TCNQ microcrystals immobilized on gold (f_o = 10 MHz) in contact with 12 mol dm^{-3} LiCl solutions. The potential was stepped from 400 mV to −100 mV

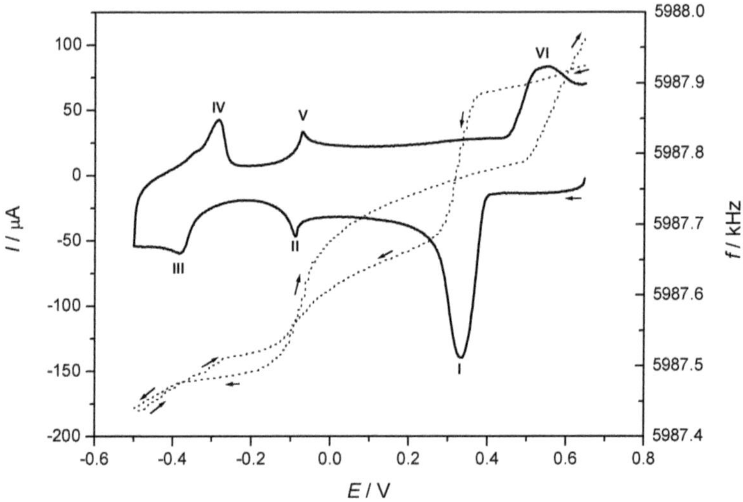

Fig. II.10.10 Cyclic voltammograms (*continuous line*) and the respective EQCM frequency responses (*dotted line*) for $RuCl_3$ microcrystals attached to a gold electrode (f_o = 6 MHz) in the presence of 0.5 mol dm^{-3} solution of CsCl. Scan rate: 20 mV s^{-1}

A knowledge of the mass changes accompanying the redox processes is especially useful concerning the elucidation of the reaction mechanism of systems showing more complicated redox and structural transformations, such as $RuCl_3$ microcrystals (Fig. II.10.10). From the results of these measurements, a conclusion

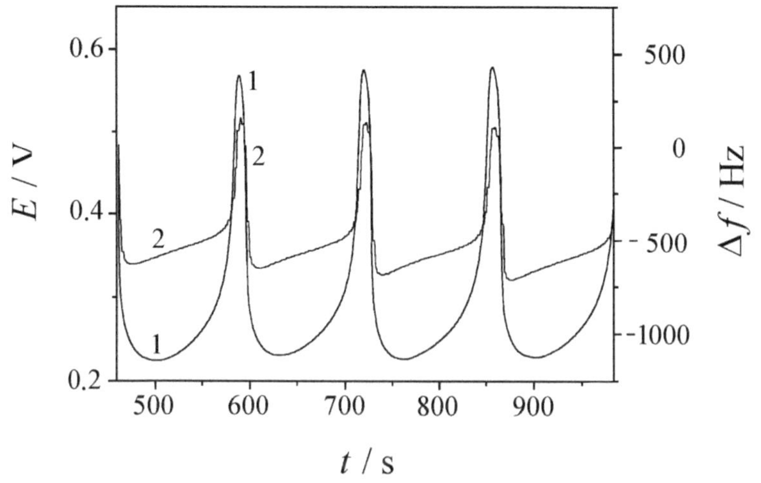

Fig. II.10.11 *1* Potential oscillations and *2* the simultaneously measured changes of the EQCM frequency response, i.e., the variation of the surface mass, in the course of galvanostatic electrooxidation of formic acid at a platinized Pt electrode ($A = 0.3\ \mathrm{cm}^2$). $f_o = 10$ MHz, $I = 0.45$ mA. Solution: 0.5 mol dm^{-3} HCOOH + 1 mol dm^{-3} $HClO_4$

can be drawn regarding the composition of the mixed valence intercalation phases formed [26].

II.10.3.5 Electrochemical Oscillations

With the help of EQCN measurements, it has been proved that the accumulation and consumption of strongly bound species produces the potential oscillations in the course of galvanostatic oxidation of small organic compound, such as formic acid [27]. Figure II.10.11 shows the simultaneous changes of the potential and the frequency as a function of time.

References

1. Buttry DA (1991) Applications of the quartz crystal microbalance to electrochemistry. In: Bard AJ (ed) Electroanalytical chemistry, vol 17. Marcel Dekker, New York, pp 1–85
2. Buttry DA, Ward MD (1992) Chem Rev 92: 1355
3. Grate JW, Martin SJ, White RM (1993) Anal Chem 65: 940A; 987A
4. Hepel M (1999) In: Wieczkowski A (ed), Interfacial electrochemistry. Marcel Dekker, New York, pp 599–630
5. Hillman AR (2003) The electrochemical quartz crystal microbalance. In: Bard AJ, Stratmann M (eds) Encyclopedia of electrochemistry, vol 3. Unwin PR (ed), Wiley-VHC, Weinheim pp 230–289
6. Buck RP, Lindner E, Kutner W, Inzelt G (2004) Pure Appl Chem 76: 1139–1160
7. Tsionsky V, Daikhin L, Urbakh M, Gileadi E (2004) Looking at the metal/solution interface with electrochemical quartz-crystal microbalance: theory and experiment. In: Bard AJ, Rubinstein I (eds) Electroanalytical chemistry, vol 22. Marcel Dekker, New York, pp 2–94

8. Lu C, Czanderna AW (eds) (1984) Applications of piezoelectric quartz crystal microbalances. Elsevier, New York
9. O'Sullivan CK, Guilbault (1999) Biosens Bioelectron 14: 663
10. Calvo EJ, Etchenique RA (1999) Kinetic applications of the electrochemical quartz crystal microbalance (EQCM). In: Compton RG, Hancock G (eds) Comprehensive chemical kinetics, vol 37. Elsevier, Amsterdam, pp 461–487
11. Kern P, Landolt D (2000) J Electrochem Soc 147: 318
12. Kim JM, Chang SM, Muramatsu H (1999) J Electrochem Soc 146: 4544
13. Henderson MJ, Hillman AR, Vieil E (1998) J Electroanal Chem 454: 1
14. Inzelt G, Horányi G (1989) J Electrochem Soc 136: 1747
15. Friedt J-M, Choi KH, Frederix F, Campitelli A (2003) J Electrochem Soc 150: H229
16. Gollas B, Bartlett PN, Denuault (2000) Anal Chem 72: 349
17. Gabrielli C, Keddam M, Perrot H, Torresi R (1994) J Electroanal Chem 378: 85
18. Sauerbrey G (1959) Z Phys 155: 206
19. Bácskai J, Láng G, Inzelt G (1991) J Electroanal Chem 319: 55
20. Lee WW, White HS, Ward MD (1993) Anal Chem 65: 3232
21. Smalley JF, Guy L, Feldberg SW, Rogers LC, Leddy J (1993) J Electroanal Chem 356: 181
22. Róka A, Varga I, Inzelt G (2006) Electrochim Acta 51: 6243
23. Wasberg M, Bácskai J, Inzelt G, Horányi G (1996) J Electroanal Chem 418: 195
24. Inzelt G (2008) Conducting polymers. A new era in electrochemistry. In: Scholz F (ed) Monographs in electrochemistry. Springer, Berlin Heidelberg
25. Inzelt G, Németh K, Róka A (2007) Electrochim Acta 52: 4015
26. Inzelt G, Puskás Z, Németh K, Varga I (2005) J Solid State Electrochem 9: 823
27. Inzelt G, Kertész V (1993) Electrochim Acta 38: 2385

Part III
Electrodes and Electrolytes

Chapter III.1
Working Electrodes

Šebojka Komorsky-Lovrić

III.1.1 Introduction

In electrochemistry an electrode is an electronic conductor in contact with an ionic conductor. The electronic conductor can be a metal, or a semiconductor, or a mixed electronic and ionic conductor. The ionic conductor is usually an electrolyte solution; however, solid electrolytes and ionic melts can be used as well. The term "electrode" is also used in a technical sense, meaning the electronic conductor only. If not specified otherwise, this meaning of the term "electrode" is the subject of the present chapter. In the simplest case the electrode is a metallic conductor immersed in an electrolyte solution. At the surface of the electrode, dissolved electroactive ions change their charges by exchanging one or more electrons with the conductor. In this electrochemical reaction both the reduced and oxidized ions remain in solution, while the conductor is chemically inert and serves only as a source and sink of electrons. The technical term "electrode" usually also includes all mechanical parts supporting the conductor (e.g., a rotating disk electrode or a static mercury drop electrode). Furthermore, it includes all chemical and physical modifications of the conductor, or its surface (e.g., a mercury film electrode, an enzyme electrode, and a carbon paste electrode). However, this term does not cover the electrolyte solution and the ionic part of a double layer at the electrode/solution interface. Ion-selective electrodes, which are used in potentiometry, will not be considered in this chapter. Theoretical and practical aspects of electrodes are covered in various books and reviews [1–9].

Electrodes for voltammetry are usually made of solid or liquid metals [2, 3, 6], or from carbon [10]. Less frequently, metal oxides or polymers are used [11–15]. The primary metallic conductor may be covered with a thin film of a secondary conductor (e.g., mercury, or oxides and polymers) [9, 13], or a monolayer of covalently bound foreign atoms or molecules such as thiols on gold substrate [16]. These are called chemically modified electrodes. The chemical preparation of the electrode

Š. Komorsky-Lovrić (✉)
Rudjer Boskovic Institute, 10001 Zagreb, Croatia
e-mail: slovric@irb.hr

F. Scholz (ed.), *Electroanalytical Methods*, 2nd ed.,
DOI 10.1007/978-3-642-02915-8_14,

surface can be performed either before the measurement, in a separate procedure, or in situ, as an integral part of the analytical protocol.

Despite some rare exceptions, the material used as an electrode is not supposed to react with the solvent and the supporting electrolyte. This requirement is best satisfied by the noble metals, glassy carbon, and graphite. Solid metal electrodes are made primarily of platinum and gold. Mercury satisfies the above requirement only partly, but it is widely used because it is liquid and possesses a large overvoltage for hydrogen evolution.

The potential range within which the electrode can be polarized is called the working window of the electrode, because only this range can be used for measurements. The electrode is polarized if a change of potential does not cause any current flow other than that for double-layer charging. This means that no faradaic processes occur. Except for thermodynamic reasons, i.e., because of the absence of faradaic processes, polarization may have kinetic reasons. An example is the reduction of hydrogen ions. At mercury electrodes this reaction is very slow and occurs at potentials much more negative than required by thermodynamics. This phenomenon is called hydrogen overvoltage. This makes mercury a very useful cathode, although it is easily oxidized at positive potentials. At platinum and gold, the hydrogen overvoltage is much lower. These metals are poor cathodes, but excellent anodes. The working window of glassy carbon electrodes is wide because of a rather high hydrogen overvoltage and a fair stability under anodic polarization.

Within its working window, an electrode can be depolarized by electroactive substances which are dissolved in the electrolyte. The electrochemical reaction on the electrode surface causes concentration gradients perpendicular to the electrode surface. The current is proportional to these concentration gradients. This relationship depends on the electrode geometry, on the hydrodynamic conditions in the solution (whether it is stirred, or not) and on the voltammetric technique. However, in all cases, the current reaches a maximum, or a limiting value, which is proportional to the bulk concentration of the reactant. This is called the concentration polarization of the working electrode. It is the basis of all analytical applications of voltammetry.

The electrode reaction is rarely as simple as described above. In many cases the product is either insoluble, or partly adsorbed at the electrode surface. Besides, the reactants of many reactions are also surface active. Furthermore, the electrode reaction can be either preceded or followed by chemical reactions. Hence, the choice of the working electrode also depends on the reaction mechanism. For instance, the reduction of lead ions on a platinum electrode is complicated by nucleation and growth of lead micro-crystals, while on a mercury electrode lead atoms are dissolved in mercury and the reduction is fast and reversible. Similarly, the well-known pigment alizarin red S and the product of its reduction are both strongly adsorbed on the surface of mercury and carbon electrodes [17]. In this case, the liquid mercury electrode is analytically much more useful because the adsorptive accumulation on the fresh electrode surface can be easily repeated by creating a new mercury drop. However, on the solid electrode, the film of irreversibly adsorbed substance is so stable that it can be formed in one solution and then transferred into another electrolyte for the measurement of the kinetics of the electrode reaction. After each experiment of this type, the surface of the carbon electrode needs careful cleaning and polishing. Finally, micro-particles of insoluble solids can be mechanically transferred to the surface of paraffin-impregnated graphite electrodes and electrochemically analyzed without prior decomposition and dissolution [18] (see Chap. II.9).

Electrodes for voltammetry can be classified according to the electron conducting material, the geometry and size, the hydrodynamic conditions under which they operate and chemical modifications of their active surfaces.

III.1.2 Electrode Materials

Electronic conductors used as electrodes are metals, rarely metal oxides, various forms of carbon and also rarely some polymers. In aqueous solutions, platinum and gold are used for electrode reactions with positive standard potentials, while mercury is useful for reactions with negative standard potentials. However, in nonaqueous aprotic media, under extremely dry conditions and in the absence of dissolved oxygen, these restrictions do not apply and the working windows are determined by the decomposition of supporting electrolytes or solvents. In organic solvents, platinum, carefully polished to a mirror finish, is a chemically inert electrode, often with very favorable electron-transfer kinetics. In these media the adsorption of reactants and products of the electrode reactions of organic compounds is usually negligible, but, in some cases, oligomers and polymers can spoil the electrode surface if the charge transfer is followed by electropolymerization. In acidic aqueous electrolytes (e.g., 0.5 mol/L H_2SO_4), platinum forms surface oxides at about 1 V vs. SCE, which can be reductively dissolved at about 0.8 V [12]. At potentials lower than 0.4 V, a platinum electrode is covered with adsorbed hydrogen. These surface reactions cause high background currents in a wide potential range. They also change the characteristics of the electrode surface. Hence, in many cases, a platinum electrode must be re-polished after each electrochemical measurement. Both platinum and gold electrodes are activated by polishing with alumina slurry on a smooth glass plate. The sizes of alumina particles are between 1 μm and 0.05 μm. The polishing starts with the largest size and continues using successively smaller sizes. After a mirror-like finish is obtained, the electrode is cleaned by rinsing with deionized water and ultrasonication in water for 15 min. For the best results the electrode should be used immediately after polishing [19].

Electrochemical cleaning of the surface of platinum immediately prior to the measurement can be performed by polarizing the electrode with a series of cyclic or square-wave voltammetric pulses in the potential range from the formation of the oxide layer to the potential of evolution of hydrogen. The procedure is particularly useful if the final potential prior to commencement of the measurements lies in the double-layer region that is within a narrow range of potentials where the electrode is not covered by either an oxide layer or adsorbed hydrogen. It is possible to prepare a monocrystal of platinum in order to use a single, well-defined face as the working electrode [20]; however, in electroanalytical chemistry, polycrystalline platinum is generally used. Thus, the working electrode surface is heterogeneous with respect to its electrochemical activity [21]. This may influence the measurements. Obviously, platinum is not an ideally inert electrode in all media. It is always recommendable to test the background current and to establish the limits of the working window in the electrolyte before an analyte is added.

Gold electrode also forms stable oxide layer on the anodic polarization prior to oxygen evolution that can be fully reduced to the bare metal during the cathodic sweep. However, gold is susceptible to corrosion in the presence of Cl^- and CN^- anions, which restricts the application of electrochemical cleaning method in these media [22]. An example of electroanalytical application of gold electrodes is the determination of traces of mercury [6, 9]. Silver electrodes can be used for the stripping voltammetry of halogens and sulfide [6]. Other metals that are only sporadically employed in electroanalysis are rhodium, palladium, germanium, gallium, and lead [9].

Mercury is a versatile working electrode because of the following physical and chemical properties: (i) it is a liquid at a room temperature, (ii) its interfacial tension is high and the surface is hydrophobic, (iii) it forms amalgams with many heavy metals, and (iv) Hg(I) ions form sparingly soluble salts with halides, sulfide, sulfate, and some other anions [23]. Most importantly, as mentioned above, the reduction of hydrogen ions on mercury requires an extraordinary high overvoltage. The cathodic limits of the working window in aqueous solutions are between -1.1 V vs. SCE at pH 1 and -2.8 V at pH 13. The anodic limit in aqueous nitrate and perchlorate solutions is about 0.2 V due to the oxidation of mercury. In chloride solution the limit is -0.1 V because of the formation and precipitation of calomel, Hg_2Cl_2. The limit in sulfide solution is -0.6 V due to the formation and precipitation of HgS. The limits in solutions of all other anions are within these boundaries. Mercury is a useful cathode for reduction of ions of amalgam-forming heavy metals, such as Bi(III), In(III), Sb(III), Cu(II), Pb(II), Sn(II), Cd(II), Zn(II), Pd(II), Ni(II), Co(II), and Tl(I). Provided that no intermetallic compounds are formed, the activities of metal atoms dissolved in mercury are proportional to their concentrations, unlike metal deposits on solid electrodes. Metals accumulated in a mercury electrode by prolonged reduction under hydrodynamic conditions can be reoxidized, and this is the basis of anodic stripping voltammetry and potentiometric stripping analysis [9, 12]. Furthermore, many simple and complex ions can be partly reduced at mercury electrodes, e.g., BrO_3^- to Br^-, Ce(IV) to Ce(III), Cr(III) to Cr(II), Cr(VI) to Cr(III), Eu(III) to Eu(II), Fe(III) to Fe(II), Mo(VI) to Mo(V), Sn(IV) to Sn(II), Ti(IV) to Ti(III), and UO_2^{2+} to UO_2^+. In addition, mercury can be used for the measurement of electrode reactions of numerous organic substances containing reducible functional groups, such as azo, carbonyl, disulfide, nitro, quinone, conjugated double and triple bonds, and aromatic rings [3, 24, 25].

The traditional mercury electrode is a drop that is periodically created and dispatched at the tip of a glass capillary that is immersed in an electrolyte solution. This is the so-called dropping mercury electrode (dme). Mercury flows from a reservoir through a capillary and forms drops because of its high inter-facial tension. The electrode can operate in the free dropping mode, or a mechanical knocker can

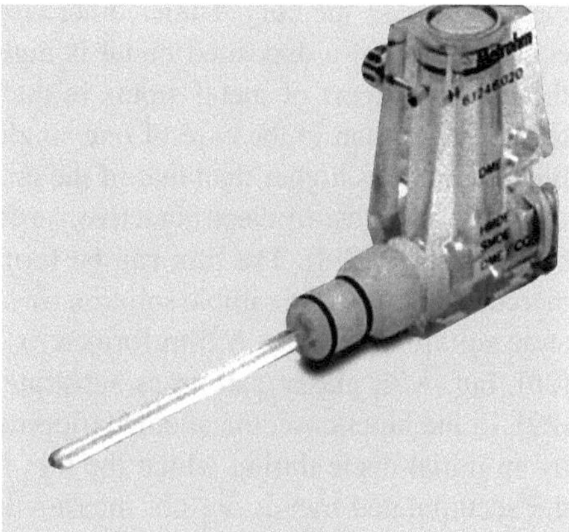

Fig. III.1.1 Metrohm multi-mode mercury electrode (courtesy of Metrohm Ltd., Herisau)

control the lifetime of the drop, i.e., the so-called drop time. The main advantage of the dropping mercury electrode is its continually renewed and clean surface, which is isotropic with respect to its physicochemical properties. Further, its surface area renews with constant frequency. The reproducible formation of a clean surface is exploited in adsorptive accumulation of surface-active and electroactive compounds and for the surface precipitation of mercuric and mercurous salts with various anions. Controlling the flow of mercury through the capillary can control the drop time within a wide range. A stable hanging mercury drop is realized either by manual control, or by electronic control. In the first case, mercury is pushed manually from a hermetically closed reservoir by a microscrew. Such device is called a *hanging mercury drop electrode* (HMDE). In a later design an electrically driven valve mechanically controls the flow (see Fig. III.1.1). This is called a *static mercury drop electrode* (SMDE). It can be synchronized with the potentiostat, and for this reason it is widely applied in all modern voltammetric measurements. Since a SMDE can always be used also as a HMDE, the manually operated electrodes are nowadays obsolete. The drop is formed with a SMDE by opening the valve for 0.1–0.3 s, depending on the drop size. After closing the valve, the volume of the drop is constant. Finally, the drop is mechanically knocked off, and a new one is created. By definition, in staircase, pulse, and differential pulse polarography, the electrode operates as a series of drops, while in staircase, pulse, differential pulse, and square-wave voltammetry, the entire potential scan is performed using one single drop (for details, see http://www.metrohm.ch).

Another type of mercury electrodes are mercury film electrodes. They are prepared by cathodic deposition of mercury on gold, glassy carbon, or iridium substrates [12]. The best film is formed on iridium because of its low solubility in mercury and its good wettability by mercury [26]. Gold is much more soluble in mercury, while, on glassy carbon, mercury is deposited as microdroplets [27]. Mercury film electrodes are used in anodic stripping voltammetry of amalgam-forming metals [9]. The stripping peak is proportional to the concentration of the metal atoms in mercury. Under otherwise identical accumulation conditions, the concentration of a deposited metal is higher in the film than in the drop. Further, the mass transport of metal atoms in the film toward the mercury solution interface is faster than in the case of one single, large drop. Hence, the response of the film electrode is higher than that of the drop electrode. However, the film electrode is highly sensitive to electroinactive, surface-active substances that can block the charge transfer [28]. The film can be formed either by adding 2×10^{-5} mol/L of mercuric ions into the sample solution (in situ) or by mercury deposition from a separate solution (ex situ). A film formed ex situ on iridium is stable for several days [26], but, with glassy carbon as substrate, an in situ film formation is preferable [29]. In the latter case, the anodic stripping voltammetric measurement is preceded by an initial cycle during which the thin film is formed, but the concentrations of the accumulated metals are not measured. In the following cycles the film grows parallel to the reduction of the analyte metal ions and this establishes reproducible conditions. In acidic perchlorate and nitrate solutions it is possible to strip off the deposited mercury by electrooxidation at 0.2 V, but in seawater this procedure would lead to calomel precipitation on the electrode surface. In the former case the measurements can be performed with approximately constant film thickness because the film is formed before each measuring cycle, and destroyed after the stripping phase. In chloride solutions all stripping scans must finish at −0.1 V to prevent electrooxidation of the deposited mercury, and the film continuously grows. After a series of measurements in the particular sample solution, the deposited mercury is electrooxidized in acidic nitrate solution.

Carbon electrodes are made of various materials, such as graphite of spectral purity, glassy carbon, graphite powder with liquid or solid binders, carbon fibers, highly oriented pyrolytic graphite (HOPG), carbon nanotubes, boron-doped diamond and titanium carbide [10, 30] (see also Chap. I.1). Graphite rods must be impregnated with paraffin to fill the pores and thus decrease the background current. The circular surface of the rod is polished and used as a disk electrode. A pencil-lead graphite can be used as a substrate for bismuth-film electrode for the determination of trace metals [31]. HOPG is prepared by exposing pyrolytic graphite to high pressure and high temperature. It is composed of condensed polyaromatic planes, or graphene sheets, which are held together by weak van der Waals forces. Within the sheet the carbon atoms are well ordered, each being bonded to three others in the plane. They create a hexagonal surface, which is atomically smooth over micrometer dimensions and is called a basal plane. The perpendicular surface is referred to as an edge plane. HOPG has metallic conductivity along basal plane and semiconducting along the edge plane. The double-layer capacitance of defect-free basal plane is very low: $C_{dl} < 5\ \mu F\ cm^{-2}$. The electrode can be cleaned by cleaving the layer planes using a sharp knife blade [32]. The glassy, or vitreous, carbon is manufactured by very slow carbonization of a premodeled polymeric resin body in an inert atmosphere at temperatures rising from 300°C and finishing at 1200°C. The material is macroscopically isotropic and seemingly poreless because the existing pores are tightly closed. It consists of cross-linked graphite-like sheets. The base of a glassy carbon cylinder is polished to a mirror finish and the cylinder is sealed in a tightly fitting tube of an insulating material to construct a stationary, or rotating, disk electrode [12]. A new type of carbon electrode material is electrically conducting diamond doped with boron, which is grown as a thin film on a conducting substrate [33]. Boron-doped diamond electrode is very hard and can be used in the combination with ultrasonic mixing [34]. Furthermore, it is characterized by the wide working window, low and stable background current and corrosion resistance. Contaminants adsorbed on the electrode surface can be removed by the exposure to clean organic solvents for half an hour [35], or by heating the electrode to about 500°C under high-vacuum conditions for 30 min [36]. Electrochemical activation is achieved by potential cycling between −0.5 V and 1.5 V vs. SCE in 0.5 M KNO_3 at 50 mV/s [37]. The carbon fibers are produced by high-temperature pyrolysis of polymer textiles. The fiber of 5–20 μm diameter is sealed by epoxy resin in a glass capillary and used as a microelectrode. The set of microelectrodes can be constructed by using a bunch of fibers that are insulated from each other by the sealant [9]. Carbon nanotubes can be synthesized by arc discharge, laser ablation, or chemical vapor deposition methods [38]. They appear either as multi-walled, or as single-walled. The first are composed of concentric and closed graphene tubules, each with a rolled-up graphene sheet. Their diameters are between 10 and 30 nm. Single-walled nanotube is made of a single graphite sheet rolled seamlessly with a diameter of 1–2 nm. The nanotubes can be immobilized on the surface of glassy carbon electrode to increase its surface area [39]. A carbon paste consists of graphite powder mixed with a liquid binder, such as the mineral oil Nujol. Electronic conductivity is achieved by the contact between graphite microparticles, while the oil ensures the compactness of the mixture. The paste is pressed into an isolating tube equipped with a piston to obtain a disk electrode with a renewable surface [6]. At the orifice of the tube the paste is leveled. Aqueous electrolytes cannot penetrate into the paste. The electrode reactions occur at the contact of graphite microparticles with the electrolyte solution. For trace metal analysis, ion exchangers can be added to the carbon paste electrode. The metal ions to be measured are firstly accumulated in the exchanger particles on the electrode surface, then reduced at the graphite particles and finally reoxidized by the stripping scan [40]. The additives to the carbon paste are called the modifiers. The choice of modifier depends on the redox reaction mechanism [41]. For instance, enzymes are added so as to develop biosensors, or insoluble solids are analyzed by grinding and adding to the paste [6, 9, 12].

Optically transparent electrodes are used in spectroelectrochemistry They can be made by evaporating 10–100 nm thick layers of platinum, gold, tin dioxide, silver, copper, mercury and carbon onto glass, or quartz substrates [42] (for more details see Chap. II.7).

Polypyrrole, polythiophene, polyaniline, polyfuran, polyacetylene, and polymethylthiophene may exhibit a mixed electronic and ionic conductivity, similar to inorganic intercalation, or insertion compounds [43]. In the conductive form these polymers are partly oxidized and these positive charges are equilibrated by inorganic anions, which can diffuse through the polymer net. Alternatively, the conductive polymers can be partly reduced, with cations as counter ions. The electronic conductivity originates from partial oxidation of conjugated π-bonds. The positive charge is mobile along the polymer chain by rearrangement of double bonds. A film of conductive polymer on a metal electrode can be deposited by oxidative electropolymerization of a monomer. Charging and discharging of the film is accompanied by insertion and expulsion of anions of the supporting electrolyte. Polymer electrodes are used in batteries, optical displays, and electroanalytical sensors for anions [44].

III.1.3 Electrode Geometry

The common forms of electrode are an inlaid disk, a sphere, a cylinder, a sheet, a net, a spiral wire, a sponge, an inlaid ring, and an inlaid rectangular plate. In analytical electrochemistry, the working electrodes most frequently appear in the first two forms, while the next four are forms of auxiliary electrodes. The ring appears in the rotating ring-disk electrode, and the plate is a frequent form of electrodes in the flow-through cells. The disk is the simplest form of solid metal and carbon electrodes. It is a cross section of a cylinder. Usually, the cylinder is sealed in an insulating material, exhibiting only the circular surface to the solution. The inlaid disk is the most convenient form for cleaning and polishing.

Mercury electrodes are usually of spherical shape. In chronoamperometry of a simple, reversible electrode reaction: $Ox^{n+} + ne^- \leftrightarrows Red$, using a hanging mercury drop electrode in an unstirred solution, the current density depends on time and electrode radius [5]:

$$\frac{I}{nFSc^*D^{1/2}} = \frac{1}{1+e^{\varphi}}\left[\frac{1}{\sqrt{\pi t}} + \frac{\sqrt{D}}{r_0}\right] \tag{III.1.1}$$

where $\varphi = nF(E - E^{\ominus})/RT$ is the dimensionless electrode potential, I is the current, n is the number of electrons, F is the Faraday constant, S is the electrode surface area, c^* is the bulk concentration of the reactant, D is the diffusion coefficient, E is the constant electrode potential, $E^{\ominus}$ is the standard potential of the redox reaction, R is the gas constant, T is the temperature, t is the measurement time, and r_0 is the radius of the drop. If the time is very long, the current tends to approach a limiting value:

$$\lim_{t\to\infty} I = \frac{4\pi nFc^*Dr_0}{1+e^{\varphi}} \tag{III.1.2}$$

considering that the area of the sphere is $S = 4\pi r_0^2$. This is called the spherical effect. It is a consequence of the spherical expansion of the diffusion layer. The limiting current appears if

$$\sqrt{\pi Dt} \gg r_0 \tag{III.1.3}$$

i.e., if the diffusion layer is much thicker than the electrode radius. Under this condition, the material is diffusing from the periphery of a large sphere toward its center. The concentration of the material near the center maintains the constant flux at the electrode surface. The theory shows that the spherical effect appears on all electrodes of finite dimensions. On an inlaid disk of radius r_0, the current density in chronoamperometry of a reversible electrode reaction is given by the formula [45]:

$$\frac{I}{nFSc^*D^{1/2}} = \frac{1}{1+e^{\varphi}}\left[\frac{1}{\sqrt{\pi t}} + \frac{4\sqrt{D}}{\pi r_0}\right] \quad \text{(III.1.4)}$$

The second term in the brackets of Eq. (III.1.4) is caused by the edge of the disk. The material lying above the insulating plane of the electrode is diffusing radially toward the edge of the conducting disk and thus contributes to the flux. Hence, neglecting the second term means neglecting the edge effect. It is justified in the case of rather large disks (e.g., $r_0 = 0.2$ cm) because the influence of the second term in Eq. (III.1.4) is smaller than 10% if the measurement time is shorter than 8 s (for $D = 10^{-5}$ cm^2/s):

$$t < \frac{\pi r_0^2}{16 \cdot 10^2 \cdot D} \quad \text{(III.1.5)}$$

However, the edge effect is maximized on microelectrodes. If the radius of the inlaid disk is only 10^{-3} cm, the influence of the first term in the brackets of Eq. (III.1.4) is smaller than 10% if the measurement time is longer than 2 s (for $D = 10^{-5}$ cm^2/s):

$$t > \frac{10^2 \pi r_0^2}{16D} \quad \text{(III.1.6)}$$

That means that the chronoamperometric current on the microdisk electrode quickly acquires a limiting value:

$$\lim_{t\to\infty} I = \frac{4nFc^*Dr_0}{1+e^{\varphi}} \quad \text{(III.1.7)}$$

In practice this means that the response to a very slow voltammetric scan appears in a form similar to a polarographic wave, with the limiting current [46]:

$$\lim_{E<<E^{\ominus}}\left(\lim_{t\to\infty} I\right) = 4nFc^*Dr_0 \quad \text{(III.1.8)}$$

This is called a steady-state response. If $r_0 = 10^{-3}$ cm, $D = 10^{-5}$ cm^2/s, and $c^* = 10^{-4}$ mol/L, the limiting current of the steady-state response is only 4×10^{-10} A. Hence, working with microelectrodes requires the use of a special current amplifier and a faradaic cage to minimize the noise. However, the voltage drop due to the solution resistance is minimal in this current range. Hence, microelectrodes are suitable for use in highly resistive media, such as ice [47]. Besides, the capacitive currents on these electrodes of extremely small surface areas are negligible [48]. For these two reasons they can be utilized for the measurement of electrode reaction kinetics by using scan rates as high as 10^6 V/s [49], or a square-wave frequency of 2000 Hz [50]. Alternatively, the electrode kinetics can be measured from the steady-state responses [45, 48, 51]. Different forms of microelectrodes include a microcylinder, a microhemisphere, an inlaid microband, an inlaid microring, a recessed microdisk, a microhole, a microhemispheroid, and a micro-blunt-cone [6, 45]. The steady-state response depends on the geometry of the particular microelectrode [45], but all of them originate either from the spherical effect (e.g., the microhemispheroid), or the edge effect (e.g., the microcylinder and the inlaid microring) [52]. Arrays of microelectrodes (e.g., the inlaid microdisks array, or an interdigitated microbands array) respond as a sum of microelectrodes only if the distance between the individual microelectrodes well exceeds their critical dimensions [53]. If this condition is not satisfied, the diffusion layers of individual micro electrodes overlap and the response of the array equals the response of a planar macroelectrode [54] (see Eq. III.1.4). For this reason a set of mercury microdroplets deposited on a glassy carbon disk works like a uniform film macroelectrode. Numerous methods for fabricating microelectrode arrays have been reviewed recently [55].

III.1.4 Hydrodynamic Conditions

In chronoamperometry, the response of a simple, reversible electrode reaction on any type of electrode is given by the equation:

$$I = \frac{nFSc^*D}{(1 + e^{\varphi}) \cdot \delta} \tag{III.1.9}$$

where δ is the so-called thickness of the diffusion layer at the electrode surface. In fact, it is the ratio c^*/δ that is equal to the concentration gradient at the electrode surface if $E \ll E^{\ominus}$ and $e^{\varphi} \to 0$, but the diffusion layer itself is significantly thicker than δ [42]. This is shown in Fig. III.1.2. However, the concept of the distance δ is useful for analyzing the transport phenomena in the cell. If the solution is not stirred, and neglecting the edge effect, the thickness δ at the surface of a large planar inlaid disk electrode depends on the measurement time: $\delta = \sqrt{\pi D t}$ (cf. Eqs. III.1.4 and III.1.9). In the case of a spherical electrode, δ is a function of its radius:

$$\delta = \frac{r_0\sqrt{\pi D t}}{r_0 + \sqrt{\pi D t}} \tag{III.1.10}$$

Equation (III.1.10) is characterized by two limiting values:

$$\lim_{r_0 \to \infty} \delta = \sqrt{\pi D t} \tag{III.1.11}$$

and

$$\lim_{t \to \infty} \delta = r_0 \tag{III.1.12}$$

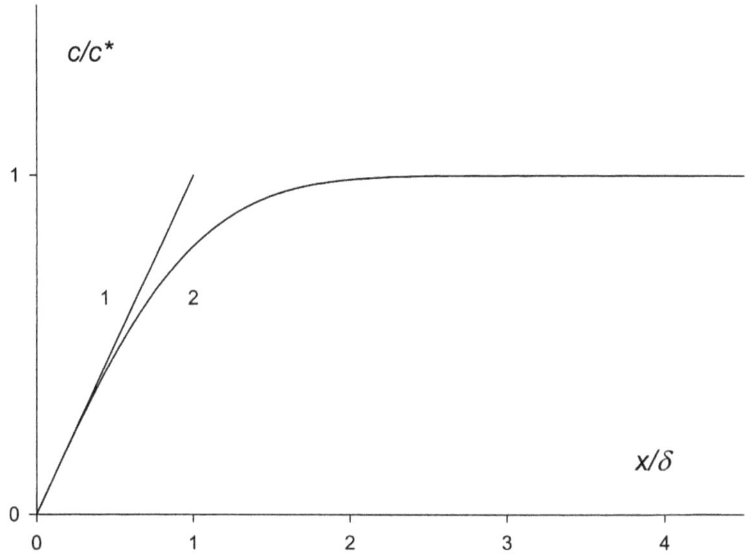

Fig. III.1.2 *1* Gradient of dimensionless concentration c/c^* at the stationary planar electrode surface in a calm solution, and *2* the dependence of the ratio c/c^* on the dimensionless distance x/δ, where $\delta = (\pi D t)^{1/2}$, $c/c^* = \mathrm{erf}\left(x\pi^{1/2}/2\delta\right)$ and $\left[\partial\left(c/c^*\right)/\partial\left(x/\delta\right)\right]_{x=0} = 1$

The second limit corresponds to the steady-state conditions under which the distance δ and the current do not depend on time. This constant current is significant only on microelectrodes, while, on macroelectrodes, practically useful steady-state conditions can be established only by stirring the solution. The stirring can be achieved either by rotating a disk electrode, or by using a stationary electrode in the agitated solution [42, 56]. In electroanalysis, the main purpose of stirring is to increase the efficacy of accumulation of analytes on the electrode surface.

The rotating-disk electrode is a small metal disk inlaid into an insulating cylinder having a large base [57–60]. The disk is situated in the center of the base. The cylinder is mounted on a metallic axle that is connected to an electromotor. The axle is perpendicular to the base and lies in the axis of the cylinder. The axle is connected to the metal disk by a wire, and bears a metallic bell that rotates in a mercury pool to obtain noiseless electrical contact with the electrode. The contact achieved by graphite brushes may appear noisy. When the cylinder rotates, the solution near its base moves angularly, radially, and axially. The coordinate system is shown in Fig. III.1.3. Because of the friction, a thin layer of solution that is in contact with the base follows the rotation of the cylinder. By viscosity this angular movement extends into a thicker layer of solution, but its rate decreases exponentially with the distance from the surface of the cylinder. The rotation of solution induces a centrifugal force that causes a radial movement of the solution. To preserve the pressure, the latter is compensated by the axial flux from the bulk of the solution. The components of the fluid velocity depend on the angular velocity of the disk ($\omega = \mathrm{d}\varphi/\mathrm{d}t$), on

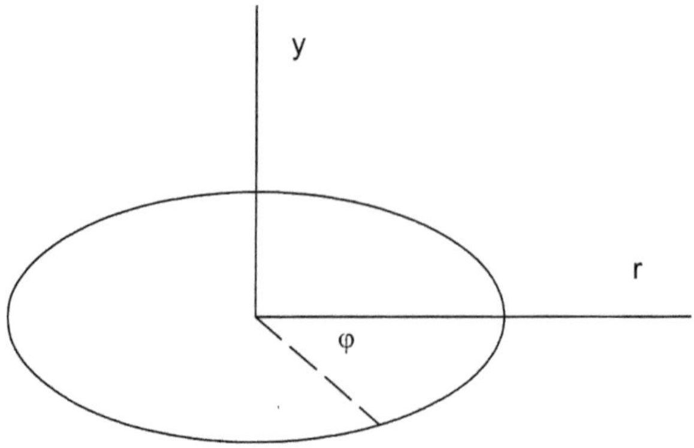

Fig. III.1.3 Rotating disk coordinate system

the radial distance from the center of the disk (r), on the coefficient of kinematic viscosity of the fluid (ν), and on the dimensionless distance from the surface of the disk $\left(\gamma = (\omega/\nu)^{1/2} y\right) : v_r = r\omega F(\gamma), v_\varphi = r\omega G(\gamma)$, and $v_y = (\omega\nu)^{1/2} H(\gamma)$ [59]. The functions $F(\gamma)$, $G(\gamma)$, and $H(\gamma)$ are shown in Fig. III.1.4. Around the center of the base of the cylinder, where the metal disk lies, the axial component of the solution velocity is most important, since the electroactive material is transported

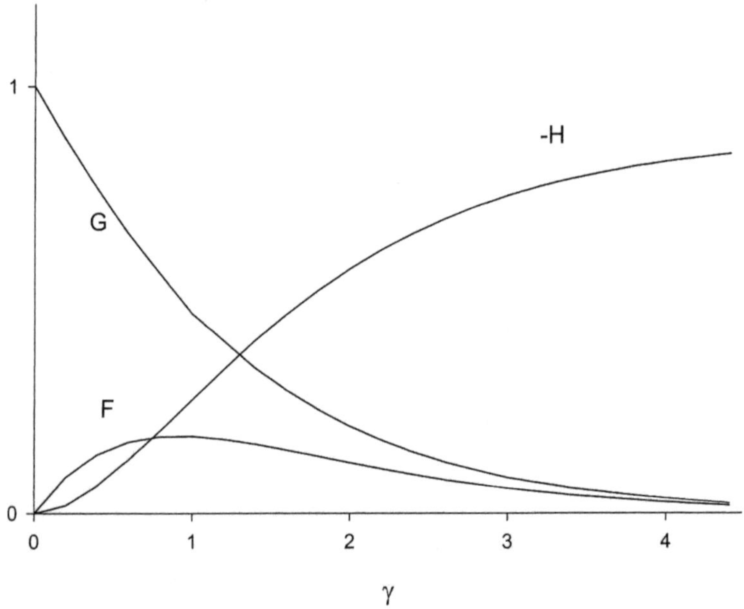

Fig. III.1.4 Dependence of functions F, G, and H on dimensionless distance $\gamma = (\omega/\nu)^{1/2} y$ from the rotating disk [58]

toward the surface in this direction only. Under chronoamperometric conditions, a diffusion layer develops at the electrode surface and extends as far into the solution as the flux at the surface is not equal to the rate of mass transport in the bulk of the solution. Under steady-state conditions, the distance δ depends on the electrode rotation rate [57]:

$$\delta = 1.61 D^{1/3} \nu^{1/6} \omega^{-1/2} \quad \text{(III.1.13)}$$

where D is the diffusion coefficient and ω is the electrode rotation rate in radians per second. Considering Eq. (III.1.9), the limiting current on the rotating disk electrode is proportional to the square root of the rotation rate. This relationship is valid only for the laminar flow of the solution. The flow pattern changes from laminar to turbulent if the Reynolds number $Re = \omega r^2/\nu$ is larger than 2×10^{-5} (where r is the radius of the cylinder). In a turbulent flow the distance δ is much smaller than in a laminar flow, but the equation for the flux is not as simple as Eq. (III.1.13) [42]. In cyclic voltammetry, the rotating disk electrode may operate either under steady-state or transient conditions. If the scan rate is very slow, the response is a wave with a limiting current determined by Eqs. (III.1.9) and (III.1.13). At fast scan rates, a voltammetric peak appears during the forward, but not the reverse scan [61]. The rotating disk electrode serves primarily for electrokinetic measurements [57, 60, 62]. For the analysis of reaction mechanisms the rotating-ring-disk electrode is useful [63]. It consists of a thin metallic ring inlaid around the metallic disk that is situated at the center of the base of the insulating cylinder. Because of the radial component of the solution movement, the products of the electrode reaction formed on the disk electrode are carried toward the ring electrode where they can be detected and characterized [42].

In electroanalytical chemistry, a thin mercury film covered rotating disk electrode and a bare glassy carbon rotating disk electrode are used for the trace determination of heavy metals [9, 29, 64]. The accumulation is always performed under steady-state conditions, but during the stripping scan it is sometimes better to stop the electrode rotation, especially if the contacts produce interfering noise [29, 64]. If linear scan voltammetry is used, the response is higher and sharper when the electrode rotates during the stripping scan [64], but, on stationary electrodes, the responses in pulse techniques are enhanced by re-plating the stripped material [29].

The current density on a stationary electrode immersed in a flowing solution is not uniform over the whole electrode surface. At an inlaid plate over which the solution flows, the diffusion layer thickness and the steady-state distance δ increase downstream [57, 65, 66]:

$$\delta = 2.94 D^{1/3} \nu^{1/6} \mathrm{v}^{-1/2} x^{1/2} \quad \text{(III.1.14)}$$

Here v is the solution velocity and x is the distance from the electrode edge in the direction of the flow. Similarly, the distance δ at a tubular electrode is

$$\delta = 1.49 D^{1/3} R^{1/3} v_0^{-1/3} x^{1/3} \quad \text{(III.1.15)}$$

where R is the inner radius of the tube, v_0 is the solution velocity at the center of the tube, and x is the distance from the edge in the axial direction. In both cases the highest current density appears at the electrode edge and decreases downstream [42].

The flow of solution can be maintained by centrifugal and peristaltic pumps that are placed outside the cell [62]. Homemade micro-pumps with a rotating smooth-disk impeller and nozzle, or with a vibrating conically perforated disk, are situated in the cell, closely below the working electrode [67]. The most favorable relationship between the average distance δ and the flow rate of the solution is found in the wall-jet systems in which a flux of the solution induced by a centrifugal pump is directed perpendicularly to the surface of a stationary disk electrode. In these systems, the current is linearly proportional to the flow rate of the solution [68].

The simplest stirrer is a magnetic bar that rotates at the bottom of the cell and causes a vertical flow of the solution. It is used in combination with the hanging, or static, mercury drop electrode. The downward flow of the solution tangentially passes by the drop. Hence, the current density is the highest near the equator of the drop and decreases toward the pole opposite to the capillary, as well as in the capillary shadow, i.e., near the orifice [5].

The mass transport can also be enhanced by ultrasound [62]. An ultrasonic source immersed in the solution produces a radial flux and the formation of bubbles [69]. When a bubble collapses at an electrode surface, a microjet of solution is formed [70]. This form of transport is called microstreaming. It can be utilized in ultrasound-enhanced electroanalysis [71–73].

Cells in which the electrochemical and hydrodynamic functions are separated are less expensive and easier to maintain. Although some of these systems operate under turbulent flow, this is not important for anodic stripping voltammetric measurements as long as the hydrodynamic conditions are reproducible [74].

III.1.5 Chemically Modified Electrodes

It is difficult to give a definition of a chemically modified electrode, as there are so many different experimental approaches to prepare electrodes with modified surfaces. Most widespread are methods based on attaching a certain compound, or a specific chemical group, to the surface of a metallic conductor [75]. These compounds are called modifiers [76]. Their role is to react chemically with an analyte and to modify its electrode reaction. The attachment can be achieved in various ways, e.g., by adsorption [77], by chemical reaction [78, 79], or by formation of a polymer film [80]. Some examples are adsorption of quinhydrone, trioctylphosphine oxide, polyacrylamide oxime, malachite green and oxine on glassy carbon electrodes, covalent bindings of substituted silanes, or metal porphyrins to the chemically oxidized surface of carbon electrodes, a film of Nafion containing dicyclohexyl-18-crown-6 ether, and a film of polypyrrole-*N*-carbodithionate on glassy carbon electrode. Modifying coating may serve either to preconcentrate a compound on the electrode [81] or to achieve electrocatalysis [82], or permselectivity [83]. Modified electrodes for preconcentration bind analytes either by complex-formation reactions, or by ion exchange, or by extraction. In all cases, information about the analyte element concentration is furnished by the reduction or oxidation currents of the complex, ionic associate, or extract formed during the preconcentration stage [9]. The electrode modification can be performed either in a preliminary step, or in situ. In the latter case, a surface-active ligand is added to the sample and the complex is adsorbed on the electrode surface. Cathodic stripping voltammetry with adsorptive accumulation was discussed in the previous chapter. Preliminary modified electrodes can be used repeatedly, but they must be regenerated after each measurement. They are prepared in the modifier solution, then washed and immersed in the sample solution to collect the analyte. After that they are transferred to the voltammetric cell to perform the measurement. In the simplest case the electrode process leads to the release of analyte from the electrode surface without destruction of the modifier film [75]. Hence, the electrode is ready for the next collection step. However, in the majority of cases, it is necessary to regenerate the electrode by extraction of the analyte from the film into a separate solution. Moreover, in the worst scenario, the electrode must be polished, and a new modification is necessary before each measurement. This procedure may restrict analytical application of solid modified electrodes. Instead, the modifier can be added to the carbon paste because the active surface of this type of electrode can be mechanically renewed [84, 85].

Electrocatalytic modified electrodes contain attached electron-transfer mediators. They accelerate electrode reactions that are kinetically hindered on the bare electrode surface. A charge transfer occurs between the attached mediator and the dissolved analyte in an unhindered redox reaction. The catalyst is readily regenerated by the fast and reversible electrode reaction. For this purpose it must be in close contact with the electrode surface. If the mediator is adsorbed as a monolayer its stability is poor due to dissolution. It is better to incorporate it into a polymer film. Alternatively, microcrystals of insoluble catalysts, such as cobalt phthalocyanine, can be mechanically transferred to the surface of a paraffin-impregnated graphite electrode [86, 87]. Electrocatalytic electrodes are used in direct voltammetric mode for the measurement of higher concentrations of inorganic and organic substances, such as H_2O_2, NO_2^-, SCN^-, As^{3+}, $S_2O_8^{2-}$, Fe(III) and Fe(II), hydrazine, cysteine, glutathione, thiols, alcohols, phenols, ascorbic acid, dopamine, NADH, ferrodoxine, hemoglobin, and insulin [75]. These electrodes are not used in the stripping mode. Their stability and repetitive application depend on whether the mediator can be completely regenerated, or not.

Polymer-modified electrodes are prepared either by casting polymer solution on the solid electrode surface and allowing the solvent to evaporate or by electropolymerization of a monomer [13]. The polymer film is either an electronic insulator, but an ionic conductor, or a mixed electronic and ionic conductor. The first type of films is used for the preparation of permselective coatings that serve to prevent unwanted matrix constituents reaching the electrode surface. Electrochemical reactions in mixed conductor films depend on several physicochemical processes, such as the partitioning of solvent and ions between the polymer and the electrolyte, the transport of ions within the polymer phase, the distribution of redox states in the polymer and electronic charge transfers between them, as well as heterogeneous charge transfer between the electrode and the polymer film [13]. Details of these processes can be found in the specialized literature [13–15, 43, 75, 88].

References

1. Parsons R (1985) The single electrode potential: its significance and calculation. In: Bard AJ, Parsons R, Jordan J (eds) Standard potentials in aqueous solution. Marcel Dekker, New York, p 13
2. Bockris JO'M, Khan SUM (1993) Surface electrochemistry. Plenum Press, New York
3. Heyrovsky J, Kuta J (1966) Principles of polarography. Academic Press, New York
4. Weppner W (1995) Electrode performance. In: Bruce PG (ed) Solid state electrochemistry. Cambridge University Press, Cambridge, p 199
5. Galus Z (1994) Fundamentals of electrochemical analysis. Ellis Horwood, New York and Polish Scientific Publishers PWN, Warsaw
6. Bond AM, Scholz F (1990) Z Chemie 30: 117
7. Cattral RW (1997) Chemical sensors. Oxford University Press, Oxford
8. Tinner U (1989) Elektroden in der Potentiometrie. Metrohm, Herisau, p 16
9. Brainina H, Neyman E (1993) Electroanalytical stripping methods. John Wiley, New York
10. Kinoshita K (1988) Carbon. John Wiley, New York
11. Gerischer H (1997) Principles of electrochemistry. In: Gellings PJ, Bouwmeester HJM (eds) CRC handbook of solid state electrochemistry. CRC Press, Boca Raton, p 9
12. Wang J (1994) Analytical electrochemistry. VCH, Weinheim
13. Doblhofer K (1994) Thin polymer films on electrodes: a physicochemical approach. In: Lipkowski J, Ross PN (eds) The electrochemistry of novel materials. VCH, Weinheim, p 141
14. Murray RW (1992) Molecular design of electrode surfaces. John Wiley, New York
15. Burgmayer P, Murray RW (1986) Polymer film covered electrodes. In: Skotheim TA (ed) Handbook of conducting polymers. Marcel Dekker, New York, p 507
16. Finklea HO (1996) Electrochemistry of organized monolayers of thiols and related molecules on electrodes. In: Bard AJ, Rubinstein I (eds) Electroanalytical chemistry, vol 19. Marcel Dekker, New York, p 109
17. Komorsky-Lovrić Š (1996) Fresenius J Anal Chem 356: 306
18. Scholz F, Meyer B (1998) Voltammetry of solid microparticles immobilized on electrode surfaces. In: Bard AJ, Rubinstein I (eds) Electroanalytical chemistry, vol 20. Marcel Dekker, New York, p 1
19. Swain GM (2007) Solid electrode materials: pretreatment and activation. In: Zoski CG (ed) Handbook of electrochemistry. Elsevier, Amsterdam, p 111
20. Budevski E, Staikov G, Lorenz WJ (1996) Electrochemical phase formation and growth. VCH, Weinheim
21. Engstrom RC (1984) Anal Chem 56: 890
22. Schmid GM, Curley-Fiorino ME (1975) Gold. In: Bard AJ (ed) Encyclopedia of electrochemistry of the elements, vol IV. Marcel Dekker, New York, p 87
23. Wrona PK, Galus Z (1977) Mercury. In: Bard AJ (ed) Encyclopedia of electrochemistry of the elements, vol IXA. Marcel Dekker, New York, p 1
24. Wang J (1988) Electroanalytical techniques in clinical chemistry and laboratory medicine. VCH, Weinheim
25. Zuman P (2000) Electroanalysis 12: 1187
26. Kounaves SP, Buffle J (1986) J Electrochem Soc 133: 2495
27. Štulikova M (1973) J Electroanal Chem 48: 33

28. Komorsky-Lovrić Š, Branica M (1994) Fresenius J Anal Chem 349: 633
29. Mlakar M, Lovrić M (1990) Analyst 115: 45
30. Baś B, Piech R, Niewiara E, Ziemnicka M, Stobierski L, Kubiak WW (2008) Electroanalysis 20: 1655
31. Demetriades D, Economou A, Voulgaropoulos A (2004) Anal Chim Acta 519: 167
32. McCreery RL (1999) Electrochemical properties of carbon surfaces. In: Wieckowski A (ed) Interfacial electrochemistry. Marcel Dekker, New York, p 631
33. Swain GM (2004) Electrically conducting diamond thin-films: advanced electrode materials for electrochemical technologies. In Bard AJ, Rubinstein I (eds) Electroanalytical chemistry, vol 22. Marcel Dekker, New York, p 182
34. Compton RG, Foord JS, Marken F (2003) Electroanalysis 15: 1349
35. Ranganathan S, Kuo TC, McCreery RL (1999) Anal Chem 71: 3574
36. Fagan DT, Hu IF, Kuwana T (1985) Anal Chem 57: 2759
37. McDermott CA, Kneten KR, McCreery RL (1993) J Electrochem Soc 140: 2593
38. Dai H (2002) Acc Chem Res 35: 1035
39. Futaba DN, Hata K, Yamada T, Mizuno K, Yumura M, Iijima S (2005) Phys Rev Lett 95: 56104
40. Navrátilová Z, Kula P (2000) J Solid State Electrochem 4: 342
41. Hočevar SB, Švancara I, Vytřas K, Ogorevc B (2005) Electrochim Acta 51: 706
42. Southampton Electrochemistry Group (1985) Instrumental methods in electrochemistry. John Wiley, New York, p 317
43. Scrosati B (1995) Polymer electrodes. In: Bruce PG (ed) Solid state electrochemistry. Cambridge University Press, Cambridge, p 229
44. Teasdale P, Wallace G (1993) Analyst 118: 329
45. Zoski CG (1996) Steady-state voltammetry at microelectrodes. In: Vanysek P (ed) Modern techniques in electroanalysis. John Wiley, New York, p 241
46. Bond AM, Oldham KB, Zoski CG (1989) Anal Chim Acta 216: 177
47. Bond AM, Pfund VB (1992) J Electroanal Chem 335: 281
48. Oldham KB, Myland JC, Zoski CG, Bond AM (1989) J Electroanal Chem 270: 79
49. Amatore C, Bonhomme F, Bruneel J-L, Servant L, Thouin L (2000) J Electroanal Chem 484: 1
50. Komorsky-Lovrić Š, Lovrić M, Bond AM (1993) Electroanalysis 5: 29
51. Oldham KB, Zoski CG (1988) J Electroanal Chem 256: 11
52. Amatore C, Fosset B (1996) Anal Chem 68: 4377
53. Feeney R, Kounaves SP (2000) Electroanalysis 12: 677
54. Szabo A, Zwanzig R (1991) J Electroanal Chem 314: 307
55. Szunerits S, Thouin L (2007) Microelectrode arrays. In: Zoski CG (ed) Handbook of electrochemistry. Elsevier, Amsterdam, p 391
56. Bard AJ, Faulkner LR (1980) Electrochemical methods. John Wiley, New York
57. Opekar F, Beran P (1976) J Electroanal Chem 69: 1
58. Pleskov YuV, Filinovskii VYu (1976) The rotating disk electrode. Consultants Bureau, New York
59. Levich V (1962) Physicochemical hydrodynamics. Prentice-Hall, Englewood Cliffs, NJ
60. Newman JS (1973) Electrochemical systems. Prentice-Hall, Englewood Cliffs, NJ
61. Brett CMA, Oliveira Brett AM (1998) Electroanalysis. Oxford University Press, Oxford
62. Alden JA, Compton RG (2000) Anal Chem 72: 199A
63. Albery WJ, Bruckenstein S (1966) Trans Faraday Soc 62: 1920
64. Komorsky-Lovrić Š (1988) Anal Chim Acta 204: 161
65. Tait RJ, Bury PC, Finnin BC, Reed BL, Bond AM (1993) J Electroanal Chem 356: 25
66. Alden JA, Compton RG (1996) J Electroanal Chem 404: 27
67. Magjer T, Branica M (1977) Croat Chem Acta 49: L1
68. Omanović D, Peharec Ž, Magjer T, Lovrić M, Branica M (1994) Electroanalysis 6: 1029
69. Mikkelsen O, Schroder KH (2000) Electroanalysis 12: 1201
70. Birkin PR, Silva-Martinez S (1996) J Electroanal Chem 416: 127

71. Agra Gutierrez C, Hardcastle JL, Ball JC, Compton RG (1999) Analyst 124: 1053
72. Saterlay AJ, Compton RG (2000) Fresenius J Anal Chem 367: 308
73. Tomčik P, Banks CE, Compton RG (2003) Electroanalysis 15: 1661
74. Macpherson JV (2000) Electroanalysis 12: 1001
75. Labuda J (1992) Select Electrode Rev 14: 33
76. Downard AJ (2000) Electroanalysis 12: 1085
77. Gong J, Lin X (2004) Electrochim Acta 49: 4351
78. Love JC, Estroff LA, Kriebel JK, Nuzzo RG, Whitesides GM (2005) Chem Rev 105: 1103
79. Anariba F, DuVall SH, McCreery RL (2003) Anal Chem 75: 3837
80. Vidal JC, Garcia-Ruiz E, Castillo JR (2003) Microchim Acta 143: 93
81. Toghill KE, Wildgoose GG, Moshar A, Mulcahy C, Compton RG (2008) Electroanalysis 20: 1731
82. Vasantha VS, Thangamuthu R, Chen SM (2008) Electroanalysis 20: 1754
83. Blanco-Lopez MC, Gutierrez-Fernandez S, Lobo-Castañon MJ, Miranda-Ordieres AJ, Tuñon-Blanco P (2004) Anal Bioanal Chem 378: 1922
84. Navratilova Z, Kula P (2003) Electroanalysis 15: 837
85. Ojani R, Raoof JB, Fathi S (2008) Electroanalysis 20: 1825
86. Komorsky-Lovrić Š (1995) J Electroanal Chem 397: 211
87. Ravi Shankaran D, Sriman Narayanan S (1999) Fresenius J Anal Chem 364: 686
88. Abruna HD (1988) Conducting polymer film on metallic substrates. In: Skotheim TA (ed) Electroresponsive molecular and polymeric systems, vol 1. Marcel Dekker, New York, p 98

Chapter III.2
Reference Electrodes

Heike Kahlert

III.2.1 Introduction

In most electrochemical measurements, it is necessary to keep one of the electrodes in an electrochemical cell at a constant potential. This so-called reference electrode allows control of the potential of a working electrode (e.g. in voltammetry) or the measurement of an indicator electrode (e.g. in potentiometry, see Chap. II.9). The standard hydrogen electrode plays the role of a basic reference element in electrochemical devices; however, in practice, it is difficult to handle. Therefore, secondary reference electrodes are preferred in most experiments. A secondary reference electrode must fulfil the following criteria: (i) it should be chemically and electrochemically reversible, i.e. its potential is governed by the Nernst equation and does not change in time; (ii) the potential must remain almost constant when a small current passes through the electrode and reverse to its original value after such small current flow (i.e. a non-polarisable electrode); and (iii) the thermal coefficient of potential should be small. Whereas there is no reference electrode that offers all these properties to the same extent, some electrodes are very close to that ideal behaviour. Here, the most important and most widespread reference electrodes will be described. In general, secondary reference electrodes are electrodes of the second kind, i.e. metal electrodes coupled to a solubility equilibrium of a salt of this metal and an electrolyte solution containing a fixed concentration of the anion of the sparingly soluble metal salt. In addition to this kind of electrodes, for some special cases, other reference systems exist. It would be ideal if the reference electrode could be placed in the same electrolyte solution that is used in the electrochemical system. In the classical two-electrode configuration of Heyrovský dc-polarography, a large non-polarisable mercury pool electrode serves as a counter electrode with a constant reference potential (Fig. III.2.1a). In the two-electrode configuration used for potentiometric measurements and in three-electrode configurations in voltammetric and in modern polarographic experiments (Fig. III.2.1b), the reference electrode can be

H. Kahlert (✉)
Institut für Biochemie, Universität Greifswald, 17487 Greifswald, Germany
e-mail: hkahlert@uni-greifswald.de

F. Scholz (ed.), *Electroanalytical Methods*, 2nd ed.,
DOI 10.1007/978-3-642-02915-8_15, © Springer-Verlag Berlin Heidelberg 2010

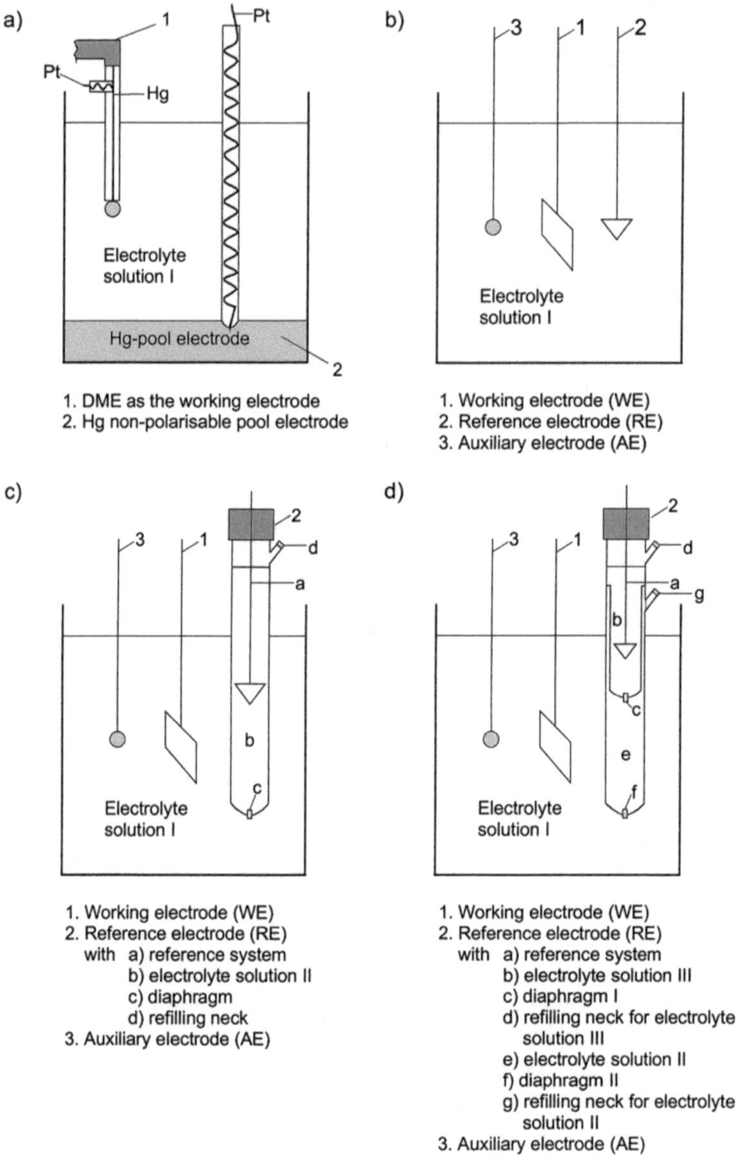

Fig. III.2.1a–d. **(a)** Scheme of the classical two-electrode configuration of the Heyrovský dc-polarography. **(b)** Scheme of a three-electrode configuration without a salt bridge between the measuring solution (electrolyte solution I) and the reference system. **(c)** Scheme of a three-electrode configuration with a salt bridge between the reference system and the measuring solution (electrolyte solution I). **(d)** Scheme of a three-electrode configuration with two salt bridges between the reference system and the measuring solution (electrolyte solution I)

either an electrode of the second kind or a metal wire (e.g. platinum). In the latter case, the potential of the metal wire may be fairly constant during an experiment; however, it is neither calculable nor will it follow the above-mentioned criteria of a real reference electrode. Therefore, such an electrode is referred to as a *pseudo reference electrode*. In voltammetry such pseudo reference electrodes are employed whenever the use of a three-electrode potentiostat is necessary. In that case, and also when a normal reference electrode is used, but separated from the electrolyte of the voltammetric cell in a way that unknown liquid junction potentials can arise (e.g. when an aqueous reference electrode is separated by a salt bridge from a non-aqueous electrolyte in the voltammetric cell), it is very useful to have a "potential standard system", e.g. ferrocene, in the solution under study. For reasons of interference, it is always safest to separate the reference electrode with its electrolyte solution from the electrochemical cell. This can be done with a diaphragm, see also Fig. III.2.1c. To prevent the electrolyte solution in the electrochemical cell from contamination by the reference electrolyte solution, a second salt bridge can be used as depicted in Fig III.2.1d. This second salt bridge may contain in some cases the same electrolyte solution as the reference electrode compartment and the electrochemical cell; mostly it will differ from one of these half-cells or both.

III.2.2 The Standard Hydrogen Electrode

The standard hydrogen electrode is the primary standard in electrochemistry. It is based on the following reversible equilibrium:

$$H_2 + 2H_2O \rightleftarrows 2H_3O^+ + 2e^- \qquad \text{(III.2.1)}$$

The standard hydrogen electrode consists of a platinum wire or a platinum sheet covered with platinum black (i.e. platinised) and an electrolyte solution containing hydronium ions. The hydrogen gas is usually continuously supplied. The Galvani potential difference of such electrode is (cf. Chap. II.9)

$$\Delta\phi = \Delta\phi^{\ominus} + \frac{RT}{2F}\ln a^2_{H_3O^+} + \frac{RT}{2F}\ln\frac{1}{f_{H_2}} = \Delta\phi^{\ominus} + \frac{RT}{F}\ln\frac{a_{H_3O^+}}{\sqrt{f_{H_2}}} \qquad \text{(III.2.2)}$$

The standard conditions are defined as: $a_{H_3O^+} = 1$ (i.e., $a_{H_3O^+} = \gamma_{H_3O^+}\frac{m_{H_3O^+}}{m^{\ominus}}$ with $\gamma_{H_3O^+} = 1$, $m_{H_3O^+} = 1\,\text{mol kg}^{-1}$, and $m^{\ominus} = 1\,\text{mol kg}^{-1}$) and $f_{H_2} = 1$ (i.e. the fugacity of hydrogen is calculated according to $f_{H_2} = \gamma_{H_2}\frac{p_{H_2}}{p^{\ominus}}$ with $\gamma_{H_2} = 1$, $p_{H_2} = 1$ standard atmosphere $= 101.325$ kPa, and $p^{\ominus} = 1$ standard atmosphere $=$ 101.325 kPa). By definition, the potential of this electrode is zero at all temperatures. As it is difficult to adjust the activity of hydronium ions to 1, for practical purpose hydrochloric acid with concentrations of 10^{-2} or 10^{-3} mol L^{-1} is used. The potential of the electrode can be calculated by using well-known activity coefficients

Table III.2.1 Mean ionic activity coefficients and paH[a] for hydrochloric acid at 25°C [1]

m_{HCl}/mol kg^{-1}	$\gamma_\pm$	paH
0.001	0.9650	3.0154
0.002	0.9519	2.7204
0.005	0.9280	2.3335
0.01	0.9040	2.0434

[a]Negative common logarithm of hydronium ion activities

of the hydrochloric acid (see Table III.2.1). In the literature, several ingenious constructions of hydrogen electrodes can be found [2]. The standard hydrogen electrode allows very precise measurements to be made; however, the demanding handling restricts its use. For instance, the hydrogen gas must be of highest purity, especially with respect to oxygen, H_2S, AsH_3, SO_2, CO and HCN, because these gases poison the platinum electrode. In solution, volatile substances, such as HCl, can be purged from the solution by the hydrogen gas, metals can be reduced at the electrode, redox systems may influence the electrode potential, etc. [3]. This will suffice to understand that the standard hydrogen electrode lost its practical importance, but maintained its role as primary standard.

III.2.3 Electrodes of the Second Kind as Reference Electrodes

III.2.3.1 Mercury-Based Reference Electrodes

III.2.3.1.1 The Calomel Electrode

The calomel electrode was introduced by Ostwald in 1890. It is an electrode of the second kind (cf. Chap. II.9). As a reference electrode of fixed, well-known and very reproducible potential, it is still important today. In the simplest case, a single drop of mercury is placed in a small tube and is covered by mercury(I) chloride (calomel Hg_2Cl_2) (see Fig. III.2.2a). Another possibility is to fill a small glass tube with a paste of mercury, mercury(I) chloride and potassium chloride solution (Fig. III.2.2b). The paste is in contact with a potassium chloride solution of constant activity. Mostly, a saturated potassium chloride solution is used and the paste additionally contains solid potassium chloride. The electrode net reaction can be formulated in the following way:

$$Hg^0 + Cl^- \rightleftarrows 1/2Hg_2Cl_2 + e^- \qquad \text{(III.2.3)}$$

Thus, the potential of this electrode against the standard hydrogen electrode is given by the equation:

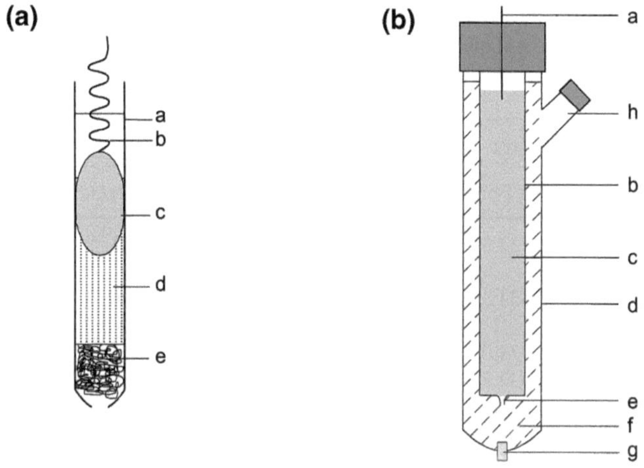

a) Glass tube
b) Platinum wire
c) Mercury droplet
d) Paste of mercury, mercury(I) chloride and potassium chloride
e) Glass wool

a) Platinum wire
b) Inner glass tube
c) Paste of mercury, mercury(I) chloride and potassium chloride solution
d) Outer glass tube
e) Small gap
f) Potassium chloride solution
g) Diaphragm
h) Refilling neck

Fig. III.2.2a–b. (**a**) Simple calomel reference electrode. (**b**) Typical construction of a calomel reference electrode for practical use in laboratories

$$E = E_{\mathrm{c}}^{\ominus'}(\mathrm{Hg,Hg_2Cl_2}) - \frac{RT}{F}\ln a_{\mathrm{Cl^-}} \qquad \text{(III.2.4)}$$

[cf. Chap. II.9 to see the meaning of the formal potential $E_{\mathrm{c}}^{\ominus'}(\mathrm{Hg,Hg_2Cl_2})$]. Table III.2.2 gives the potentials of the calomel electrode for different concentrations of potassium chloride at different temperatures [2]. The problem is that, in a saturated potassium chloride solution already at temperatures above 35°C a disproportion reaction takes place according to

$$\mathrm{Hg_2Cl_2} \rightleftarrows \mathrm{Hg} + \mathrm{HgCl_2} \qquad \text{(III.2.5)}$$

The back reaction by cooling down the electrode is very slow so that a hysteresis of the electrode potential occurs. This is the reason why it is recommended that the calomel electrode only be used at lower temperatures, in maximum up to 70°C. The thermal coefficient is smallest for a calomel electrode with 0.1 M KCl, but it is easier to handle the saturated calomel electrode.

Table III.2.2 Electrode potentials of the calomel electrode at different temperatures and different concentrations of KCl (the concentrations are related to 25°C) [2]

	E/mV				
	m_{KCl}	c_{KCl}			
T/°C	1 mol kg^{-1}	0.1 mol L^{-1}	1 mol L^{-1}	3.5 mol L^{-1}	Saturated
0		333.8	285.4		260.2
5	272.90				
10	271.94	334.3	283.9		254.1
15	270.87				250.9
20	269.62	334.0	281.5		247.7
25	268.23	333.7	280.1		244.4
30	266.61	333.2	278.6		241.1
35	264.90				237.7
40	263.06	331.6	275.3	246.6	234.3
45	261.04				
50		329.6	271.6	242.8	227.2
60		322.9	267.3	237.7	219.9
70			262.2	233.1	212.4

III.2.3.1.2 The Mercury/Mercury(I) Sulphate Electrode

Another mercury-containing reference electrode of the second kind is the mercury/mercury(I) sulphate electrode. In principle, the construction is the same as for the calomel electrode. The electrolyte solution consist either of potassium sulphate in a certain concentration or sulphuric acid. The electrode net reaction is

$$2Hg^0 + SO_4^{2-} \rightleftarrows Hg_2SO_4 + 2e^- \qquad \text{(III.2.6)}$$

The electrode potential follows as

$$E = E_c^{\ominus'}(Hg,Hg_2SO_4) - \frac{RT}{F}\ln a_{SO_4^{2-}} \qquad \text{(III.2.7)}$$

The electrode potentials of this electrode at different temperatures are given in Table III.2.3. The mercury/mercury(I) sulphate electrode with sulphuric acid is useful as a reference electrode in solutions containing sulphuric acid.

III.2.3.1.3 The Mercury/Mercuric Oxide Electrode

In general, metal/metal oxide electrodes are used in systems of high alkalinity. Ideally, these electrodes respond to the pH of the electrolyte solution in a similar way as the hydrogen electrode. They may be regarded as a special sort of electrodes of the second kind, because the oxide ions are in equilibrium with hydroxyl ions of

Table III.2.3 Electrode potentials of the mercury–mercurous sulphate electrode at different temperatures [1]

T/°C	E/mV	
	$c_{H_2SO_4}$ 1 mol L^{-1} (at 25°C)	$c_{K_2SO_4}$ Saturated
0	634.95	671.8
5	630.97	667.6
10	627.04	663.5
15	623.07	659.4
20	619.30	655.3
25	615.15	651.3
30	611.07	647.3
35	607.01	643.3
40	603.05	639.2
45	599.00	635.1
50	594.87	630.9
55	590.51	626.6
60	586.59	622.6
65		617.7
70		613.3
75		608.4
80		603.4
85		598.4
90		593.1

the solvent. In the case of mercury/mercury oxide electrodes we can formulate the following equilibria:

$$Hg \rightleftarrows Hg^{2+} + 2e^- \quad \text{(III.2.8)}$$

$$Hg^{2+} + O^{2-} \rightleftarrows HgO \quad \text{(III.2.9)}$$

$$O^{2-} + H_2O \rightleftarrows 2OH^- \quad \text{(III.2.10)}$$

$$2OH^- + 2H^+ \rightleftarrows 2H_2O \quad \text{(III.2.11)}$$

The potential of this electrode is

$$E = E_c^{\ominus'}\left(Hg,Hg^{2+}\right) + \frac{RT}{2F}\ln a_{Hg^{2+}} \quad \text{(III.2.12)}$$

The solubility of the oxide is defined by the solubility constant; however, the oxide ions are also involved in reactions (III.2.10) and (III.2.11), so that

$$K_s = a_{Hg^{2+}} \times \left(a_{OH^-}\right)^2 \quad \text{(III.2.13)}$$

where K_s is the solubility product of the hydroxide. Equation (III.2.12) can then be formulated as

$$E = E_c^{\ominus'}\left(\mathrm{Hg,Hg^{2+}}\right) + \frac{RT}{2F}\ln K_s - \frac{RT}{F}\ln a_{\mathrm{OH^-}} = E_c^{\ominus'}(\mathrm{Hg,HgO}) - \frac{RT}{F}\ln a_{\mathrm{OH^-}} \tag{III.2.14}$$

The formal potential of this electrode, $E_c^{\ominus'}$(Hg,HgO) is 0.9258 V [2]. Because of its solubility properties, the use of the mercury/mercury oxide electrode is confined to strong alkaline solutions. According to Ives and Janz [2], the mercuric oxide is best prepared by gentle ignition of carefully crystallised mercuric nitrate. The construction is similar to the calomel electrode with an alkaline solution [e.g. saturated $Ca(OH)_2$] instead of the potassium chloride as the electrolyte solution.

III.2.3.2 The Silver/Silver Chloride Electrode

This electrode of the second kind is the most frequently used reference electrode in practical measurements, because the construction is very simple, the potential is very well reproducible, and last, but not least, this electrode is free of mercury! Normally, a silver wire is covered with silver chloride, which can be achieved electrochemically or thermally [1]. Electrochemically produced films are thinner than thermally produced films. The construction of a commercially available silver/silver chloride electrode is similar to the calomel electrode (see Fig. III.2.3). A very simple method for preparing a silver/silver chloride electrode has been described by Thomas [4]. Because reference systems based on silver/silver chloride can be produced in a very small size, they are often used in microsystems [5–9]. The electrolyte solution in these reference systems is normally a potassium chloride solution (mostly saturated or 3 M), and only seldom sodium or lithium chloride. The electrode net reaction is

$$\mathrm{Ag^0 + Cl^- \rightleftarrows AgCl + e^-} \tag{III.2.15}$$

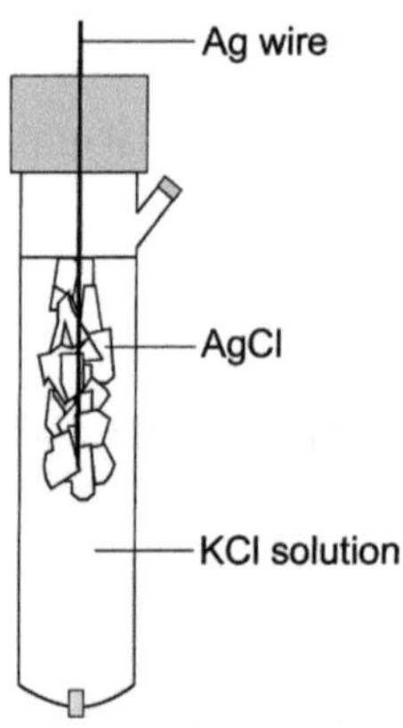

Fig. III.2.3 Typical construction of a silver/silver chloride reference electrode

Table III.2.4 Electrode potentials of the silver–silver chloride electrode at different temperatures and different concentrations of KCl (the concentrations are related to 25°C) [1]

	E/mV				
	m_{KCl}	c_{KCl}			
T/°C	1 mol kg^{-1}	1 mol L^{-1}	3 mol L^{-1}	3.5 mol L^{-1}	Saturated
0	236.6	249.3	224.2	222.1	220.5
5	234.1	246.9	220.9	218.7	216.1
10	231.4	244.4	217.4	215.2	211.5
15	228.6	241.8	214.0	211.5	206.8
20	225.6	239.6	210.5	207.6	201.9
25	222.3	236.3	207.0	203.7	197.0
30	219.0	233.4	203.4	199.6	191.9
35	215.7	230.4	199.8	195.4	186.7
40	212.1	227.3	196.1	191.2	181.4
45	208.4	224.1	192.3	186.8	176.1
50	204.5	220.8	188.4	182.4	170.7
55	200.6	217.4	184.4	178.0	165.3
60	196.5	213.9	180.3	173.5	159.8
65		210.4	176.4	169.0	154.3
70	187.8	206.9	172.1	164.5	148.8
75		203.4	167.7	160.0	143.3
80	178.7	199.9	163.1	155.6	137.8
85		196.3	158.3	151.1	132.3
90	169.5	192.7	153.3	146.8	126.9
95	165.1	189.1	148.1	142.5	121.5

The electrode potential depends on the activity of chloride ions in the electrolyte solution according to (see also Chap. II.9)

$$E = E_{c}^{\ominus\prime}(\mathrm{Ag,AgCl}) - \frac{RT}{F}\ln a_{\mathrm{Cl}^-} \qquad \text{(III.2.16)}$$

In Table III.2.4 electrode potentials for different chloride concentrations at different temperatures are given. Compared to the calomel electrode, the silver/silver chloride reference system has the great advantage that measurements at elevated temperatures are possible. Special devices have been developed based on the silver/silver chloride reference systems for measurements in high-temperature aqueous solutions [10] and under changing pressure conditions [11].

III.2.4 pH-Based Reference Electrodes

pH-sensitive electrodes can also be used as reference systems with a buffer solution of constant and well-known pH. The glass electrode is rarely used as a reference electrode, because it requires two reference electrodes itself and the potential does not maintain a constant value because of a changing asymmetry potential. This prompts frequent standardisations of such reference electrode.

An alternative is the quinhydrone electrode, which is based on the electrochemically reversible oxidation–reduction system of p-benzoquinone (quinone) and hydroquinone in which hydrogen ions participate (see also Chap. II.9). The construction is very simple, in that a noble metal wire, usually platinum, is introduced into a solution containing some crystals of quinhydrone. For pH < 9, the potential of this electrode depends on the pH of the adjacent solution according to

$$E(25°\mathrm{C}) = E_c^{\ominus'}(\text{quinhydrone}, \mathrm{H_3O^+}) - 0.0592[\mathrm{V}] \times \mathrm{pH} \qquad \text{(III.2.17)}$$

At higher pH values it is of increasing significance that hydroquinone is a dibasic

Table III.2.5 Formal potentials of the quinhydrone electrode at different temperatures [2] and electrode potentials at different temperatures and at different pH values

		E/mV			
T/°C	$E_c^{\ominus'}$/mV	pH = 1	pH = 3	pH = 5	pH = 7
0	717.98	663.78	555.38	446.98	338.58
5	714.37	659.17	548.77	438.37	327.97
10	710.73	654.53	542.13	429.73	317.33
15	707.09	649.89	535.49	421.09	306.69
20	703.43	645.23	528.83	412.43	296.03
25	699.76	640.56	522.16	403.76	285.36
30	696.07	635.87	515.47	395.07	274.67
35	692.37	631.17	508.77	386.37	263.97
40	688.65	626.55	502.35	378.15	253.95

acid and the first and second dissociation constants influence the electrode potential. On the other hand, hydroquinone is a reducing agent and undergoes aerial oxidation at a rate that increases quite abruptly with rising pH at around pH 8. This limits the pH of the buffer solution. Under well-defined conditions, the potential response of this electrode is reproducible, fast (within a few seconds) and it reaches the theoretical value to a few microvolts [2]. Table III.2.5 give the formal potentials of the quinhydrone electrode at different temperatures and the electrode potentials at different temperatures and at different pH values. This electrode has been used up to now in solutions containing hydrofluoric acid and in some organic solvents (methanol, ethanol, n-butanol, acetone and formic acid). Application of the quinhydrone electrode to organic solutions needs consideration of the different solubilities of the two components.

III.2.5 Inner Potential Standards

The standard hydrogen electrode is the universally accepted reference electrode in aqueous solutions. Unfortunately, such a universal reference electrode does not exist for non-aqueous solutions. By using reference electrodes such as the calomel electrode or the silver/silver chloride electrode with aqueous reference electrolyte

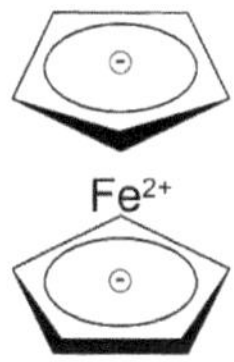

Fig. III.2.4 Ferrocene molecule

solutions, an unknown liquid-junction potential is introduced into the measurements. There has been great interest in finding a reference redox couple, the potential of which is independent of the solvent [12–19]. Most successful were complex compounds containing large ligands and a transition metal ion, e.g. ferrocene (fc) (see Fig. III.2.4) and its derivatives. Gritzner and Kuta [20] suggested the ferrocene (fc)/ferrocenium (fc^+) couple as a reference system. The well-defined one-electron system fc/fc^+ is now widely used in cyclic voltammetric studies in highly resistive organic solvents as a reference potential. The typical concentration of ferrocene is between 0.5 mM and 1 mM [19]. Bond et al. have shown that ferrocene can be used also in aqueous media under carefully controlled conditions [16]. Through semi-integration of cyclic voltammograms it is possible to use the ferrocene/ferrocenium system in high-resistance organic media not only to calibrate the potential scale but also to determine the uncompensated resistance [19]. Formal potentials of ferrocene in different solvents can be found in the literature [24]. When an inner potential standard such as ferrocene is used, the procedure in cyclic voltammetry is that the formal potential of the fc/fc^+ system is measured as the mid-peak potential (cf. Chap. I.2) and it is given the tabulated value. The characteristic potentials of all other voltammetric signals are then related to the formal potential of fc/fc^+. Of course, the inner standard should be well removed from all other voltammetric systems in the solution studied. Decamethylferrocene is much more hydrophobic than ferrocene, and its formal potential in different solvents depends much less on the solvent properties. Therefore, it has been proposed as a very useful potential standard [26].

III.2.6 Solid-State Reference Electrodes

In recent years, the so-called solid-state reference electrodes have been developed for application in solution electrochemistry. In these electrodes, an electron conducting element (e.g. a silver wire or a silver layer) is first covered with a solid salt layer (e.g. silver chloride) and then with a second solid salt/polymer layer (e.g. a mixture of silver chloride and potassium chloride in PVC) Finally, the electrode is covered with a polymer, for instance cellulose acetate is deposited (cf. Fig. III.2.5) [21]. The influence of the polymer on the properties of such electrode was investigated by Lee et al. [22]. Because these electrodes can be fabricated in a very small size and because only small amounts of the electrolyte are dissolved, these electrode systems are often used in microelectrode arrangements for medicinal measurements.

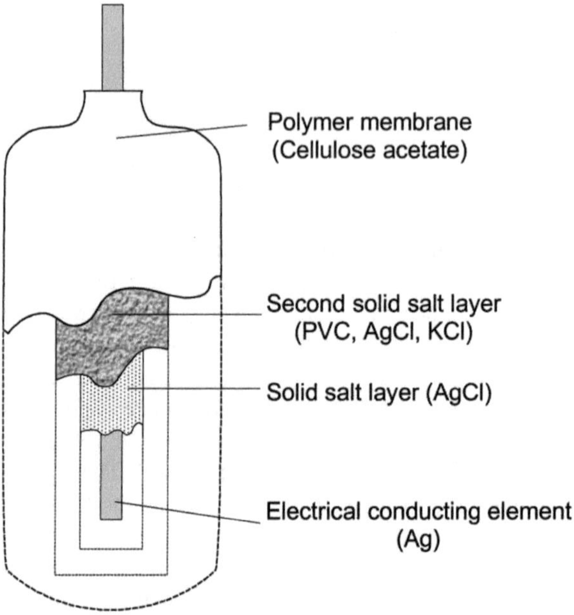

Fig. III.2.5 Solid-state silver/silver chloride reference electrode

Recently, Vonau et. al [27] have suggested an all-solid reference electrode consisting of a sintered Ag/AgCl mixture, which is embedded in a solid melt of KCl inside a cylindrical hollow body of porous alumina oxide ceramics. The outer surface of the ceramics is coated with a chemical-resistant insulating layer. At the bottom end, a circular area is held free from the coating and acts as diaphragm.

The solid-contact reference systems used in all-solid-state membrane based ion selective electrodes are discussed in Sect. II.9.3.4.3.

III.2.7 Pseudo Reference Electrodes

Pseudo reference electrodes are electrodes that are used as reference electrodes, especially in three-electrode potentiostatic measurements, but do not possess the properties of "real" reference electrodes, i.e. they are *not* non-polarisable, they do *not* possess a thermodynamically calculable potential and they do *not* have a potential, which is independent of the electrolyte in the cell. Nevertheless, such electrodes, normally simple metal wires of platinum or gold, can be used provided that the potential scale is calibrated with an inner standard. A somewhat worse alternative is to measure the potential of the pseudo reference electrode versus a conventional reference electrode in a separate experiment. This is reliable only when one can be sure that the potential of the pseudo reference electrode is the same in the calibration and the application experiment. Since the pseudo reference electrode is directly placed into the cell electrolyte, the impedance is usually small (no salt bridge, no liquid junction potentials etc.), which is desirable for the potentiostat to function properly.

III.2.8 Practical Problems

III.2.8.1 The Electrolyte of Reference Electrodes

The electrolyte solution of reference electrodes may serve two functions: (a) to provide a constant potential of the reference electrode, and (b) to serve as the electrolyte bridge to the analyte solution, ideally with negligible diffusion potentials. The possibilities to contact the reference electrode to the adjacent solution vary greatly. The most common arrangements are shown in Fig. III.2.6. A detailed description of the diaphragms is given below. When such arrangement is used, the levels of the analyte and the reference electrode electrolytes should be balanced to prevent any contamination, either of the analyte or the reference electrode compartment.

Very often it is imperative to prevent any contact between the analyte solution and the reference electrolyte solution, and this can be achieved by an additional salt bridge (see Fig. III.2.7). The electrolyte solution in the salt bridge has to be chosen so that it will not influence the measurement. It must be tolerated by both the reference electrolyte solution and the analyte solution. The most important electrolytes for salt bridges are potassium nitrate, potassium chloride, sodium sulphate, or ammonium nitrate solutions.

At a diaphragm two solutions are separated from each other. When these two solutions consist of different electrolytes or electrolytes in different concentrations, a diffusion of the constituent of the solutions occurs. This leads to a potential difference, which is called a liquid-junction potential, the magnitude of which depends strongly on the composition of the solutions. Ideally, this liquid-junction potential should be very small and constant to minimise errors. To estimate liquid-junction potentials, the Henderson equation is applicable [23]:

$$\Delta\phi_{\text{diff}} = -\frac{RT}{F}\frac{\sum\limits_i \frac{(a_i^{\text{I}}-a_i^{\text{II}})u_i|z_i|}{z_i}}{\sum\limits_i \left(a_i^{\text{I}}-a_i^{\text{II}}\right)u_i|z_i|}\ln\frac{\sum\limits_i a_i^{\text{I}}u\,|z_i|}{\sum\limits_i a_i^{\text{II}}u\,|z_i|} \qquad \text{(III.2.18)}$$

The most important reference systems (calomel electrode, silver/silver chloride electrode) contain chloride ions. Here, mostly potassium chloride solutions are used, because the mobility of potassium ions (76.2×10^5 $\text{cm}^2\text{V}^{-1}\text{s}^{-1}$ in infinitely diluted aqueous solution at 25°C) and chloride ions (79.1×10^5 $\text{cm}^2\text{V}^{-1}\text{s}^{-1}$ in infinitely diluted aqueous solution at 25°C) are quite similar [1]. Considering two solutions containing potassium chloride of different concentrations, Eq. III.2.18 can simplified to

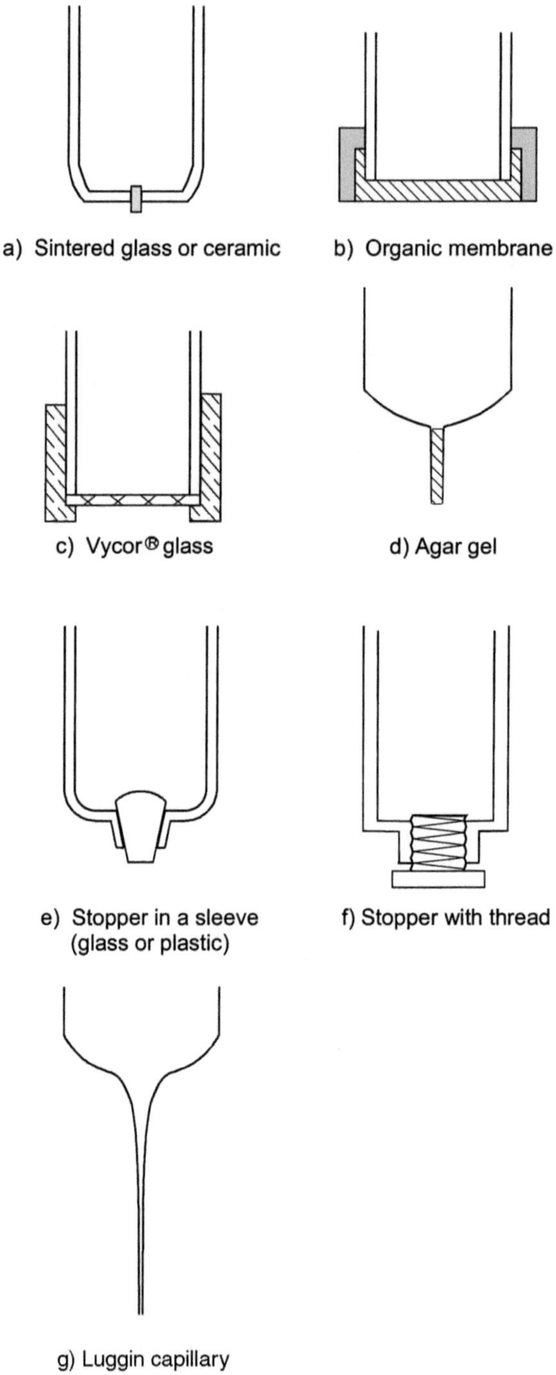

Fig. III.2.6 Most common arrangements of electrolyte junctions

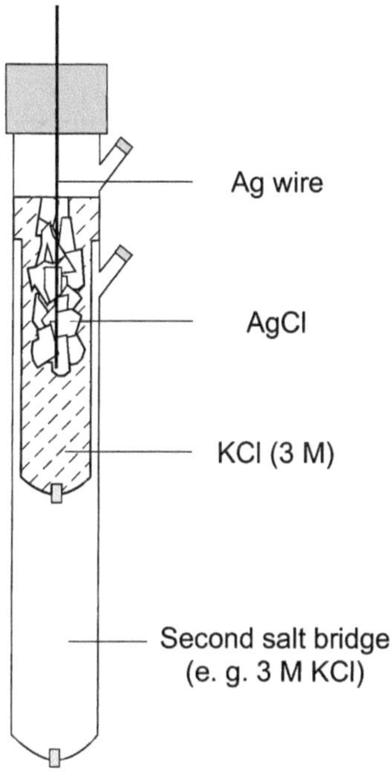

Fig. III.2.7 Construction of a silver/silver chloride reference electrode with a second salt bridge

$$\Delta\phi_{\text{diff}} = -\frac{RT}{F}\left(t^{\text{K}^+} - t^{\text{Cl}^-}\right)\ln\frac{a^{\text{II}}_{\text{KCl}}}{a^{\text{I}}_{\text{KCl}}} \tag{III.2.19}$$

and, with $t^{\text{K}^+} = 0.4906$ and $t^{\text{Cl}^-} = 0.5094$ the liquid-junction potential for different concentrations can be calculated, as it was done in Table III.2.6. If a solution of any electrolyte (e.g. hydrochloric acid) is in contact with an electrolyte of comparable mobilities of anions and cations (e.g. potassium chloride), then the first term in Eq. (III.2.18) is small when the concentration of the potassium chloride solution is high compared to the concentration of HCl. The change in the second term is not so dramatic because of the logarithm and the liquid-junction potential becomes very small. This is one reason why the concentration of the reference electrolyte should be as high as possible, for instance saturated. Saturated means that at each possible temperature a certain amount of the solid salt must be present. However, in the case of reference electrodes of the second kind the thermal coefficient is usually high due to the strong temperature dependence of the solubility and of the solubility product. On the other hand, electrodes with a saturated reference electrolyte solution show a strong hysteresis and a precipitation of the electrolyte salt can lead to a blocking of the contact area to the analyte solution. For reference systems such as the calomel electrode and the silver/silver chloride electrode, an optimal concentration of potassium chloride is between 3 mol L^{-1} and 3.5 mol L^{-1}. To minimise the

Table III.2.6 Diffusion potential between two solutions of KCl with different concentrations at 25°C

$c_{\text{Electrolyte I}}$/mol L^{-1}	$c_{\text{Electrolyte II}}$/mol L^{-1}	$\Delta\phi_{\text{diff}}$/mV
Saturated	0.1	$-1.8^{(a)}$
3.5	1.0	$-0.2^{(a)}$
3.5	0.1	$-0.6^{(a)}$
3.5	0.01	$-1.0^{(a)}$
0.06	0.04	$-0.2^{(b)}$
0.04	0.02	$-0.34^{(b)}$
0.04	0.005	$-1.0^{(b)}$
0.02	0.01	$-0.34^{(b)}$
0.01	0.005	$-0.33^{(b)}$

[a]From [1].
[b]From [25].

pressure sensitivity and required maintenance efforts, nowadays liquid electrolyte solutions are replaced by gel-solidified reference electrolytes with sealers such as agar-agar, aerosil™, methyl and hydrooxyethyl cellulose and others. Advantages and drawbacks of gel-solidified reference electrolytes are summarised in [27].

III.2.8.2 The Diaphragm

The diaphragm of a reference electrode must fulfil two contradictory tasks, i.e. to provide the electrolytic conductivity between the electrodes, and to prevent the free mixing of the electrolytes. It loses this function when impurities plug the diaphragm and when the pores of the diaphragm are too large. Proteins are potential diaphragm blockers, whereas some solutions, e.g. strongly alkaline solutions, may easily dissolve glass diaphragms. A frequently encountered pitfall is the precipitation of potassium perchlorate when perchlorate ions are present in the analyte solution and KCl in the electrolyte bridge. This problem can be easily avoided by using NaCl in the electrolyte bridge. Common diaphragms (cf. Fig. III.2.6) are made of sintered glass, porcelain, or of the so-called Vycor® glass (Vycor® is the registered trademark of Corning, Inc. It is a porous glass of sufficient conductivity, and it is very well suited to be used as diaphragm between aqueous and non-aqueous solutions. For this purpose thin plates of e.g. 2 mm thickness are cut from a rod and fitted to the end of a glass tube with the help of a shrink tube). Ground-in stoppers, of course without grease, but having instead a thin electrolyte film, can be used as well. In some cases a capillary filled with an electrolyte is used as a salt bridge. This so-called Luggin capillary is useful when the solution in the electrochemical cell has an appreciable electric resistance. The mouth of the capillary can be placed very near to the surface of the working electrode, thus minimising the *IR* drop in the electrolyte layer between the working electrode and the capillary. The diaphragm of commercial available reference electrodes can be cleaned from time to time. In case of proteins a cleaning solution consisting of hydrochloric acid (0.1 M) and pepsin ($w = 1\%$) can be used; in the case of silver sulphide a cleaning solution consisting of hydrochloric acid (0.1 M) and thiourea ($w = 7.5\%$) is recommended [1]. Before the washing procedure, the reference electrolyte solution must be removed. After the cleaning, the reference tube must be rinsed carefully with the electrolyte solution to remove traces of the cleaning solution.

III.2.8.3 Refilling of the Reference Solution

Because of the diffusion between the electrolytes in the reference electrode half-cell and the main cell, the reference electrolyte must be replaced by fresh solution from time to time. Two things are important: (i) the new solution must have exactly the right composition and concentration as necessary for the reference electrode and (ii) air bubbles must not be allowed to be present in the tubes. The refilling has to be done very carefully, so that no air bubbles are left in the system.

III.2.8.4 Maintenance of Reference Electrodes

The most significant problem is that reference electrodes must be stored between the measurements very carefully so that they do not dry out. It is recommended that the refilling necks and the diaphragm be closed with a plastic cap. It is also recommended that the reference electrode be stored in the reference electrolyte solution as it is possible for some commercially available electrodes (see Fig. III.2.8).

It is very important to check the reference electrode potential from time to time to exclude erroneous measurements. Normally, the potentials of reference systems are given versus the standard hydrogen electrode. Because of the difficult handling of a hydrogen electrode, it is recommended to measure the potential of the reference electrode against a second (new) reference electrode, for instance a calomel electrode. In practice, this electrode should be used exclusively for this purpose and

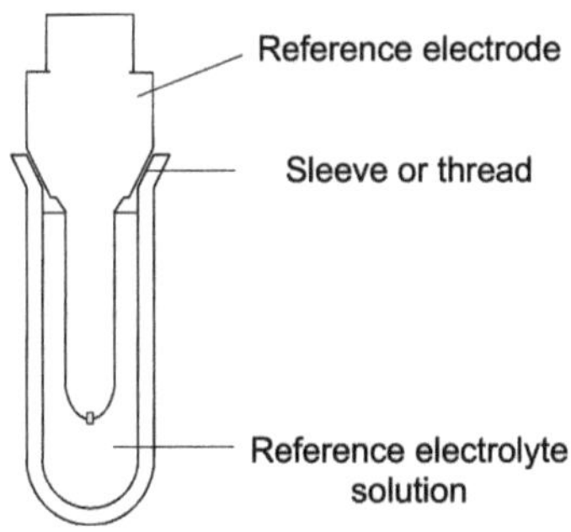

Fig. III.2.8 Storage of a commercially available reference electrode

then stored in saturated potassium chloride solution. For measuring the potential, both electrodes are introduced into a saturated potassium chloride solution and are connected electrically through a high-impedance voltmeter. The temperature should be kept constant; the deviation should not exceed ± 0.5°C. Then, the formal potential of the second electrode is added to the measured potential to obtain the potential of the tested reference electrode:

$$E_{\mathrm{c}}^{\ominus\prime}(1\text{st ref.}) = E + E_{\mathrm{c}}^{\ominus\prime}(2\text{nd ref.}) \tag{III.2.20}$$

References

1. Galster H (1990) pH-Messung. VCH, Weinheim
2. Ives DJG, Janz GJ (1961) Reference electrodes, theory and practice. Academic Press, New York
3. Schwabe K (1976) pH-Meßtechnik. Verlag Theodor Steinkopf, Dresden
4. Thomas JM (1999) J Chem Ed 76: 97
5. Hassel AW, Fushimi K, Seo M (1999) Electrochem Commun 1: 180
6. Sinsabaugh SL, Fu CW, Fung CD (1986) Proc Electrochem Soc 86: 66
7. Arquinth P, van den Berg A, van der Schoot BH, de Rooij NF, Bühler H, Morff WE, Dürselen LFJ (1993) Sens Actuators B B13–14: 340
8. Mroz A, Borchardt M, Dükmann C, Cammann K, Knoll M, Dumschat C (1998) Analyst 123: 1373
9. Suzuki H, Hirakawa T, Sasaki S, Karube I, (1998) Sens Actuators B B46: 146
10. Lvov SN, Gao H, Macdonald DD (1998) J Electroanal Chem 443: 186
11. Peters G(1997) Anal Chem 69: 2362
12. Koepp WM, Wendt H, Strehlow H (1960) Z Elektrochem 64: 483
13. Kolthoff IM, Thomas FG (1965) J Phys Chem 69: 3049
14. Tanaka N, Ogata T (1974) Inorg Nucl Chem Lett 10: 511
15. Bond AM, Lay PA (1986) J Electroanal Chem 199: 285
16. Bond AM, McLennan EA, Stojanovic RS, Thomas FG (1987) Anal Chem 59: 2853
17. Stojanovic RS, Bond AM (1993) Anal Chem 65: 56
18. Podlaha J, Štěpnička P, Gyepes R, Mareček V, Lhotský A, Polášek M, Kubišta J, Nejezchleba M (1997) Collect Czech Chem Commun 62: 185
19. Bond AM, Oldham KB, Snook GA (2000) Anal Chem 72: 3492
20. Gritzner G, Kuta J (1982) Pure Appl Chem 54: 1527
21. Nagy K, Eine K, Syverud K, Aune O (1997) J Electrochem Soc 144: L1
22. Lee HJ, Hong US, Lee DK, Shin JH, Nam H, Cha GS (1998) Anal Chem 70: 3377
23. Hamann CH, Hamnett A, Vielstich W (1998) Electrochemistry. Wiley-VCH, Weinheim
24. Togni A, Hayashi T (1995) Ferrocenes. VCH, Weinheim
25. Yu TR, Ji GL (1993) Electrochemical methods in soil and water research. Pergamon Press, Oxford
26. Noviandri I, Brown KN, Fleming DS, Gulyas PT, Lay PA, Masters AF, Phillips L (1999) J Phys Chem B 103: 6713
27. Guth U, Gerlach F, Decker M, Oelßner W, Vonau W (2009) J Solid State Electrochem 13: 27

Chapter III.3
Electrolytes

Šebojka Komorsky-Lovrić

III.3.1 Introduction

An electrochemical cell consists of two electronic conductors (electrodes) connected via the external circuit (metallic conductor) and separated by an ionic conductor that is called the electrolyte. While the electrodes can be either pure metallic conductors, or mixed electronic and ionic conductors, the separator must be an electronic insulator to prevent a short circuit between the electrodes. In principle, electrolytes can be used in all three physical states: solid, liquid, and gas. Solid electrolytes, e.g., $RbAg_4I_5$, are confined to special studies using solid-state electrochemical cells and sensors for gases such as oxygen, hydrogen, sulfur dioxide, and carbon dioxide as well as for ion-selective electrodes [1]. The most common solid electrolyte sensor is a pH electrode in which a glass membrane is an ionic conductor with sodium ions as charge carriers [2]. Liquid electrolytes are either solutions of ionic salts, acids and bases, or molten ionic salts. Gaseous electrolytes are some ionized vapors, but the gases that exhibit mixed electronic and ionic conduction are not good separators. In analytical electrochemistry the most common electrolytes are solutions.

The basis of ionic conduction is the mobility of ions [3]. In liquid electrolytes it is the consequence of a three-dimensional random movement of ions. The characteristic of the random walk is that the mean distance $<x>$ traveled by the ion is zero, but the mean square distance $<x^2>$ is proportional to time. Because of this movement, the concentration of ions is uniform throughout the volume of the electrolyte in the absence of an electric field. Under the influence of a certain force, e.g., in an electric field, the ions acquire a nonrandom component of velocity in the direction of the force. The velocity developed under unit applied force is called the absolute mobility of the ion. The conventional, or electrochemical, mobility is the velocity of ions in a unit electric field. The relationship between the absolute and conventional mobility is

$$u_{conv} = u_{abs} z_i e_0 \tag{III.3.1}$$

Š. Komorsky-Lovrić (✉)
Rudjer Boskovic Institute, 10001 Zagreb, Croatia
e-mail: slovric@irb.hr

F. Scholz (ed.), *Electroanalytical Methods*, 2nd ed.,
DOI 10.1007/978-3-642-02915-8_16,

where $z_i e_0$ is the charge on the ion [3]. The electrolyte contains at least two types of ions with opposite charge. In liquids, all ions are mobile and contribute to the conductivity, provided that no ion pairing occurs. In solid electrolytes often only one of the ions is mobile. Ion conductivity of solid materials can result from a variety of different intrinsic and extrinsic defects. Details of these processes can be found in the specialized literature [1].

III.3.2 Ionic Transport

The ionic conductivity is achieved by a nonrandom, direct movement of ions, which results in the transport of matter and a flow of charge [4]. The rate of the transport density is called the flux. It is the number of moles of a certain species crossing a unit area of a reference plane in 1 s. The migration or conduction is a flow of charge produced by an electric field between the electrodes [3]. Cations move to the cathode and anions to the anode. In a simple z:z-valent electrolyte the current density, or the flux of charge, is proportional to the concentration of ions ($c = c^+ = c^-$ also: $z = z^+ = z^-$), their conventional mobility (u^+_{conv} and u^-_{conv}) and the gradient of the electrostatic potential, i.e., the electric field $X = -\frac{\mathrm{d}\psi}{\mathrm{dx}}$:

$$\vec{j} = zFc(u^+_{\mathrm{conv}} + u^-_{\mathrm{conv}})X \tag{III.3.2}$$

The product:

$$\sigma = zFc(u^+_{\mathrm{conv}} + u^-_{\mathrm{conv}}) \tag{III.3.3}$$

is called the specific conductivity of the solution. It is the conductance of a cube of electrolyte solution 1 cm long and 1 cm^2 in area. Its unit is siemens per centimeter. The units of other variables in Eq. (III.3.2) are: F (C mol^{-1}), c (mol cm^{-3}), u_{conv} (cm^2 V^{-1} s^{-1}), X (V cm^{-1}), and $\vec{j}$ (A cm^{-2}). The current flowing through the cell is

$$I = S\sigma X \tag{III.3.4}$$

where S is the area of the reference plane. Equation (III.3.4) shows that an ionic conductor obeys Ohm's law for a dc current under steady-state conditions. The ratios

$$\Lambda_{\mathrm{m}} = \frac{\sigma}{c} \tag{III.3.5}$$

and

$$\Lambda = \frac{\sigma}{zc} \tag{III.3.6}$$

Table III.3.1 Equivalent conductivities Λ^o and Λ (S cm^2 mol^{-1}) of aqueous electrolytes at 25°C

		Concentrations (mol/L)		
Compound	Infinite dilution	0.001	0.01	0.1
HCl	425.95	421.15	411.80	391.13
KCl	149.79	146.88	141.20	128.90
$KClO_4$	139.97	137.80	131.39	115.14
KNO_3	144.89	141.77	132.75	120.34
KOH	271.5	234.0	228.0	213.0
LiCl	114.97	112.34	107.27	95.81
$LiClO_4$	105.93	103.39	98.56	88.52
1/2 $MgCl_2$	129.34	124.15	114.49	97.05
NH_4Cl	149.6	146.7	141.21	128.69
NaCl	126.39	123.68	118.45	106.69
$NaClO_4$	117.42	114.82	109.54	98.38
$NaOOCCH_3$	91.0	88.5	83.72	72.76
NaOH	247.7	244.6	237.9	–
1/2 Na_2SO_4	129.8	124.09	112.38	89.94

are called the molar conductivity and the equivalent conductivity of the electrolyte, respectively. At infinite dilution the equivalent conductivity of the electrolyte can be separated into the equivalent conductivities of individual ions:

$$\Lambda^o = \lambda_+^o + \lambda_-^o \tag{III.3.7}$$

where $\lambda_+^o = F(u_{conv}^+)^o$ and $\lambda_-^o = F(u_{conv}^-)^o$. This is the law of the independent migration of ions. However, the conventional mobility of ions depends on the concentration of the electrolyte because of interionic interactions. Hence, the ions can migrate independently only at infinite dilution. At increasing concentration, the migration of ions is coupled in a way that the drift of positive ions hinders the drift of negative ions. For this reason the equivalent conductivity of an electrolyte is diminished as follows:

$$\Lambda = \Lambda^0 - Ac^{1/2} \tag{III.3.8}$$

where A is an empirical constant. Equation (III.3.8) applies to concentrations up to 10^{-2} mol/L. Equivalent conductivities of many electrolytes at infinite dilution and various concentrations are reported in the literature [5]. Equivalent ionic conductivities extrapolated to infinite dilution are also reported. Selected data are presented in Tables III.3.1 and III.3.2.

In a solution containing several electrolytes, the total current density is made up of the contributions of all the ionic species in the solution.

$$\vec{j} = \sum_i z_i F c_i (u_{conv})_i X \tag{III.3.9}$$

Table III.3.2 Equivalent ionic conductivities λ_+^o and λ_-^o (S cm^2 mol^{-1}) in aqueous solution at infinite dilution at 25°C

Cations	λ_+^o	Anions	λ_-^o
1/3 Al^{3+}	61	Br^-	78.1
Cs^+	77.2	CN^-	78
H^+	349.65	Cl^-	76.31
K^+	73.48	ClO_4^-	67.3
Li^+	38.66	F^-	55.4
1/2 Mg^{2+}	53.0	HCO_3^-	44.5
NH_4^+	73.5	1/2 HPO_4^{2-}	33
Na^+	50.08	NO_3^-	71.42
Rb^+	77.8	OH^-	198
Tl^+	74.7	1/2 SO_4^{2-}	80.0
Diethylammonium$^+$	42.0	Acetate$^-$	40.9
Dimethylammonium$^+$	51.0	Benzoate$^-$	32.4
Methylammonium$^+$	58.7	1/3 citrate^{3-}	70.2
Octadecylpyridinium$^+$	20	Formate$^-$	54.6
Octadecyltributylammonium$^+$	16.6	1/2 maleate^{2-}	58.8
Octadecyltripropylammonium$^+$	17.2	1/2 oxalate^{2-}	74.11
Tetra-n-butylammonium$^+$	19.5	Picrate$^-$	30.37
Tetraethylammonium$^+$	32.6	Propionate$^-$	35.8
Tetramethylammonium$^+$	44.9	Salicylate$^-$	36
Trimethylammonium$^+$	47.23	Trichloroacetate$^-$	36.6

The fraction of the total current carried by a particular ionic species is known as the transport number:

$$t_i = \frac{z_i c_i (u_{\text{conv}})_i}{\sum\limits_i z_i c_i (u_{\text{conv}})_i} \qquad \text{(III.3.10)}$$

In the case of a 1:1-valent electrolyte at infinite dilution, the transport numbers are

$$t_+ = \frac{\lambda_+^o}{\lambda_+^o + \lambda_-^o} \qquad \text{(III.3.11)}$$

and

$$t_- = \frac{\lambda_-^o}{\lambda_+^o + \lambda_-^o} \qquad \text{(III.3.12)}$$

The transport number depends on the relative concentration of the respective ion. In electroanalytical chemistry, the electrochemical reaction of an analyte ion A^{z+} is measured in the presence of a supporting electrolyte MX. Usually, the concentration of the electrolyte MX is at least 1000 times higher than the concentration of the ion A^{z+}. Under this condition the transport number of the analyte is virtually zero:

$$t_A \approx \frac{z_A c_A}{z_M c_M} \times \frac{u_A}{u_M + u_X} \approx 10^{-3} \qquad \text{(III.3.13)}$$

which means that the migration of the analyte ion A^{z+} can be neglected.

The supporting electrolyte primarily serves to minimize the solution resistance and the resulting so-called *IR* drop. In voltammetry, the potential of the working electrode is controlled versus a reference electrode using a potentiostat, which allows the current to pass between the working and auxiliary electrodes while ensuring that none passes through the reference electrode. The potential drop caused by the resistance of the electrolyte solution is determined by Eq. (III.3.4). Usually, the reference electrode is separated by a salt bridge from the electrochemical cell; however, when the *IR* drop in the analyte solution is not negligible, it is necessary to give the salt bridge the form of a capillary (a so-called Luggin capillary), the tip of which is positioned very close to the working electrode surface. This arrangement leaves a small, uncompensated solution resistance (R_u) which depends on the electrode geometry and the distance between the electrode surface and the capillary tip. The potentiostat maintains a constant potential between the working and the reference electrodes, but the actual potential across the electrolyte–electrode interface is in error by a variable amount which depends on the current flowing through the cell: $V_{IR} = IR_u$. The manifestation of this IR_u drop in cyclic voltammetry consists of decreasing peak heights and increasing peak separations, which is similar to the influence of a slow electron transfer step. Obviously, the error can be minimized by increasing the specific conductivity of the solution. This is achieved by the addition of the supporting electrolyte in concentrations between 0.1 and 1 mol/L. The alternative is to apply an *IR* compensation (cf. Chap. II.1).

The second mode of transport of ions is diffusion, a process caused by a concentration gradient [6]. A diffusional flux appears near the electrode surface where the charge-transfer reaction causes a depletion of ions and the development of the concentration gradient. The flux is determined by Ficks's first law:

$$J_D = -D(\mathrm{d}c/\mathrm{d}x) \tag{III.3.14}$$

while the concentration of ions at a certain distance from the electrode surface is determined by Fick's second law:

$$\frac{\partial c}{\partial t} = D\frac{\partial^2 c}{\partial x^2} \tag{III.3.15}$$

If the solution is not mechanically stirred, the diffusion is the only mode of transport for uncharged electroactive molecules and the dominant mode for ions in traces, the migration of which is negligible in the excess of a supporting electrolyte. This is particularly important in analytical electrochemistry. The relationship between the current and the bulk concentration of the analyte can be exactly determined by solving the partial differential Eq. (III.3.15) only if the migration of analyzed ions can be neglected. The supporting electrolyte ensures electroneutrality in the diffusion layer and suppresses the gradient of the diffusion electric potential (see below). Besides, it conducts the current through the cell and prevents the concentration polarization of the auxiliary electrode, the surface area of which is usually much larger than the area of the working electrode surface. In a ceramic plug, which separates the solution under study from the salt bridge to the reference electrode, a liquid-junction potential with restrained diffusion may arise unless a high concentration of an indifferent electrolyte is added to both the solution and the bridge. Finally, the supporting electrolyte determines the value of the ionic strength of the solution. The activities of all the ions in the solution are proportional to their concentrations, the activity coefficients being a function of the concentration of the excess electrolyte alone.

The origin of both the diffusion and the migration is the random walk of ions. It was shown by Einstein and Smoluchowski that the diffusion coefficient is proportional to the mean square distance of the random movements of ions:

$$D = \frac{\langle x^2 \rangle}{2t} \tag{III.3.16}$$

where t is the duration of measurement. In the case of semi-infinite planar diffusion of ions from an instantaneous source, the square root of the mean square distance $((\langle x^2 \rangle)^{1/2} = (2Dt)^{1/2})$ is the limit of the layer inside which 68% of initially introduced ions can be found, while the remaining 32% of ions have advanced beyond this distance. Equations (III.3.14) and (III.3.16) show that the diffusion appears because it is more probable that an ion moves from the region of higher concentration to that of lower concentration, than vice versa. When the movement of ions is discontinuous, i.e., if it consists of a series of jumps between the fixed sites, as in solid conductors, the diffusion coefficient is proportional to the square of the mean jump distance and to the jump frequency, which depends on the energy barrier between the sites [7].

From a thermodynamic point of view, the driving force for diffusion is the gradient of chemical potential:

$$J_{\mathrm{D}} = -Bc\mathrm{d}(\mu^{\ominus} + RT \ln a)/\mathrm{d}x \tag{III.3.17}$$

where $a = fc$ is the activity of ions, f is the activity coefficient, and B is a constant of proportionality. Considering that $\mathrm{d}(\ln fc)/\mathrm{d}x = f^{-1}c^{-1}(f\mathrm{d}c/\mathrm{d}x + c\mathrm{d}f/\mathrm{d}x)$, Eq. (III.3.17) can be transformed into a form of Fick's first law (Eq. III.3.14), with a concentration-dependent diffusion coefficient:

$$D = BRT(1 + \mathrm{d}(\ln f)/\mathrm{d}(\ln c)) \tag{III.3.18}$$

The second term of Eq. (III.3.18), which is called the thermodynamic factor [6, 7], can be changed significantly in solid electrolytes, but in solutions its variation is negligible, and hence the diffusion coefficient is practically independent of the concentration.

Discussing the situation in which the influence of the concentration gradient of ions is balanced by the electric field acting in the opposite direction of the gradient, Einstein found that the diffusion coefficient is proportional to the absolute mobility of the ion:

$$D = u_{\mathrm{abs}} \cdot kT \tag{III.3.19}$$

where k is the Boltzmann constant. Using Eq. (III.3.1), this relationship can be transformed into

$$D = \frac{RT}{zF} u_{\text{conv}} \tag{III.3.20}$$

where $R = kN_A$, $F = e_0 N_A$ and N_A is Avogadro's number. Furthermore, at infinite dilution, the conventional mobility of ions is related to the equivalent ion conductivity (see Eq. III.3.7), so that the equivalent conductivity of a 1:1-valent electrolyte can be related to the diffusion coefficients of cations and anions:

$$\Lambda^{\text{o}} = \frac{zF^2}{RT}(D_+^{\text{o}} + D_-^{\text{o}}) \tag{III.3.21}$$

Equation (III.3.21) does not apply to higher concentrations. The equivalent conductivity decreases proportionally to the square root of concentration (see Eq. III.3.8), but the diffusion coefficient depends on the ionic strength of the solution, according to Eq. (III.3.18). Using the Debye–Hückel limiting law for a 1:1-valent electrolyte [3]:

$$\ln f = -A^* c^{1/2} \tag{III.3.22}$$

where A^* is the constant, and introducing it into Eq. (III.3.18), considering that $\mathrm{d}(\ln f)/\mathrm{d}(\ln c) = c\mathrm{d}(\ln f)/\mathrm{d}c$, one obtains

$$D = D^{\text{o}}\left(1 - \frac{1}{2}A^* c^{1/2}\right) \tag{III.3.23}$$

The difference between Eqs. (III.3.8) and (III.3.23) shows that the influences of concentration on D and Λ are not identical.

However, if the concentration gradient of ions can be balanced by the electric field, it can also be balanced by some other force, such as the Stokes viscous force: $F_s = 6\,\pi r \eta v$, where r is a radius of an ion, η is the viscosity of the solution, and v is a velocity of an ion. This consideration shows the relationship between the diffusion coefficient and the viscosity of the medium:

$$D = \frac{kT}{6\pi r \eta} \tag{III.3.24}$$

Equation (III.3.24) indicates that ions with different radii must have different diffusion coefficients in the same medium and, consequently, different mobility (Eqs. III.3.19 and III.3.20) and conductivity (Eq. III.3.21). Moreover, Eqs. (III.3.21) and (III.3.24) show that the product of the equivalent conductivity of a certain electrolyte and the viscosity of medium is approximately constant. This is the empirical Walden's rule, which applies if the radii of solvated ions in various media are not significantly different.

Finally, as both concentration gradient and the electric field induce fluxes of ions that can be added or subtracted as vectors, a unified driving force for the flux, the gradient of electrochemical potential, is introduced:

$$J_{+} = -\frac{D_{+}c_{+}}{RT} \cdot \frac{\mathrm{d}}{\mathrm{d}x}(\mu^{\ominus} + RT \ln c_{+} + z_{+}F\psi) \tag{III.3.25}$$

$$J_{+} = -D_{+}\frac{\mathrm{d}c_{+}}{\mathrm{d}x} - c_{+}u_{\mathrm{conv}}^{+}\frac{\mathrm{d}\psi}{\mathrm{d}x} \tag{III.3.26}$$

$$\frac{\partial c_{+}}{\partial t} = D_{+}\frac{\partial^{2} c_{+}}{\partial x^{2}} + u_{conv}^{+}\frac{\partial c_{+}}{\partial x} \cdot \frac{\mathrm{d}\psi}{\mathrm{d}x} \tag{III.3.27}$$

Introducing Eq. (III.3.20) into Eq. (III.3.25) derives Eq. (III.3.26), while the last equation is the consequence of Fick's second law [6]:

$$\frac{\mathrm{d}c}{\mathrm{d}t} = -\frac{\mathrm{d}J}{\mathrm{d}x} \tag{III.3.28}$$

and the assumption that $\mathrm{d}^2\psi/\mathrm{d}x^2 = 0$, which is valid if the electroneutrality condition

$$\sum_{i} z_i c_i = 0 \tag{III.3.29}$$

is satisfied [4]. The diffusion and migration components of the flux of a particular ion (Eq. III.3.26) may be in the same or opposite directions, depending on the direction of the electric field and the charge on the ion. The directions are opposite if negatively charged complex ions (e.g., $\mathrm{Cu(CN)}_4^{2-}$) are reduced and when cations are oxidized. In the bulk solution, the concentration gradients are generally small, and the total current is carried mainly by migration. However, if the transport number (Eq. III.3.10) of the particular ion is close to zero, as in the presence of the supporting electrolyte (see Eq. III.3.13), the migration component of its flux is negligible both in the bulk solution and near the electrode. Hence, this ion is transported solely by diffusion. Besides, each ion of the supporting electrolyte develops a concentration gradient in the electrode vicinity, inducing the diffusion component of the flux that opposes the migration component. So, its net flux is zero at the electrode surface because it is neither reduced nor oxidized. These gradients also preserve the electroneutrality in the diffusion layer.

The flux of cations is coupled with the flux of anions. The total current density is given by Faraday's law:

$$\vec{j} = \sum_{i} z_i F J_i \tag{III.3.30}$$

(see Eqs. III.3.2 and III.3.9). By introducing Eq. (III.3.26) into Eq. (III.3.30), one obtains

$$\vec{j} = -F\sum_{i} z_i D_i \frac{\mathrm{d}c_i}{\mathrm{d}x} - \sigma\frac{\mathrm{d}\psi}{\mathrm{d}x} \tag{III.3.31}$$

where σ is given by Eq. (III.3.3). The electric potential gradient consists of the ohmic and the diffusional components:

$$\frac{\mathrm{d}\psi}{\mathrm{d}x} = -\frac{j}{\sigma} - \frac{F}{\sigma}\sum_i z_i D_i \frac{\mathrm{d}c_i}{\mathrm{d}x}. \tag{III.3.32}$$

The electric current passing through the solution causes the ohmic potential gradient, while the diffusional component is a consequence of different mobility of ions. It will not be formed if all mobilities are identical, but if they are not, this component may exist even in the absence of the current. If there is no external electric field (i.e., if $\vec{j} = 0$), the diffusion of a 1:1-valent electrolyte with different diffusion coefficients of cations and anions causes the separation of charges in the diffusion layer and the development of a diffusion potential which opposes further charge separation:

$$\frac{\mathrm{d}\psi}{\mathrm{d}x} = -\frac{RT}{F}\sum_i \frac{t_i}{z_i} \cdot \frac{\mathrm{d}(\ln c_i)}{\mathrm{d}x} \tag{III.3.33}$$

where t_i is given by Eq. (III.3.10). This is called the electroneutrality field. It is slowing down the faster ion and accelerating the slower one. Hence, both ions acquire the common diffusion coefficient:

$$D_{\pm} = \frac{2D_+ D_-}{D_+ + D_-} \tag{III.3.34}$$

Generally, the flux of one ion depends on the fluxes of all other ions. The solutions of Eqs. (III.3.27) and (III.3.31) for particular problems can be found in the literature [3, 4, 6–8].

III.3.3 Ionic Solutions

In electroanalytical chemistry, the most frequently used electrolytes are ionic solutions. Dissolved substances are true or potential electrolytes. True electrolytes are compounds that in the pure solid state appear as ionic crystals. When melted, true electrolytes in pure liquid form are ionic conductors. All salts belong to this class. Potential electrolytes show little conductivity in the pure liquid state. They consist of separate neutral molecules and the bonding of the atoms in these molecules is essentially covalent. They form ions by acid–base reactions with solvents [9]. There are two broad categories of solvents: amphiprotic, which are capable of both accepting and donating protons, and aprotic, which do not accept or donate protons. The solvents can be further classified according to their protogenic or protophilic properties, and the relative permittivity or polarity [10]. Molecules of polar solvents contain a permanent dipole moment. In an electric field, the dipoles orient against it and cause a counter electric field. In certain solvent molecules the electric field can induce a dipole moment. Hence, the net electric field in the solvent is less than it is in vacuum. This decreasing of the field is proportional to an empirical constant that is called the dielectric constant, or the permittivity of the solvent. Table III.3.3 reports

Table III.3.3 Permittivity of pure liquids at 20°C

Substance	ε	Substance	ε
Acetic acid	6.15	Formamide	109.5
Acetic anhydride	20.7	Formic acid	57.9
Acetone	20.7	Glycerol	42.5
Acetonitrile	37.5	Isobutyronitrile	20.4
Acrylonitrile	38	Methanol	32.6
Ammonia	25[a]	2-Methoxyethanol	16
Benzonitrile	25.6	Morpholine	7.3
1-Butanol	17.8	Nitromethane	36.7
Butyronitrile	20.3	*N*-Methylacetamide	179
Dichloromethane	8.9	*N*-Methylformamide	182.4
Diethyl ether	4.3	1-Pentanol	15.3
1,2-Dimethoxyethane	3.5	Phenylacetonitrile	18.7
Dimethylacetamide	37.8	1-Propanol	20.1
Dimethylformamide	36.7	2-Propanol	18.3
Dimethylsulfone	37	Propionitrile	27.2
Dimethyl sulfoxide	47	Propylene carbonate	64.4
1,4-Dioxane	2.2	Pyridine	12.3
Ethanol	24.3	Sulfolane	44
Ethylenediamine	12.5	Tetrahydrofuran	7.6
Ethylene glycol	37.7	Tetramethylurea	23

[a] $T = -78°C$.

the dielectric constants of some pure liquids used as solvents in electrochemistry. The values of dipole moments of these molecules in the gas phase are presented in Table III.3.4. The permittivity is diminished by increasing temperature, as can be seen for water in Table III.3.5. The heat accelerates the random movement of dipole

Table III.3.4 Dipole moments of molecules in the gas phase

Substance	μ (D[a])	Substance	μ (D)
Acetic acid	1.70	Ethylene glycol	2.28
Acetic anhydride	2.8	Formamide	3.73
Acetone	2.88	Formic acid	1.41
Acetonitrile	3.92	Glycerol	2.56
Ammonia	1.47	Isobutyronitrile	4.29
Benzonitrile	4.18	Methanol	1.7
1-Butanol	1.66	2-Methoxyethanol	2.36
Butyronitrile	4.07	Morpholine	1.55
Dichloromethane	1.6	Nitromethane	3.46
Diethyl ether	1.15	*N*-Methylformamide	3.8
Dimethylacetamide	3.79	1-Pentanol	1.8
Dimethylformamide	3.82	Phenylacetonitrile	3.51
Dimethyl sulfoxide	3.96	1-Propanol	1.55
1,4-Dioxane	0.45	2-Propanol	1.58
Ethanol	1.68	Propionitrile	4.05
Ethylenediamine	1.99	Pyridine	2.22

[a] $D = 3.33564 \times 10^{-30}$ C cm.

Table III.3.5 Permittivity of water between 0°C and 100°C

T (°C)	ε
0	87.90
20	80.20
40	73.17
60	66.73
80	60.86
100	55.51

molecules and disturbs their orientation in the electric field. The permittivity of solvents determines the solubility of true electrolytes. The process of solvation in which the ions are surrounded by the dipoles of the solvent and enticed out of the lattice and into the solution dissolves the crystal lattice. Solvation by water molecules is called hydration. The stability of solvated, or hydrated, ions depends on the energy of the ion–dipole interaction.

The energy of ion-pair formation is the sum of attraction and repulsion energies. The attraction energy is the work of the Coulomb force from infinity to a distance d. The repulsive force results from the penetration of the electron clouds. For a pair of singly charged ions, the formation energy is [11]

$$E = -\frac{1}{4\pi\varepsilon_0} \cdot \frac{e^2}{d_{\mathrm{eq}}} \left(1 - \frac{1}{n}\right) \qquad \text{(III.3.35)}$$

where $+e$ and $-e$ are the charges of ions, ε_0 is the permittivity of vacuum, $d_{\mathrm{eq}} = r_+ + r_-$ is the equilibrium distance between the centers of ions, r_+ and r_- are the crystallographic radii of ions, and $6 < n < 12$ is a constant which can be estimated from the compressibility of ionic crystals. A small correction $1/n$ means that only a minor deformation of the electron clouds takes place when an ion pair is formed. A three-dimensional aggregate of positively and negatively charged ions is a crystal lattice. The energy per ion pair of NaCl crystal formation is

$$E = -\frac{1.748}{4\pi\varepsilon_0} \cdot \frac{e^2}{d_{\mathrm{eq}}} \left(1 - \frac{1}{n}\right) \qquad \text{(III.3.36)}$$

where 1.748 is the Madelung constant, which depends on the geometry of the crystal. An ionic crystal, such as NaCl, can be dissolved if the sum of solvation energies of Na^+ and Cl^- ions is higher than the crystal formation energy. The energy of a single ion–dipole pair formation is

$$E = -\frac{e\mu}{4\pi\varepsilon_0 (r_{\mathrm{i}} + r_{\mathrm{s}})^2} \qquad \text{(III.3.37)}$$

where $\mu = r_{\mathrm{s}}Q$ is the dipole moment and r_{i} and r_{s} are the radii of the ion and solvent molecule, respectively. The solvation energy depends mainly on the charge and the radius of the ion and on the dipole moment and the number of dipoles in the primary solvation sphere. In addition, there is a small contribution of solvent molecules from the secondary solvation sheath. The number of solvent dipoles in the first coordination shell can be estimated by measuring the mobility and the partial molar volumes of ions, the compressibility and the permittivity of ionic solution, the entropy of

Table III.3.6 Hydration numbers of some cations and anions

Ion	Hydration number	Ion	Hydration number
Li^+	5	Ca^{2+}	9
Na^+	5	Zn^{2+}	12
K^+	5	Cd^{2+}	12
F^-	3	Fe^{2+}	12
Cl^-	2	Co^{2+}	13
Br^-	2	Ni^{2+}	13
Mg^{2+}	13	Pb^{2+}	11
Sr^{2+}	11	Cu^{2+}	12
Ba^{2+}	11	Sn^{2+}	13

solvation and by several kinds of spectroscopy, of the electrolytic solutions. The results obtained by various methods may significantly differ, which partly depends on the models used for the interpretation of data. A distinction between coordination number and solvation number has been proposed [3]. The first number refers to the total number of solvent molecules in the first layer around the ion when it rests. The second number corresponds to the solvent molecules that remain with the ion while it moves. These solvent molecules have strictly oriented dipoles to interact maximally with the ion. When the ion stops, the number of solvent molecules in the first layer may increase, but their dipoles are oriented at 90° and do not interact with the ion. Average hydration numbers of some ions are compiled in Table III.3.6. In the secondary solvation shell, which includes several molecular layers, the dipoles have no permanent orientation. In these layers the dielectric constant suddenly increases from a small value near the ion to the permittivity of the pure solvent. This diminishes the Coulomb force between the dissolved ions.

Ionic solutions are also prepared by dissolution of potential electrolytes in amphiprotic solvents. Pure amphiprotic solvents undergo self-dissociation, or autoprotolysis:

$$\mathrm{HSo + HSo \rightleftarrows H_2So^+ + So^-} \quad \text{(III.3.38)}$$

This reaction is characterized by an equilibrium constant (autoprotolysis constant K_{AP}):

$$K_{\mathrm{AP}} = [\mathrm{H_2So^+}][\mathrm{So^-}] \quad \text{(III.3.39)}$$

The constants of several solvents are listed in Table III.3.7. Some examples of autoprotolysis are the reactions of water, methanol, acetic acid, and liquid ammonia:

$$\mathrm{2H_2O \rightleftarrows H_3O^+ + OH^-} \quad \text{(III.3.40)}$$

$$\mathrm{2H_3COH \rightleftarrows H_3COH_2^+ + H_3CO^-} \quad \text{(III.3.41)}$$

$$\mathrm{2H_3CCOOH \rightleftarrows H_3CCOOH_2^+ + H_3CCOO^-} \quad \text{(III.3.42)}$$

$$\mathrm{2NH_3 \rightleftarrows NH_4^+ + NH_2^-} \quad \text{(III.3.43)}$$

Table III.3.7 Equilibrium constants of autoprotolysis of solvents at 25°C

Solvent	K_{AP}
Water	1.00×10^{-14}
Methanol	2.0×10^{-17}
Ethanol	8×10^{-20}
Ammonia[a]	2×10^{-38}
Ethylenediamine	5×10^{-16}
Formic acid	6.3×10^{-7}
Acetic acid	3.5×10^{-15}
Sulfuric acid	2.5×10^{-4}

[a] $T = -50°C$.

The products of autoprotolysis are the conjugate acid H_2So^+ and base So^-. A potential electrolyte may react with the solvent either as an acid

$$HA + HSo \rightleftarrows A^- + H_2So^+ \tag{III.3.44}$$

or as a base:

$$B + HSo \rightleftarrows HB^+ + So^- \tag{III.3.45}$$

These reactions depend on the permittivity of the solvent. In water, strong acids and bases are completely dissociated, while weak acids and bases are only partly dissociated, which is characterized by the equilibrium constants of the reactions (III.3.44) and (III.3.45):

$$K_{dis,ac} = \frac{[A^-][H_3O^+]}{[HA]} \tag{III.3.46}$$

$$K_{dis,b} = \frac{[HB^+][OH^-]}{[B]} \tag{III.3.47}$$

Hence, in water, which has a very high dielectric constant, the so-called strong electrolytes include all salts and strong acids and bases. The weak electrolytes are almost all organic acids and bases. However, this distinction depends on the characteristics of the solvent. Some solvents are more acidic than water (e.g., formic and acetic acid), while the others are more basic (e.g., ethylenediamine). Besides, the permittivity of organic solvents is generally smaller than the permittivity of water. So, the dissolution of a potential electrolyte in a nonaqueous medium proceeds in two steps: ionization and dissociation.

$$HA + HSo \rightleftarrows A^-H_2So^+ \tag{III.3.48}$$

$$A^-H_2So^+ \rightleftarrows A^- + H_2So^+ \tag{III.3.49}$$

For example, in acetic acid ($\varepsilon = 6$), perchloric acid is completely ionized, but the dissociation constant of the ion pair $H_3CCOOH_2^+ClO_4^-$ is only about 10^{-5} mol/L and $HClO_4$ can be classified as a weak acid in this medium. The ionization of acids is enhanced in basic solvents and diminished in acidic solvents, but the dissociation of the ion pair depends only on the dielectric constant of the solvent. Nevertheless, the acidity of weak acids is enhanced in basic solvents, and the basicity of weak bases is enhanced in acidic solvents. Examples are solutions of phenol in ethylenediamine:

$$C_6H_5-OH+En \rightleftarrows C_6H_5-O^-HEn^+ \quad \text{(III.3.50)}$$

$$C_6H_5-O^-HEn^+ \rightleftarrows C_6H_5-O^- + HEn^+ \quad \text{(III.3.51)}$$

and of pyridine in acetic acid:

$$C_5H_5N+H_3CCOOH \rightleftarrows C_5H_5NH^+H_3CCOO^- \quad \text{(III.3.52)}$$

$$C_5H_5NH^+H_3CCOO^- \rightleftarrows C_5H_5NH^+ + H_3CCOO^- \quad \text{(III.3.53)}$$

The base H_3CCOO^- can be partly neutralized by the addition of the acid $H_3CCOOH_2^+ClO_4^-$ to prepare a buffer solution.

In aprotic solvents, such as acetonitrile (H_3CCN), dimethyl sulfoxide (H_3CSOCH_3), or methylisobutyl ketone ($H_3CCOCH(CH_3)_2$), the potential electrolytes can be dissolved, but not ionized. These solvents have moderate permittivity, and they support the dissociation of true electrolytes. Dissolved acids (e.g., C_6H_5OH or H_2O) may act as proton donors if a certain proton acceptor is created in the electrode reaction.

The choice of solvent is dictated primarily by the solubility of the analyte. Both the solvent and the supporting electrolyte should not react with the analyte and should not undergo electrochemical reactions over a wide potential range [12]. Aprotic solvents, carefully purified of proton donors (e.g., H_2O), are used for the analysis of electrochemically generated radicals. Dissolved oxygen must be removed from all solvents because it interferes with the majority of electrode reactions occurring at potentials negative to 0 V versus a saturated calomel electrode. The concentration of oxygen in water is about 5×10^{-4} mol/L. Oxygen is reduced in two steps to peroxide and hydroxide ions or water, depending on the pH of solution. Oxygen and peroxide can react with the products of electroreductions. Oxygen can be removed from aqueous solutions by purging them with high-purity nitrogen gas for about 5–10 min prior to the measurement and by maintaining a nitrogen blanket over the solution thereafter. It is recommendable to repeat the bubbling in-between the measurements as often as possible. Oxygen is best removed from some organic solvents by distillation under argon, followed by preventing any contact of the purified solvent with air, and by additional purging with argon or nitrogen immediately before the measurements.

The simplest approach to the use of solvents in electrochemistry is to divide them according to their chemical structure.

III.3.3.1 Aqueous Electrolyte Solutions

III.3.3.1.1 Synthetic Aqueous Electrolyte Solutions

Double-distilled water can be used in the majority of analyses, but for trace analysis additional distillations of water in all-quartz equipment is required. Traces of organic compounds, which may severely interfere in stripping analysis, can be very efficiently removed when the first distillation is made by the addition of potassium permanganate and some KOH pellets. In this case, it is necessary to electrically heat the glass tubes through which the steam is transported to the cooler. This is the only way to prevent permanganate solution creeping into the product vessel. Nowadays, deionized water is frequently used. It is purified by passing through cation- and anion-exchange columns until the specific resistance is 18.2 MΩ cm. In addition, special columns, e.g., filled with charcoal, for the removal of organic substances may be applied.

Water is a very good solvent for many inorganic compounds. According to the empirical solubility rules, all alkali metal and ammonium ion salts are soluble, except LiF, $KClO_4$, NH_4ClO_4, and alkali metal fluorosilicates. In addition, all alkaline earth metal chlorides, bromides, iodides, oxides, sulfides, selenides, tellurides, acetates, chlorates, nitrates, nitrites, perchlorates, permanganates, sulfates, fluorosilicates, thiocyanates, and thiosulfates are very soluble, except BeO, MgO, CaO, $CaSO_4$, $SrSO_4$, $BaSO_4$, and barium fluorosilicate. Regarding other metal salts, all chlorides, bromides, and iodides are also very soluble, except AgCl, Hg_2Cl_2, $PbCl_2$, $HgBr_2$, HgI_2, BiI_3, and SnI_4. Moreover, all acetates, chlorates, nitrates, nitrites, perchlorates, and permanganates possess good solubility, except silver acetate and nitrite. Finally, all sulfates are very soluble, except $PbSO_4$ and Ag_2SO_4 [13].

Water can dissolve nonionic compounds that are capable of forming hydrogen bonds, such as alcohols, phenols, ammonia, amines, amides, and carboxylic acids.

Supporting electrolytes which satisfy general requirements are inorganic salts (e.g., KCl, KNO_3, K_2SO_4, NaCl, $NaClO_4$, $NaNO_3$, and NH_4Cl), mineral acids and bases (e.g., HCl, $HClO_4$, H_2SO_4, HNO_3, KOH, NaOH, and NH_3) and buffers (e.g., borate, citrate, dihydrogen, and hydrogen phosphate). The ability of anions to complex metal ions may affect the selectivity of voltammetric measurements. To enhance the effect, a masking agent (e.g., a polydentate ligand) may be added to the electrolyte. The hydrolysis of highly acidic cations may be controlled by the addition of acids or bases to the solution. Hence, the composition of the electrolyte is tailored to meet the goals of the particular experiment.

The potential window of the working electrode is the potential range within which the working electrode is polarized. In pure water the window is limited by reductive and oxidative decompositions:

$$2H_2O + 2e^- \rightleftarrows H_2 + 2OH^- \qquad (III.3.54)$$

$$2H_2O - 4e^- \rightleftarrows O_2 + 4H^+ \qquad (III.3.55)$$

Table III.3.8 Utilizable ranges of voltammetric working electrodes in aqueous electrolytes

Electrode	Electrolyte	Anodic limit (V vs. SCE[a])	Cathodic limit (V vs. SCE)
Pt	1 M $HClO_4$	1.3	−0.25
Pt	1 M NaOH	0.5	−1.15
C	1 M $HClO_4$	1.4	−1.0
C	1 M NaOH	0.9	−1.5
Hg	1 M $HClO_4$	0.5	−1.1
Hg	1 M NaOH	0.0	−2.8

[a]SCE ≡ saturated calomel electrode, $E_{SCE} = 0.241$ V vs. SHE ($T = 25°C$).

If reactions (III.3.54) and (III.3.55) are reversible, the corresponding Nernst equations define the limits of the potential window:

$$E_{H^+/H_2} = -0.059\ \text{pH (V vs. SHE)} \tag{III.3.56}$$

$$E_{O_2/H_2O} = 1.23 - 0.059\ \text{pH (V vs. SHE)} \tag{III.3.57}$$

where SHE is a standard hydrogen electrode. So, in the presence of an electrochemically inert buffer and assuming that thermodynamics controls the electrode reactions, the theoretical window in water is 1.23 V regardless of the pH. In the absence of a buffer, the working window is wider, approaching 2 V, because the solution in the vicinity of the working electrode is basic if it acts as a cathode, and acidic if it acts as an anode. Moreover, the electrode reactions (III.3.54) and (III.3.55) are not reversible and require overpotentials to overcome kinetic polarization. On a smooth platinum electrode in 1 mol/L H_2SO_4 the overpotential for the reduction of hydrogen ions is −0.08 V. The overpotential for the oxidation of hydroxide ions on the same electrode in 1 mol/L KOH is 0.44 V [9]. Hence, the working window depends on the electrode material and the electrolyte. Some examples are listed in Table III.3.8. The high over-voltages of hydrogen evolution on carbon and mercury electrodes were discussed earlier. The window of a mercury electrode is limited by its anodic dissolution.

Generally, the potential range of electrode polarization depends on the solvent, the electrolyte decomposition, and the electrode material.

III.3.3.1.2 Seawater, a Natural Aqueous Electrolyte Solution

Seawater is the most abundant natural and, of course, *aqueous* electrolyte. Because of its very special properties, it is appropriate to discuss seawater here in a separate section. It is an aqueous solution of almost all elements, but there are eleven ions that account for more than 99.5% of the total dissolved solids in seawater. They are listed in Table III.3.9. The total mass of all salts dissolved in 1 kg of seawater is called the salinity. On average, the salinity of seawater is 35‰ (i.e., 35 g per 1 kg

Table III.3.9 Composition of seawater at 35‰ salinity and 25°C

Constituent	Concentration (mol/L)
Na^+	0.47912
Mg^{2+}	0.05449
Ca^{2+}	0.01051
K^+	0.01045
Sr^{2+}	0.00009
Cl^-	0.55865
SO_4^{2-}	0.02890
Br^-	0.00086
HCO_3^-	0.00238
$B(OH)_3$	0.00043
F^-	0.00006

of the solution). The major components of seawater are conservative, which means that their concentrations are linearly proportional to the salinity. This is because they have been uniformly mixed throughout the world's oceans. The chlorinity (*Cl*) is the chlorine equivalent (in grams) of the total mass of halides that can be precipitated from 1 kg of seawater by the addition of silver nitrate. The relationship between the chlorinity and salinity is

$$S = 1.80655Cl \tag{III.3.58}$$

The concentrations of the major ionic components in seawater of any salinity can be calculated by using the data in Table III.3.9 because their concentration ratios are essentially constant. However, minor variations (± 10%) in Ca^{2+}, Sr^{2+}, and HCO_3^- concentrations are possible. The density of seawater is a complex function of the salinity and temperature. The ionic strength of seawater $I = 0.5 \sum c_i z_i^2$ depends on the density and the salinity:

$$I = 1.9927 \times 10^{-2} d_{sw} S \tag{IIII.3.59}$$

If the salinity is 35‰ and the temperature is 25°C, the ionic strength is 0.71374 mol/L.

The permittivity of seawater decreases with the chlorinity:

$$\varepsilon_{sw} = (\varepsilon_w - A_1 Cl)/(1 + A_1 Cl) \tag{III.3.60}$$

where $A_1 = 6.00 \times 10^{-3} + 5.33 \times 10^{-6} T$, and T is the temperature in °C. The freezing point is a function of salinity:

$$T_f = -0.0137 - 0.05199S - 7.225 \times 10^{-5} S^2 - 7.58 \times 10^{-4} h \text{ (in °C)} \tag{III.3.61}$$

where h is the depth in meters. Similar relationships are established for the viscosity, the osmotic coefficient and pressure, the vapor pressure, the surface tension, and the solubility of gases [14].

The main gases dissolved in seawater are nitrogen, oxygen, and carbon dioxide. In well-aerated waters, the concentrations of O_2 and CO_2 are about 5×10^{-4} and 10^{-5} mol/L, respectively.

The carbon dioxide system controls the pH of seawater:

$$CO_2(aq) + H_2O \rightleftarrows HCO_3^- + H^+ \quad \text{(III.3.62)}$$

$$HCO_3^- \leftrightarrow CO_3^{2-} + H^+ \quad \text{(III.3.63)}$$

where $CO_2(aq)$ is dissolved gas:

$$[CO_2]_{sw} = K \times p_{co_2} \quad \text{(III.3.64)}$$

$K = 2.91 \times 10^{-2}$M/atm (for $S = 35‰$ and $T = 25°C$) and $p_{co_2} = 3.3 \times 10^{-4}$ atm is the partial pressure of CO_2 in the atmosphere (1 atm $= 1.01325 \times 10^5$ Pa). If the total carbon dioxide content is about 2.4×10^{-3} mol/L, the pH of seawater is 8.2 ± 0.1. In the laboratory it can be altered to about 8.6 during purging the sample with an inert gas, because of the loss of dissolved CO_2. This can be avoided by using a nitrogen/carbon dioxide mixture of known composition. The mixture can be prepared by bubbling the nitrogen gas through a sodium bicarbonate buffer solution. As supporting electrolyte, the seawater is a buffered solution of four cations (Na^+, Mg^{2+}, Ca^{2+} and K^+) and four anions (Cl^-, SO_4^{2-}, HCO_3^- and Br^-), with a remarkably low concentration of impurities. Apart from Sr^{2+}, F^-, and $B(OH)_3$, there are 12 elements that appear in concentrations between 10^{-4} and 10^{-8} mol/L. Because of dissolved oxygen, they are all in their highest oxidation states. Besides, they are either fully hydrolyzed or condensed, forming free oxyanions and ion pairs with major cations, or complexes with hydroxyl ions and the major anions. These minor components of seawater appear as the following ionic species: H_4SiO_4 and $H_3SiO_4^-$ (a total concentration 7.1×10^{-5}mol/L), NO_3^- and $NaNO_3$(3.6×10^{-5}mol/L), $NaHPO_4^-$, HPO_4^{2-} and $MgHPO_4$(2×10^{-6}mol/L), Rb^+(1.4×10^{-6}mol/L), IO_3^-, $NaIO_3$ and $MgIO_3^+$(4.7×10^{-7}mol/L), Li^+(2.6×10^{-7}mol/L), Ba^{2+} and $BaCl^+$(1.5×10^{-7}mol/L), MoO_4^{2-}(1×10^{-7}mol/L), $Al(OH)_3$ and $Al(OH)_4^-$(7×10^{-8}mol/L), $HAsO_4^{2-}$(5×10^{-8}mol/L), $Fe(OH)_3$(4×10^{-8}mol/L) and Ni^{2+}, $NiCl^+$ and $NiCO_3$(3×10^{-8}mol/L). The concentrations of all other elements in seawater are below 10^{-8} mol/L.

III.3.3.2 Nonaqueous Electrolyte Solutions

III.3.3.2.1 Alcohols

The properties of alcohols are similar to that of water, but they are better solvents for organic compounds. Apart from inorganic salts and bases ($NaClO_4$, LiCl, NH_4Cl, and KOH), useful supporting electrolytes are tetraalkylammonium salts: tetra-i-amylammonium perchlorate ($TAAClO_4$), tetra-i-amylammonium tetraphenylborate (TAATPB), tetra-i-amylammonium tetra-i-amylborate (TAATAB), tetrabutylammonium perchlorate ($TBAClO_4$), tetrabutylammonium chloride (TBACl), tetrabutylammonium bromide (TBABr), tetrabutylammonium hexafluorophosphate (TBAHFP), tetrabutylammonium iodide (TBAI), tetrabutylammonium tetrafluoroborate (TBATFB), tetrabutylammonium tetraphenylborate (TBATPB), tetraethylammonium perchlorate ($TEAClO_4$), tetraethylammonium chloride (TEACl), tetraethylammonium bromide (TEABr), tetraethylammonium iodide (TEAI), tetraethylammonium nitrate ($TEANO_3$), tetraethylammonium picrate (TEAP), and tetramethylammonium bromide (TMABr). The preparation and purification of these salts, as well as the purification and drying procedures for all organic solvents listed below, are described in the literature [15–19]. The salts can be recrystallized from hot water (e.g., TEAHFP, TBATFB and $LiClO_4$), or aqueous methanol (TEATFB and $LiClO_4$), or ethanol (TBATFB and TBAHFP), or acetone (TEATPB) and dried under vacuum at 60°C (TEATPB), 70°C (TBAHFP), 80°C (TBATFB), 95°C (TEATFB), or 175°C ($LiClO_4$). In alcohols a mercury electrode is polarized up to −2 V versus a mercury pool in LiCl. The recommended reference electrode is Ag|AgCl(s), LiCl saturated in ethanol ($E = 0.143$ V vs. SHE [20]).

III.3.3.2.2 Acids

Acetic acid, acetic anhydride, formic acid, and methanesulfonic acid are sometimes used in acid–base titrations of weak bases, but rarely in polarography and voltammetry. In the presence of inorganic salts and acids (LiCl, KCl, $NaClO_4$, $Mg(ClO_4)_2$, HCl, $HClO_4$, and H_2SO_4), the cathodic limit of the accessible potential range on a mercury electrode is between –0.8 V and −1.7 V versus the aqueous saturated calomel electrode (SCE), which is determined by discharge of hydrogen. On a platinum electrode the anodic limit is about +1.5 V versus aqueous SCE. The best reference electrode is the mercury–calomel couple: Hg| Hg_2Cl_2(s), LiCl(s) | acetic acid with the potential −0.055 V versus aqueous SCE [21]. In addition, the silver/silver(I) couple can be used as a reference: Ag|AgCl(s), 1 mol/L $LiClO_4$ in acetic acid ($E = 0.350$ V vs. SHE [20]).

III.3.3.2.3 Amines

Pyridine (C_5H_5N), ethylenediamine ($H_2NCH_2CH_2NH_2$), hexamethylphosphoramide ($[(H_3C)_2N]_3PO$) and morpholine (tetrahydro-1,4-oxazine, C_4H_8NO) are basic solvents used for cathodic reductions of metal ions at mercury, platinum, and graphite electrodes. They form Lewis acid–base adducts with metallic ions and they dissolve a variety of alkali, alkaline earth, rare earth and transition metal salts, organometallic compounds and some organic compounds. The following elements have been electrodeposited from amines: Li, Na, K, Cu, Ag, Mg, Ca, Ba, Zn, Pb, and Fe [15]. The supporting electrolytes for these solvents include lithium, sodium, ammonium, and tetraalkylammonium salts (LiCl, $LiClO_4$, $LiNO_3$, $NaClO_4$, $NaNO_3$, NH_4NO_3, NH_4Cl, TMABr, $TEANO_3$, TEAP, TEACl, $TEAClO_4$, $TBAClO_4$, etc.). Useful reference electrodes are: Ag|1 mol/L $AgNO_3$,

C_5H_5N ($E = 0.09$ V vs. SCE aq. [22]), Ag|0.01 mol/L $AgNO_3$, 0.5 mol/L $NaClO_4$, $[(H_3C)_2N]_3PO$ and ZnHg(s) | $ZnCl_2$(s) | 0.25 mol/L LiCl, $(H_2NCH_2)_2$ ($E = -1.1$ V vs. SCE aq. [23]).

III.3.3.2.4 Ethers

Ethers are good solvents for organic compounds, but they have rather low dielectric constants which make solutions quite resistive. They are electrochemically inert and provide fairly wide potential windows, which are limited by the decomposition of electrolytes or the oxidation of mercury. The following ethers have been used in voltammetry: tetrahydrofuran (tetramethylene oxide, C_4H_8O), 1,2-dimethoxyethane (H_3C-O-CH_2CH_2-O-CH_3), 1,2-epoxybutane (C_4H_8O), 1,4-dioxane and diethyl ether. Suitable supporting electrolytes are $LiClO_4$, $NaClO_4$, and $TBAClO_4$. The reference electrodes are: Ag|$AgClO_4$(s)| 0.3 mol/L $LiClO_4$, C_4H_8O and Ag|$AgNO_3$(s)|0.1 mol/L $TBAClO_4$, $(H_3COCH_2)_2$($E = 0.627$ V vs. SCE aq. [24]).

III.3.3.2.5 Nitriles

Acetonitrile, H_3CCN, is the best-known member of this class (See Fig. III.3.1). It is an excellent solvent for many polar or ionized organic compounds and for some inorganic salts, but not for saturated hydrocarbons and NaCl. Acetonitrile is electrochemically inert and has a fairly high permittivity. The best electrolytes are $LiClO_4$, $NaClO_4$, TBATPB, TEATPB, TAATPB, $TBAClO_4$, $TEAClO_4$, and mineral acids. The best reference electrode is Ag|0.1 mol/L $AgNO_3$||0.1 mol/L $NaClO_4$ ($E = 0.3$ V vs. SCE aq. [15]). The working windows are very wide: on a platinum electrode in a solution of $LiClO_4$ the limits are 2.4 V and -3.5 V vs. Ag/$AgNO_3$. The limits are determined by the decomposition of the electrolyte. On a mercury electrode in a solution of $TEAClO_4$ the limits are 0.6 V and -2.8 V. Electrochemical properties of other nitriles, such as propionitrile (C_2H_5CN), phenylacetonitrile ($C_6H_5CH_2CN$), isobutyronitrile $[(H_3C)_2CHCN]$, benzonitrile, and acrylonitrile, are very similar to the properties of acetonitrile. Commercial acetonitrile is usually contaminated by water, acetamide, and ammonia. The water can be removed by shaking with silica gel followed by the treatment with calcium hydride until hydrogen evolution stops and by fractional distillation over calcium hydride [19].

III.3.3.2.6 Amides

The permittivity of low-molecular-weight *N*-methyl-substituted amides is larger than that of water. The most widely used solvent in this group is dimethylformamide $[HCON(CH_3)_2]$. It readily dissolves polar and nonpolar organic compounds and inorganic perchlorates. Useful supporting electrolytes are $NaClO_4$, $TBAClO_4$, TBATPB, $TEAClO_4$, and TBAI. The working window of a platinum electrode in $TBAClO_4$ solution spans from +1.5 V to -2.5 V vs. SCE aq. The window of a mercury electrode in this medium extends from +0.5 V to -3.0 V vs. SCE aq. The recommended reference electrode is: CdHg(s)|$CdCl_2$(s), $CdCl_2 \times H_2O$(s), NaCl(s)|$HCON(CH_3)_2$ [25]. The other members of this group are dimethylacetamide [$H_3CCON(CH_3)_2$], *N*-methylformamide ($HCONHCH_3$), *N*-methylacetamide ($H_3CCONHCH_3$), and formamide ($HCONH_2$). All amides are electrochemically similar. Dimethylformamide can be purified by drying with silica gel followed by vacuum distillation and storage over 4 Å molecular sieves [18].

III.3.3.2.7 Dimethyl Sulfoxide

Dimethyl sulfoxide is a polar liquid which dissolves many organic and inorganic compounds. Alcohols, aldehydes, ketones, esters, ethers, heterocycles, aromatic compounds, iodides, bromides, chlorides, perchlorates, and nitrates are soluble, but paraffins, higher alcohols, fluorides, sulfates, and carbonates are not. There is a wide choice of supporting electrolytes among which $NaClO_4$, LiCl, $NaNO_3$, $KClO_4$, and tetraalkylammonium perchlorates are the most frequently used. The working window of a mercury electrode in $TBAClO_4$ solution is very convenient for the measurement of reduction processes. The limits of this window are +0.3 V and −2.7 V vs. SCE aq. Alkali metal ions are reduced in the potential range between −1.8 V and −2.2 V vs. SCE aq. Because of its high boiling point, dimethyl sulfoxide can be used for high-temperature electrochemical studies. A useful reference electrode is Ag|AgCl(s), KCl(s)|$(H_3C)_2SO$ ($E = 0.3$ V vs. SCE aq. [15]).

III.3.3.2.8 Methylene Chloride

Methylene chloride (CH_2Cl_2) is a very good solvent for organometallic compounds, but its permittivity is rather low. Using $TBAClO_4$ as the supporting electrolyte, the accessible potential ranges are between +1.8 V and −1.7 V vs. SCE for a platinum electrode, and between +0.8 V and −1.9 V vs. SCE for a mercury electrode. CH_2Cl_2 is aprotic and a very weakly coordinating solvent, without Lewis base properties [16], and tetrakis(pentafluorophenyl)-borate salts are very valuable supporting electrolytes for a Pt electrode [17]. It is a convenient medium for voltammetric measurements of metalloporphyrins and cation radicals of aromatic hydrocarbons. For this purpose the content of water in the solvent must be minimized. The purification is usually performed by refluxing CH_2Cl_2 under an inert gas atmosphere over a drying agent (P_2O_5 or CaH_2) followed by distillation and storage over P_2O_5 or 4 Å molecular sieves [16]. CH_2Cl_2 should be kept in brown bottles to prevent photolysis by ambient light.

Table III.3.10 Freezing and boiling temperatures of some organic solvents

Common name	Freezing point (°C)	Boiling point (°C)
Acetonitrile	−43.8	81.7
Benzonitrile	−14.0	191.1
Dimethylformamide	−60.5	153
Dimethyl sulfoxide	17.9	189
Methanol	−97.5	64.6
Methylene chloride	−97.2	40.0
Tetrahydrofuran	−108.4	65

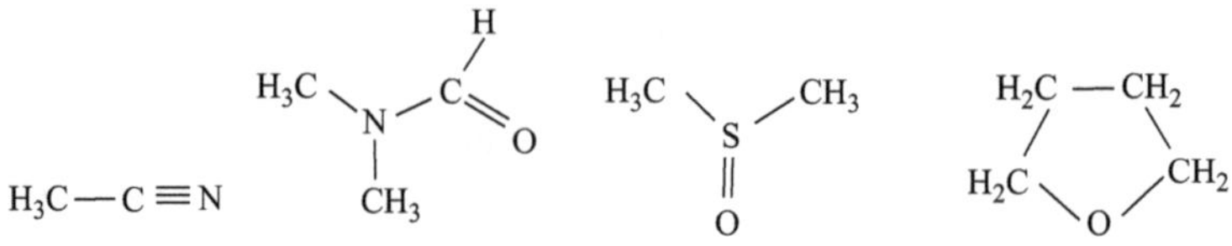

Acetonitrile Dimethylformamide Dimethyl sulfoxide Tetrahydrofuran

Fig. III.3.1 Structures of four organic solvents

References

1. Gellings PJ, Bouwmeester HJM (1997) CRC handbook of solid state electrochemistry. CRC Press, Boca Raton
2. Bach H, Baucke FGK, Krause D (2001) Electrochemistry of glasses and glass melts, including glass electrodes. Springer, Berlin Heidelberg New York
3. Bockris J O'M, Reddy AKN (1998) Modern electrochemistry: ionics, vol 1. Plenum Press, New York
4. Koryta J, Dvorak J (1987) Principles of electrochemistry. John Wiley, Chichester
5. Vanysek P (1989) Equivalent conductivities of some electrolytes in aqueous solution. In: Weast RC, Lide DR, Astle MJ, Beyer WH (eds) CRC handbook of chemistry and physics. CRC Press, Boca Raton, p D-169
6. Jost W (1960) Diffusion in solids, liquids and gases. Academic Press, New York
7. Kärger J, Heitjans P, Haberlandt R (1998) Diffusion in condensed matter. Friedrich Vieweg & Sohn, Braunschweig
8. Brett CMA, Oliveira Brett AM (1998) Electrochemistry. Oxford University Press, Oxford
9. Dahmen EAMF (1986) Electroanalysis. Elsevier, Amsterdam
10. Manahan SE (1986) Quantitative chemical analysis. Brooks/Cole Publ. Co., Monterey
11. Kuhn H, Försterling HD (1999) Principles of physical chemistry. John Wiley, Chichester
12. Wang J (1994) Analytical electrochemistry. VCH, Weinheim
13. Moeller T, O'Connor R (1972) Ions in aqueous systems. McGraw-Hill, New York
14. Whitfield M, Turner DR (1981) Seawater as an electrochemical medium. In: Whitfield M, Jagner D (eds) Marine electrochemistry. John Wiley, Chichester, p 3
15. Mann CK (1969) Nonaqueous solvents for electrochemical use. In: Bard AJ (ed) Electroanalytical chemistry, vol 3. Marcel Dekker, New York, p 57
16. Kadish KH, Anderson JE (1987) Pure Appl Chem 59: 707
17. LeSuer RJ, Geiger WE (2000) Angew Chem 112: 254
18. Armarego WLF, Chai CLL (2003) Purification of laboratory chemicals, 5th edn. Butterworth Heinemann, Amsterdam
19. Creager S (2007) Solvents and supporting electrolytes. In: Zoski CG (ed) Handbook of electrochemistry. Elsevier, Amsterdam
20. Tinner U (1989) Elektroden in der Potentiometrie. Metrohm, Herisau, p 16
21. Cihalik J, Simek J (1958) Collect Czech Chem Commun 23: 615
22. Cisak A, Elving PJ (1963) J Electrochem Soc 110: 160
23. Schaap WB, Bayer RE, Siefker JR, Kim JY, Breewster PW, Smith FC (1961) Record Chem Prog 22: 197
24. Hoffmann AK, Hodgson WG, Maricle DL, Jura HW (1964) J Am Chem Soc 86: 631
25. Marple LW (1967) Anal Chem 39: 844

Chapter III.4
Experimental Setup

Zbigniew Stojek

III.4.1 Introduction

To run a successful electrochemical experiment it is essential that the experimental setup is electrically correct and appropriate for the experiment planned. There are several points that should be carefully considered before the experiments are started. They include proper choice of the working, reference and auxiliary electrodes, proper selection of the solvent and supporting electrolyte, proper selection of the electroanalytical technique and its parameters, proper wiring of the electrochemical circuit and, finally, proper setting of the parameters of the potentiostat/voltammograph used.

III.4.2 The Working Electrode

The working electrode is the electrode at which the investigated processes occur. The investigator can choose one of the electrodes listed in Chap. III.1. The main criterion is the available potential window, which should meet the requirements of the investigation. Usually, in the range of positive potentials, platinum, gold and carbon (graphite, glassy carbon) electrodes are used. The surfaces of these materials are partially oxidised in aqueous solutions at this potential range. Thin layers of oxides are formed at gold and platinum, and various functional groups, such as –C=O and –OH, are attached to the carbon materials. In the negative range of potential, in aqueous solutions and other protic solvents, mercury electrodes are superior due to high overpotential of the reduction of hydrogen. On the other hand, many organic compounds strongly adsorb on mercury, which may complicate the analysis of voltammograms. In aprotic solvents, Pt, Au and C electrodes can be used in both positive and negative ranges of potential. Electrodes made of other noble metals, such as Ir and Ag, are less frequently used.

Z. Stojek (✉)
Department of Chemistry, University of Warsaw, Pasteura 1, 02-093, Warsaw, Poland
e-mail: stojek@chem.uw.edu.pl

F. Scholz (ed.), *Electroanalytical Methods*, 2nd ed.,
DOI 10.1007/978-3-642-02915-8_17,

A solid electrode, in comparison to a mercury drop, usually requires very careful pretreatment. The electrode surface should be clean and polished on a very wet pad to mirror gloss. This can be done using abrasive powders (or their suspensions in water), such as diamond and alumina, of various particle sizes. Depending on the actual state of the surface the polishing should start with the appropriate size of the abrasive material, followed by polishing with smaller particles. Usually, at the end of this process, 0.1 and 0.05 μm particles are used. After polishing, the electrode surface should be carefully cleaned (removal of abrasive material) with a dynamic stream of water, and dried with, e.g. methanol if a nonaqueous solvent is used in the experiments. Water-based suspensions of both diamond and alumina are available commercially. Oil-based suspensions are not recommended for polishing, since they lead to hydrophobic properties of the polished surfaces and this limits substantially the applicability of the electrode. An examination of the quality of the electrode surface with a microscope should always be carried out before the polishing procedure and before the experiments. This also helps to ensure that the abrasive material has been quantitatively removed from the surface. Traces of alumina attached to the electrode surface may alter, e.g. pH at the electrode/solution interface. For particular electrochemical purposes cheap disposable electrodes are proposed to ensure a good reproducibility of the measurements. To obtain well-defined voltammograms, it is also important that the electroactive part of the electrodes is perfectly sealed into the electrode body. Otherwise, the background for the voltammograms is usually excessive and steep.

Often, to obtain reproducible curves/waves, the solid electrode needs an electrochemical activation/regeneration. This is usually done by cycling the potential in an appropriate range while keeping the electrode in an appropriate solution. There is no universal range of potential and universal solution that can be employed for such activation. In the work with Au and Pt electrodes, often the voltammetric cycling/activation is done in a diluted sulphuric acid.

Solid electrodes covered by membranes or modified with polymers, gels and various composite materials cannot be treated by polishing. The only way to make them work reproducibly is to apply an appropriate conditioning potential (or a sequence of potentials) before the voltammetric experiments.

It is important that, before carrying out voltammetric experiments with analytes, the available potential window (range) is known and that there are no peaks of unwanted impurities in that range. The available potential window is determined by the currents of reduction/oxidation of the supporting electrolyte/solvent and, in the experiments, one should avoid entering potentials where these processes visibly occur. The products generated at the potential limits may interfere with the system under investigation and may affect the electrode surface.

After use, the working electrode should be thoroughly rinsed and dried to avoid crystallisation of the substrate on the electrode surface.

III.4.3 The Reference Electrode

The selection of a proper reference electrode is equally vital in voltammetry especially when accurate and precise data on the formal potentials of the red-ox couples under examination are needed. Traditional electrodes based on Hg and Ag ($Hg/HgCl$, Hg/Hg_2SO_4, $Ag/AgCl$, see Chap. III.2) can always be used; however, their concentrated electrolytes should be well separated from the analysed solution. In other words, everything should be done to prevent a leakage of the solution from the reference electrode to the cell, and vice versa. Therefore, first of all, a good electrolyte bridge filled with the analysed solution and well sealed with appropriate, conductive stoppers should be used. Such a bridge may well protect the solution in the cell against a leak of unwanted ions, but not the reference electrode as well. Also, it is usually advised to use the same solvent in the cell and in the reference electrode compartment.

When the experiments are performed with a two-electrode system, the current flows through the reference electrode. Under such conditions the reference electrode potential may not be stable over time. The smaller the working electrode the smaller is the risk of affecting the potential of the reference electrode. In the work with microelectrodes (electrodes with the dimensions of their active parts in the range of micrometers or less) as the working electrodes, the use of the two-electrode system is justified. If the three-electrode system is used, the reference electrode is practically charged with a very small current only; this current is in the range of pA. Such small currents cannot affect substantially the activities of the species that determine the potential of the reference electrode. So, in justified situations, when the voltammetric half-wave potential or peak potential does not have to be known precisely and what really matters is the peak or wave height, the so-called quasi-reference electrodes can be used. Most often a piece of platinum foil immersed into the examined solution is used as the quasi-reference electrode. Quasi-reference electrodes are especially useful when voltammetry at a very low ionic strength solution is performed [1].

III.4.4 The Counter Electrode

The counter (auxiliary) electrode is used in the three-electrode system only. In this system, the current flows between the working and the counter electrode. Either a piece of platinum foil or a platinum or titanium wire is usually employed as the counter electrode. Carbon rods are also used. It is recommended that the area of the counter electrode is substantially larger than that of the working electrode. If this condition is met, the counter electrode should not affect the current measurement due to, e.g. passivation, deactivation and blocking of the surface.

III.4.5 Instrumental Parameters and Wiring

What determines the quality of a potentiostat? Obviously, the ideal potentiostat should be fast, of low noise and of high input impedance. These three requirements rarely go together. Often the user has to choose between "high speed" and "high stability". High stability or low noise in voltammetry is usually achieved in instruments by inserting extra capacitors in the electronic circuit. This solution leads to distorted voltammetric signals: peaks are smaller and thicker, and waves are more sloped; however, calibration plots constructed for analytical purposes can still be linear. For a more detailed description of the potentiostat construction the reader is referred to the appropriate chapter in ref. [2].

High-input impedance (>1 × 10^{12} Ω) allows the use of electrolyte bridges of high resistance. It may even be possible to fill the bridge with just deionised water. It is important to realise that any resistance in the cell leads to an ohmic drop in the potential applied to the electrode. Fine instruments offer a possibility of positive-feedback compensation of the ohmic drop. To do this well the value of the solution resistance should be known and inserted either manually or through the appropriate dialog box of the software. The system works in such a way that an extra potential is added to the applied potential to counter the ohmic drop related to the resistance of the solution. If the solution resistance is not known, the method of "set and try" may also be satisfactory. A declaration of a too large resistance usually results in the oscillation of the system and the appearance of faulty peaks. Another possibility of eliminating the ohmic drop is to use the current interrupt method introduced by EG&G in their potentiostats. In this method, while the voltammetric curve is run, the circuit is repeatably disconnected for several microseconds to analyse the decay of the potential at the working electrode, and an appropriate correction to the potential is applied. Usually, the potentiostats can compensate up to 12 V of ohmic drop, which means that up to 12 V can be imposed between the working and the auxiliary (counter) electrodes. Some EG&G potentiostats can apply up to 100 V. It is worth noting here that a lost connection to the reference electrode during the measurement will result in the application of the maximum potential difference between the counter and working electrodes and possibly in a heavy damage to the electrode surface. Some instruments, those with an impedance-measuring circuit built in, offer an on-line compensation of the ohmic potential drop without any current interrupt.

III.4.6 Nonaqueous Media

Electrochemical investigations are often carried out in aproptic solvents such as dimethylformamide, dimethyl sulfoxide, acetonitrile, propylene carbonate, nitromethane and others. Electrode processes are different in these solvents compared to water. Often well-defined one-electron waves/peaks are obtained. However, to be sure that the results obtained are reliable, one should know the concentration of water in the applied solvent. This concentration cannot be equal to or higher than the analyte concentration to achieve real nonaqueous characteristics. An extra attention should be also paid to the ohmic potential drop, since the resistance of nonaqueous solutions is usually higher compared to aqueous solutions.

III.4.7 Elimination of Electrical Noise

Electrical noise often distorts obtained voltammograms. This may be especially troublesome when the concentration of the examined species is very low. If the frequency of the noise is sufficiently high, the experimental data can be easily smoothed with most of the commercial electrochemical and graphical software. A simple moving average method as well as the Fourier transform method will give satisfactory results. It is much more difficult to eliminate low frequency (of a few Hz) noise from the voltammograms: usually, the routines applied will affect the faradaic signals too. Even more complicated is the situation when the frequency and the magnitude of the noise change with either potential or time.

Good protection against electrical noise is provided by the application of a Faraday cage during the experiments. A Faraday box can be built quickly by covering a regular carton with aluminium foil. The entire cell with all non-screened electrical connections should be enclosed in the box. Only very small openings should be made for the potentiostat cables. For the best results the Faradaic box (in fact the aluminium foil) should be well grounded.

References

1. Ciszkowska M, Stojek Z (2000) Anal Chem 72: 754A
2. Bard AJ, Faulkner RF (2000) Electrochemical methods, 2nd edn. John Wiley, New York

Part IV
Publications in Electrochemistry

Chapter IV.1
Seminal Publications in Electrochemistry and Electroanalysis

Fritz Scholz, György Inzelt, and Zbigniew Stojek

This compilation of seminal publications in electrochemistry and electroanalysis is neither complete nor are all the listed contributions of the same importance. The authors feel that it might be of interest and very rewarding for people who use electroanalytical methods in the laboratory to go back to the roots and read some of the publications which later initiated strong developments of the science that is presented in this book. Many of the contributions to science cited here became an inherent part of textbooks and common knowledge so that the original work is usually not referred to and access to this information is difficult. The following web pages give access to much more information on the history of electrochemistry and electroanalysis:

1. http://chem.ch.huji.ac.il/history/electrochemists.htm
2. http://www.chemheritage.org/
   Short biographies of most of the listed scientists are also available in: Bard AJ, Inzelt G, Scholz F (eds) (2008) Electrochemical dictionary, Springer, Berlin. Publications listed below, which are accessible in the web as English translations and originals, respectively, are marked by (*).

1791 **L. GALVANI** (1737–1798) publishes electrical experiments with frogs: (1791) De Bononiensi Scientiarum et Artium Instituto atque Academia Commentarii VII: 363, [Ostwalds Klassiker Bd 52 (1894)]

1800 **A. VOLTA** (1745–1827) communicates the construction of an electrochemical pile in a letter to Sir Joseph Banks, published in: (1800) Philos Trans II: 405 (*)

1800 **W. CRUIKSHANKS** (1745–1800) publishes the first qualitative analysis (copper) performed with the help of electrolysis: (1800) Nicholsons Journal 4: 187

1807 **H. DAVY** (1778–1829) publishes a theory of electrolysis: (1807) Philos Trans 1

1807 **H. DAVY** (1778–1829) discovers sodium and potassium by electrolysis: (1808) Philos Trans (*)

1812 **J. J. BERZELIUS** (1779–1848) publishes an electrochemical theory of the chemical bond: (1812) J Chem Phys 6: 119

F. Scholz (✉)
Institut für Biochemie, Universität Greifswald, 17487 Greifswald, Germany
e-mail: fscholz@uni-greifswald.de

F. Scholz (ed.), *Electroanalytical Methods*, 2nd ed.,
DOI 10.1007/978-3-642-02915-8_18, © Springer-Verlag Berlin Heidelberg 2010

1826 **G. S. OHM** (1789–1854) publishes the law that became known as Ohm's law: (1826) Schweigger's J 46: 137

1833–1834 **M. FARADAY** (1791–1867) publishes numerous observations on electrochemistry, including the famous law and introduced the modern electrochemical nomenclature (e.g., ion, anion, cation, electrolyte, electrode): (1832–1834) Philos Trans, (Ostwalds Klassiker Bd. 81, 86, 87, 126, 128, 131, 134, 136, Leipzig, 1896–1903) (*); see also: "Faraday as a discoverer" by J. Tyndall (1894)

1864 **W. GIBBS** (1822–1908) publishes the first electrogravimetric analysis: (1864) Fresenius Z Anal Chem 3: 334

1873 **G. J. LIPPMANN** (1845–1921) introduces the capillary electrometer that bears his name: Ann Phys 225:546

1879 **H. L. F. v. HELMHOLTZ** (1821–1894) introduces the dropping mercury electrode, publishes a model of the electrical double layer: (1879) Wied Ann 7: 337

1881 **H. L. F. v. HELMHOLTZ** (1821–1894) reports in his Faraday Lecture the electrochemical cell with a glass membrane as the separator and build by his assistant **W. GIESE** (1847–?): (1881) J Chem Soc 39: 277. This was the first glass electrode, although neither Giese nor Helmholtz realized its significance for pH measurements (*)

1887 **S.A. ARRHENIUS** (1859–1927) publishes the theory of dissociation of electrolytes in water: (1887) Z physik Chem I: 631 (*)

1888 **F. W. OSTWALD** (1853–1932) publishes a law, which later became known as Ostwalds' dilution law: (1888) Z physik Chem 2: 36 (*)

1889 **W. NERNST** (1864–1941) publishes in his habilitation "Die elektro-motorische Wirksamkeit der Jonen", Leipzig, 1889, the fundamental equation which relates the potential to ion activities: (1889) Z phys Chem 4: 29

1893 **R. BEHREND** (1856–1926) publishes the first potentiometric precipitation titration: (1893) Z physik Chem 11: 466

1896 **E. SALOMON** (1874–?) publishes the first current-voltage curve in the form of discrete points: (1896) Z Elektrochem 3: 264 (see also his PhD thesis, Göttingen 1897)

1897 **W. BÖTTGER** (1871–1949) publishes the first potentiometric acid-base titration: (1897) Z physik Chem 24: 253

1897 **E. SALOMON** (1874–?) describes the first amperometric titration: Z Elektrochem 4:71 (see also his PhD thesis, Göttingen 1897)

1900 **F. CROTOGINO** (1878–1947) publishes the first potentiometric redox titration: (1900) Z anorg Chem 24: 225

1901 **E. H. RIESENFELD** (1877–1957) and **W. NERNST** (1864–1941) publish the first experiments involving an ion transfer between immiscible electrolyte solutions: (1901) Ann Phys 8:600

1903 **F. W. KÜSTER** (1861–1917) and **M. GRÜTERS** publish the first conducto-metric titration: (1903) Z anorg Chem 35: 54

1903 **F. G. COTTRELL** (1877–1948) publishes the equation which is known as the Cottrell equation: (1903) Z physik Chem 42: 385

1903 **B. KUČERA** (1874–1921) introduces the dropping mercury electrode for electrocapillary studies: (1903) Ann Phys 11: 529

1905 **J. TAFEL** (1862–1918) publishes his empirically discovered equation: (1905) Z phys Chem 50: 641

1906 **M. CREMER** (1865–1935) publishes the observation of a voltage drop across a glass membrane and thus led the basis for modern glass electrodes: (1906) Z Biol 29: 562

1909 **S. P. L. SØRENSEN** (1868–1939) introduces the concept of pH: (1909) Compt Rend Lab Carlsberg 8: 1, 396; (1909) Biochem Z 21: 131, 201; 22: 352 (*)

1909 **F. HABER** (1868–1934) and **Z. KLEMENSIEWICZ** (1886–1963) publish the glass electrode for pH-measurements: (1909) Z physik Chem 67: 385

1911 **F. G. DONNAN** (1870–1956) publishes a theory of membrane potentials: (1911) Z Elektrochem 17: 204; (1924) Chem Rev 1: 73

1909–1913 **L. G. GOUY** (1854–1926) [(1909) Compt Rend 149: 654 and (1910) J Phys 9: 457] and **D. L. CHAPMAN** (1869–1958) [(1913) Phil Mag 25: 475] publish a model of the electrical double layer

1921 **E. Ch. S. BIILMANN** (1873–1946) introduces the quinhydrone electrode for pH measurements: (1921) Ann Chim 15:109

1922 **J. HEYROVSKÝ** (1890–1967) publishes first results with the method which he soon called polarography: (1922) Chemicke Listy 16: 256; (1923) Philos Mag 45: 303

1923 **J. N. BRØNSTED** (1879–1947) gives a new definition of acids and bases: Rec Trav Chim Pays-Bas 42: 718 (*)

1923 **P. J. W. DEBYE** (1884–1966) and **E. A. J. HÜCKEL** (1896–1980) publish what is now known as the Debye-Hückel theory of strong electrolytes: Debye P, Hückel E (1923) Phys Z 24:185, 305

1924 **O. STERN** (1888–1969) publishes his model of the electrical double layer: (1924) Z Elektrochem 30: 508

1924 **J. A. V. BUTLER** (1899–1977) publishes his contribution to the equation which is known as the Butler-Volmer equation: (1924) Trans Faraday Soc 19: 734

1925 **J. HEYROVSKÝ** (1890–1967) and **M. SHIKATA** (1895–1965) publish the construction of the first polarograph, which was the first automatic analyzer: (1925) Rec Trav chim Pays-Bas 46: 496

1930 **J. HEYROVSKÝ** (1890–1967) and **J. BABICKA** publish the first observation of a catalytic hydrogen wave: (1930) Coll Czech Chem Commun 2: 370

1930 **T. ERDEY-GRÚZ** (1902–1976) and **M. VOLMER** (1885–1965) publish the equation which becomes later known as the Butler-Volmer equation, and they introduce the transfer coefficient: (1930) Z phys Chem A150: 203

1931 **Ch. ZBINDEN** (1904–1983) publishes a copper determination which can be termed chronopotentiometric stripping microcoulometry: (1931) Bull de la soc chim biol 13: 35. This can be regarded as the key step towards anodic stripping voltammetry.

1932 **I. SLENDYK** publishes the first example of a catalytic hydrogen wave caused by platinum traces: (1932) Coll Czech Chem Commun 4: 335

1933 **A.N. FRUMKIN** (1895–1976) describes the influence of the electrical double layer on the kinetics of electrode reactions: (1933) Z Physik Chem 164A: 121

1934 **D. ILKOVIČ** (1907–1980) publishes the equation which is known as the Ilkovic equation: (1934) Coll Czech Chem Commun 6: 498

1935 **A. O. BECKMANN** ((1900–2004) invented the first pH meter based on vacuum-tube amplifier (US Patent No. 2,058,761)

1935 **J. HORIUTI** (1901–1979) and **M POLANYI** (1891–1976) explain the meaning of the charge transfer coefficient by the help of transition-state theory: Horiuti J, Polanyi M (1935) Acta Physicochim URSS 2:505

1939 **L. SZEBELLEDY** (1901–1944) and **Z. SOMOGYI** (1915–1945) establish coulometric analysis: (1938) Fresenius Z Anal Chem 112: 313, 323, 385, 391, 395, 400

1940 **I. M. KOLTHOFF** (1894–1993) and **H. A. LAITINEN** (1915–1991) introduce the term voltammetry: (1940) Science 92: 152

1942 **V. G. LEVICH** (1917–1987) publishes the equation describing the limiting current at a rotating disk electrode: (1942) Acta Physicochim USSR 17: 257

1948 **M. KALOUSEK** (1915–1996) introduces what is later called 'Kalousek commutator': (1948) Coll Czech Chem Commun 13–105

1948 **J.E.B. RANDLES** (1912–1998) develops linear sweep voltammetry and gives the equation for the peak current of linear sweep voltammetry known as the Randles-Sevcik equation: (1948) Trans Faraday Soc 44: 327. **A. SEVCIK** (1926–2006) derives a similar equation independently: (1948) Coll Czech Chem Commun 13: 349

1952 **G. C. BARKER** (1915–2000) and **I. L. JENKINS** introduce square-wave polarography: (1952) Analyst 77: 685

1952 **V. G. LEVICH** (1917–1987) publishes his fundamental book Physico-chemical Hydrodynamics (Izd Akad Nauk SSSR, 2nd ed 1952, Engl Translation: Prentice Hall 1962)

1952 **W. KEMULA** (1902–1985) introduces polarographic detection in liquid chromatography (so-called chromatopolarography): (1952) Roczn chem 26: 281

1956–1959 **W. KEMULA** (1902–1985) and **Z. KUBLIK** (1922–2005) introduce the hanging mercury drop electrode (HMDE or Kemula electrode) and anodic stripping voltammetry: (1956) Roczn chem 30: 1005; (1958) Anal Chim Acta 18: 104; (1959/60) J Electroanal Chem 1: 123

1957 **G. MAMANTOV, P. PAPOFF** and **P. DELAHAY** publish on anodic stripping analysis with hanging mercury drop electrodes employing current and potential step methods: (1957) J Am Chem Soc 79: 4034

1959 **A. H. MAKI** and **D. H. GESKE** develop the electrochemical ESR cell: (1959) J Chem Phys 30: 1356

1960 **G. C. BARKER** (1915–2000) and **A. W. GARDNER** introduce pulse polarography: (1960) Fresenius Z. Anal. Chem 173: 79. [They published it as early as 1958, however, in a lesser known journal: (1958) At Energy Res Establ, Harwell, C/R 2297).

1961 **S. BRUCKENSTEIN** and **T. NAGAI** report the first potentiometric stripping analysis: (1961) Anal Chem 33: 1201

1961 **E. PUNGOR (1923–2007)** and **E. HOLLOS-ROKOSINYI** develop the first ion-selective electrode after the invention of the glass electrode: (1961) Acta Chim Hung 27: 63

1962 **A. Y. GOKHSHTEIN** and **Y. P. GOKHSHTEIN** (1906–1996) introduce the static mercury drop electrode: (1962) Zh fiz khim 36: 651

1963 **J. O'M. BOCKRIS, M. A. V. DEVANATHAN** and **K. MÜLLER** publish their model of the electrical double layer: (1963) Proc Royal Soc (London) A 274: 55

1964 **T. KUWANA, R.K. DARLINGTON** and **D.W. LEDDY** develop spectroelectrochemistry by using optically transparent electrodes: (1964) Anal Chem 36: 2023

1964 **R. S. NICHOLSON** and **I. SHAIN** publish the characteristic equations for cyclic voltammetry which they developed using numerical methods: Anal Chem 36:706

1965 **R. A. MARCUS** publishes his theory for electron transfer reactions: (1965) J Chem Phys 43: 679

1975 **P. R. MOSES, L. WIER** and **R. W. MURRAY** introduce chemically modified electrodes: (1975) Anal Chem 47: 1889

1976 **D. JAGNER** and **A. GRANELI** introduce potentiometric stripping analysis on a broad scale: (1976) Anal Chim Acta 83: 19

1976 **R. N. ADAMS** (1924–2002) starts to use microelectrodes: (1976) Anal Chem 48: 1126A

1978 **A. MERZ** and **A. J. BARD** [(1978) J Am Chem Soc 100: 3222] as well as **M. R. VAN DE MARK** and **L. L. MILLER** publish the first papers on polymer film electrodes: (1978) J Am Chem Soc 100: 3223

1979 **K. B. OLDHAM** publishes an analytical formula describing an entire linear scan voltammogram: J Electroanal Chem 105:373

1980 **K. AOKI, J. OSTERYOUNG** and **R. A. OSTERYOUNG** (1927–2004) introduce differential double pulse polarography: (1980) J Electroanal Chem 110: 1

1979–1985 **R. MALPAS, R. A. FREDLEIN** and **A. J. BARD** [(1979) J Electroanal Chem 98: 171]; **J. H. KAUFMAN, K. K. KANAZAWA** and **G. B. SREET** [(1984) Phys Rev Lett 53: 2461], as well as **S. BRUCKENSTEIN** and **S. SWATHIRAJAN** [(1985) Electrochim Acta 30: 851] develop the electrochemical quartz crystal microbalance (EQCM) technique.

1989 **A. J. BARD, F.-R. FAN, J. KWAK** and **O. LEV** publish the first true scanning electrochemical microscopy (SECM): (1989) Anal Chem 61: 132

Chapter IV.2
Books on Fundamental Electrochemistry and Electroanalytical Techniques

Fritz Scholz

IV.2.1 Electrochemical Dictionaries, Encyclopedias, and Reference Books

1. Bard AJ, Inzelt G, Scholz F (eds) (2008) Electrochemical dictionary. Springer, Berlin
2. Bard AJ, Stratmann M (eds) (2002–2007) Encyclopedia of electrochemistry, 11 vols. Wiley-VCH, Weinheim
3. Zoski CG (ed) (2007) Handbook of electrochemistry. Elsevier, Boston
4. Bard AJ, Parsons R, Jordan J (eds) (1985) Standard potentials in aqueous solution. Marcel Dekker, New York
5. Bard AJ (ed) vol 1 (1973) – vol 15 (1984), vol 9, part B (1986) Encyclopedia of electrochemistry of elements. Marcel Dekker, New York

IV.2.2 Books on General Electrochemistry

1. Plieth W (2008) Electrochemistry for material science. Elsevier, Amsterdam
2. Bagotsky VS (ed) (2006) Fundamentals of electrochemistry. Wiley, Hoboken
3. Bard AJ, Faulkner LR (2001) Electrochemical methods, fundamentals and applications, 2nd ed. Wiley, New York
4. Bockris JO'M, Reddy AKN, (Gamboa-Adelco ME/2A) (1998/2001) Modern electrochemistry, 2nd edn, vol 1/2A. Plenum, New York
5. Hamann CH, Hamnett A, Vielstich W (1998) Electrochemistry. John Wiley, Weinheim
6. Sato N (1998) Electrochemistry at metal and semiconductor electrodes. Elsevier, Amsterdam
7. Bockris JO'M, Khan ShUM (1993) Surface electrochemistry, a molecular level approach. Plenum, New York
8. Schmickler W (1996) Interfacial electrochemistry. Oxford University Press, New York
9. Rubinstein I (ed) (1995) Physical electrochemistry. Marcel Dekker, New York
10. Oldham HB, Myland JC (1994) Fundamentals of electrochemical science. Academic Press, San Diego
11. Galus Z (1994) Fundamentals of electrochemical analysis, 2nd ed. Ellis Horwood, New York

---

F. Scholz (✉)
Institut für Biochemie, Universität Greifswald, 17487 Greifswald, Germany
e-mail: fscholz@uni-greifswald.de

F. Scholz (ed.), *Electroanalytical Methods*, 2nd ed.,
DOI 10.1007/978-3-642-02915-8_19,

12. Gileadi E (1993) Electrode kinetics for chemists, chemical engineers and materials scientists. VCH, New York
13. Brett CMA, Oliveira Brett AM (1993) Electrochemistry. Oxford University Press, Oxford
14. Rieger PH (1987) Electrochemistry. Prentice-Hall, London
15. Vetter KJ (1961) Elektrochemische Kinetik. Springer, Berlin; Engl ed: Vetter KJ (1967) Electrochemical kinetics. Academic Press, New York

IV.2.3 Monographs on Special Techniques and Subjects

1. Doménech-Carbó A, Doménech-Carbó MT, Costa V (2009) Electrochemical methods in archeometry, conservation and restoration. In: Scholz F (ed) Monographs in electrochemistry. Springer, Berlin
2. Inzelt G (2008) Conducting polymers, a new era in electrochemistry. In: Scholz F (ed) Monographs in electrochemistry. Springer, Berlin
3. Compton RG, Banks CE (2007) Understanding voltammetry. World Science, New Jersey
4. Mirčeski V, Komorsky-Lovrić S, Lovrić M (2007) Square-wave voltammetry, theory and application. In: Scholz F (ed) Monographs in Electrochemistry, Springer, Berlin
5. Wang J (2006) Analytical Electrochemistry, 3rd ed. VCH, Weinheim
6. Scholz F, Schröder U, Gulaboski R (2005) Electrochemistry of immobilized particles and droplets, Springer, Berlin
7. Zanello P (2003) Inorganic electrochemistry, theory, practice and application. The Royal Society of Chemistry, Cambridge
8. Bond AM (2002) Broadening electrochemical horizons. Principles illustration of voltammetric and related techniques. University Press, Oxford
9. Buchberger W (1998) Elektrochemische Analysenverfahren. Spektrum, Heidelberg
10. Eggins BR (1997) Biosensors. Wiley, Chichester
11. Kissinger PT, Heineman WR (ed) (1996) Laboratory techniques in electroanalytical chemistry, 2nd ed. Marcel Dekker, New York
12. Sawyer DT, Sobkowiak A, Roberts JL (1995) Electrochemistry for chemists, 2nd ed. Wiley, Chichester
13. Gosser DK Jr (1993) Cyclic voltammetry, simulation and analysis of reaction mechanisms. VCH, New York
14. Brainina Kh, Neyman E (1993) Electroanalytical stripping methods. In: Winefordner JD (ed) Chemical Analysis, Wiley, New York
15. Noel M, Vasu KI (1990) Cyclic voltammetry and the frontiers of electrochemistry. Aspect Publications, London
16. Macdonald R (1987) Impedance spectroscopy. Wiley Interscience, New York
17. Baizer MM, Lund H (eds) (1983) Organic electrochemistry. Marcel Dekker, New York
18. Bond AM (1980) Modern polarographic methods in analytical chemistry. Marcel Dekker, New York
19. Heyrovský J, Kůta J (1966) Principles of polarography. Academic Press, New York (German edn: Heyrovský J, Kůta J (1965) Grundlagen der Polarographie. Akademie-Verlag, Berlin)
20. Breyer B, Bauer HH (1963) Alternating current polarography and tensammetry. Interscience, New York
21. Ives DJG, Janz GI (eds) (1961) Reference electrodes: theory and practice. Academic Press, New York, reprinted: NACE International, Houston, TX 1996
22. Delahay P (1954) New instrumental methods in electrochemistry. Interscience, New York
23. Kolthoff IM, Lingane JJ (1952) Polarography, 2nd ed, vol 1 and 2. Interscience Publ, New York

IV.2.4 Series Editions of Advances in Electrochemistry

1. Advances in electrochemistry and electrochemical engineering, vol 1 (1961) – vol 8, Delahay P, Tobias CW (eds), vol 9, Muller RH (ed), vol 10–13 (1984) Gerischer H, Tobias CW (eds), Wiley-Interscience, New York
2. Advances in electrochemical sciences and engineering, vol 1(1997) – vol 4, Gerischer H, Tobias CW (eds), VCH, New York 1990–1995, vol 5, Alkire RC, Gerischer H, Kolb DM, Tobias CW (eds), vol 6–8, Alkire RC, Kolb DM (eds), vol 9 (2006) Alkire RC, Kolb DM, Lipkowski J, Ross P (eds), J. Lipkowski, and P. Ross, Wiley-VCH, New York
3. Comprehensive treatise of electrochemistry, vol 1 (1980) – vol 10 (1985) Bockris J O'M, and others (eds) Plenum, New York
4. Electroanalytical chemistry – a series of advances. vol 1 (1966) Bard AJ (ed) – vol 22 (2004) Bard AJ, Rubinstein I (eds), Marcel Dekker, New York
5. Modern aspects of electrochemistry, vol 1 (1954) Bockris J O'M, Conway BE, (eds) Butterworth, Washington – vol 43 (2008) Schlesinger M (ed) Springer, New York

IV.2.5 Electrochemical Journals

1. Bioelectrochemistry, Elsevier
2. Corrosion Science, Elsevier
3. Electroanalysis, Wiley-VCH
4. Electrochemical and Solid State Letters, The Electrochemical Society,
5. Electrochemistry Communications, Elsevier
6. Electrochimica Acta, Elsevier
7. Ionics, Springer
8. Journal of Applied Electrochemistry, Springer
9. Journal of Electroanalytical Chemistry, Elsevier
10. Journal of the Electrochemical Society, The Electrochemical Society
11. Journal of Solid State Electrochemistry, Springer
12. Journal of Power Sources, Elsevier
13. Russian Journal of Electrochemistry, Springer
14. Solid State Ionics, Elsevier

IV.2.6 Journals that Regularly Publish Papers on Electrochemistry and/or Electroanalysis

1. Analytical Chemistry, The American Chemical Society
2. Journal of Physical Chemistry (A, B, and C), The American Chemical Society
3. Langmuir, The American Chemical Society
4. Journal of Colloid and Interface Science, Academic Press
5. Analytical and Bioanalytical Chemistry, Springer
6. Analytica Chimica Acta, Elsevier
7. PCCP (Physical Chemistry Chemical Physics), RSC
8. Sensors and Actuators B: Chemical. Elsevier

Index

F. Scholz (ed.), *Electroanalytical Methods*, 2nd ed.,
DOI 10.1007/978-3-642-02915-8, © Springer-Verlag Berlin Heidelberg 2010

N

Q

T

MIX
Papier aus verantwortungsvollen Quellen
Paper from responsible sources
FSC® C105338

If you have any concerns about our products,
you can contact us on
ProductSafety@springernature.com

In case Publisher is established outside the EU,
the EU authorized representative is:
Springer Nature Customer Service Center GmbH
Europaplatz 3, 69115 Heidelberg, Germany

Printed by Libri Plureos GmbH
in Hamburg, Germany